STP 1421

Outdoor Atmospheric Corrosion

Herbert E. Townsend, editor

ASTM Stock Number: STP1421

ASTM
100 Barr Harbor Drive
West Conshohocken, PA 19428-2959

INTERNATIONAL

Printed in the U.S.A.

Library of Congress Cataloging-in-Publication Data

Outdoor atmospheric corrosion / Herbert E. Townsend, editor.
p. cm.—(STP ; 1421)
"ASTM Stock Number: STP1421."
Includes bibliographical references and index.
ISBN 0-8031-2896-7
1. Corrosion and anti-corrosives—Congresses. I. Townsend, Herbert E., 1938– II. ASTM special technical publication ; 1421

TA418.74 .O88 2002
620.1′1223—dc21

2002074627

Peer Review Policy

Each paper published in this volume was evaluated by two peer reviewers and at least one editor. The authors addressed all of the reviewers' comments to the satisfaction of both the technical editor(s) and the ASTM International Committee on Publications.

To make technical information available as quickly as possible, the peer-reviewed papers in this publication were prepared "camera-ready" as submitted by the authors.

The quality of the papers in this publication reflects not only the obvious efforts of the authors and the technical editor(s), but also the work of the peer reviewers. In keeping with long-standing publication practices, ASTM International maintains the anonymity of the peer reviewers. The ASTM International Committee on Publications acknowledges with appreciation their dedication and contribution of time and effort on behalf of ASTM International.

Printed in Phila., PA
August 2002

Foreword

This publication, Outdoor Atmospheric Corrosion, contains papers presented at the symposium of the same name held in Phoenix, Arizona, on 8-9 May 2001. The symposium was sponsored by ASTM International Committee G1 on Corrosion of Metals. The symposium co-chairman was Herbert E. Townsend, Consultant, Center Valley, PA.

Dedication to Seymour K. Coburn
1917–2001

This volume is dedicated to the memory of Seymour K. Coburn, who passed away on January 4, 2001.

Sy, as he was known to many of his friends, was born in Chicago in 1917. He received a BS in Chemistry from the University of Chicago in 1940, and an MS from Illinois Institute of Technology in 1951. After initially working for Minor laboratories, Lever Brothers, and the Association of American Railroads, he began a long career as a corrosion specialist at the Applied Research Laboratories of US Steel Corporation.

Working with C. P. Larabee at US Steel, he became well known throughout the industry for pioneering their studies of the effects of alloying elements on the corrosion of steels. To do this, they studied the corrosion performance of hundreds of steel compositions exposed to rural, marine, and industrial environments, and defined the beneficial effects of copper, nickel, phosphorus, chromium, and silicon. No treatment of the subject is complete without a reference to their classic paper, "The Atmospheric Corrosion of Steels as Influenced by Changes in Chemical Composition," that was presented in 1961 to the First International Congress on Metallic Corrosion in London.

Sy went on to become one of the leading advocates of weathering steels, that is, low-alloy steels which develop a protective patina during exposure in the atmosphere so that they become corrosion-resistant without painting for use in applications such as bridges, utility towers, and buildings. He was US Steel's research consultant for the John Deere Headquarters on Moline, IL, the first building constructed with weathering steel, as well as the Chicago Civic Center, and some of the first unpainted weathering steel bridges.

In 1970, he was transferred to the Special Technical Services unit of US Steel's Metallurgical Department where he became the top promoter and trouble-shooter for bridges and other weathering steel applications. But it was not until he attended a workshop of the Steel Structures Paint Council that he achieved his real goal in life—he became a teacher.

An active member of ASTM International, Sy chaired Subcommittee G1.04 on Atmospheric Corrosion from 1964 to 1970, and was instrumental in organizing this subcommittee. He also was the prime mover in organizing and editing STP 646, "Atmospheric Factors Affecting the Corrosion of Engineering Materials," and he chaired the symposium that led to that STP, a celebration of 50 years of exposure testing at the State College, PA, ASTM International atmospheric corrosion test site in May 1976.

After retiring in 1984, he continued to teach and actively consult around the world in matters related to weathering steels and protective coatings. In addition to his ASTM International activities, Sy was also a member of the American Chemical Society, The American Society for Metals, the National Association of Corrosion Engineers, and the Steel Structures Painting Council.

Stan Lore
612 Scrubgrass Road
Pittsburgh, PA 15243

Contents

Overview

This book is a collection of papers presented at the ASTM International Symposium on Outdoor and Indoor Atmospheric Corrosion that was held in Phoenix, AZ in May 2001. With presentations from authors representing ten counties in North and South America, Europe, and Asia, the symposium was truly international.

The symposium was originally conceived as a vehicle to present results of the 1976 ASTM International outdoor atmospheric corrosion test program. During the initial scheduling, it was combined with another symposium being planned by Robert Baboian on indoor corrosion to form a joint symposium on both outdoor and indoor corrosion. Although a joint symposium was organized accordingly, contributions on the indoor topic did not materialize. Consequently, this STP is devoted entirely to the outdoor topic.

Corrosion of metals in the atmosphere has been an important topic for many years, as evidenced by the many symposium volumes previously published by ASTM International.

- *STP 67, Symposium on Atmospheric Exposure Tests on Nonferrous Metals,* 1946.
- *STP 175, Symposium on Atmospheric Corrosion of Non-Ferrous Metals,* 1956.
- *STP 290, Twenty-Year Atmospheric Investigation of Zinc-Coated and Uncoated Wire and Wire Products,* 1959.
- *STP 435, Metal Corrosion in the Atmosphere,* 1968.
- *STP 558, Corrosion in Natural Environments,* 1974.
- *STP 646, Atmospheric Factors Affecting the Corrosion of Engineering Materials,* 1978, S. K. Coburn, Editor.
- *STP 767, Atmospheric Corrosion of Metals,* 1982, S. W. Dean, Jr. and E. C. Rhea, Editors.
- *STP 965, Degradation of Metals in the Atmosphere,* 1988, S. W. Dean, Jr. and T. S. Lee, Editors.
- *STP 1239, Atmospheric Corrosion,* 1995, W. W. Kirk and Herbert H. Lawson, Editors.
- *STP 1399, Marine Corrosion in Tropical Environments,* 2000, S. W. Dean, Jr., Guillermo Hernandez-Duque Delgadillo, and James B. Bushman, Editors.

The present volume can be viewed as the most recent in a series on a topic of continuing economic and ecological significance. As previously discussed (see "Extending the Limits of Growth through Development of Corrosion-Resistant Steel Products," *Corrosion,* Vol. 55, No. 6, 1999, 547–553), controlling losses of the world's resources due to atmospheric corrosion may be an important component of continuing economic development. Four major themes are evident in this collection.

Prediction of Outdoor Corrosion Performance

One theme focuses on prediction of atmospheric corrosion performance from climatic data, particularly in relation to methods being developed by the International Standards Organization (ISO). These attempt to classify the corrosivity of a location based either on short-term exposure of standard coupons, or on local time of wetness, and deposition rates of chloride and sulfate. Many of the assumptions in developing the ISO methodology are now being reconsidered in the light of recently completed testing, and work continues to improve the models.

Laboratory and Specialized Outdoor Test Methods

A second theme considers laboratory tests related to outdoor corrosion, and specialized outdoor methods. These include methods of evaluating the results of outdoor tests, ways to predict outdoor performance based on laboratory tests, and on work to develop a seaside (salt-resistant) steel by additions of calcium and sulfur.

Effects of Corrosion Products on the Environment

A third theme examines the ecological effects of corrosion product runoff, a subject that blends corrosion science, environmental technology, analytical chemistry and politics. Contributions from the Swedish Royal Institute of Technology, and the US Department of Energy reflect a growing concern in developed countries for the ecological effects of dissolved metals.

Long-Term Outdoor Corrosion Performance of Engineering Materials

The fourth theme is the documentation of the actual long-term outdoor behavior of engineering materials. This topic includes reports of the 21-year results of the 1976 ASTM International outdoor atmospheric corrosion test program on nickel alloys, Galvalume, galvanized, and aluminum-coated steel sheet. Articles on the performance of unpainted, low-alloy weathering steel include a survey of utility poles in a wide range of environments, work to establish a lean-alloy (Cu-P) grade as an inexpensive alternative to A588A, and the development of a new ASTM G101 corrosion index for estimating relative corrosion resistance from composition.

I am indebted to many for support and to the success of the symposium and this book. These include the members of the Atmospheric Corrosion Subcommittee G1.04, symposium co-chairman Robert Baboian, a plethora of skilled reviewers, the presenters and authors of a large number of high-quality papers, and the help of ASTM International staff including Dorothy Fitzpatrick, Annette Adams, and Maria Langiewicz. This book, like the symposium, is dedicated to the memory of Seymour Coburn, a pioneer in the development of weathering steels, and an active contributor to the efforts of ASTM International in the field of outdoor atmospheric corrosion.

Herbert E. Townsend
Consultant
Center Valley, PA
symposium co-chair and editor

PREDICTION OF OUTDOOR CORROSION PERFORMANCE

Sheldon W. Dean[1] and David B. Reiser[2]

Analysis of Long-Term Atmospheric Corrosion Results from ISO CORRAG Program

Reference: Dean, S. W. and Reiser, D. B., "**Analysis of Long-Term Atmospheric Corrosion Results from ISO CORRAG Program**," *Outdoor Atmospheric Corrosion, ASTM STP 1421*, H. Townsend Ed., American Society for Testing and Materials International, West Conshohocken, PA, 2002.

Abstract: A series of regression analyses was made on the multi-year corrosion losses of panels of steel, zinc, copper, and aluminum in the ISO CORRAG program. In every case, the only sites selected for the analyses were sites with all four exposures reported and complete data sets on the time of wetness, sulfur dioxide, and chloride deposition. The regressions with significant R values were then selected for further analyses. The time exponent and one-year corrosion coefficient were regressed against the environmental variables. None of the exponent regressions showed large environmental effects. The steel exponent was increased by chloride deposition and time of wetness. The copper exponent was increased by increasing time of wetness and decreased by increasing chloride. Neither zinc nor aluminum exponents showed significant effects from the environmental data. The best environmental regressions were only able to predict the measured corrosion losses to within a factor of two for steel, zinc, and copper. The aluminum loss predictions were worse. Some other environmental variables will need to be found to improve this approach to predicting atmospheric corrosion.

Keywords: atmospheric corrosion, time of wetness, chloride deposition, sulfation, sulfur dioxide deposition, ISO CORRAG program, regression analysis, time exponent

Introduction

Atmospheric corrosion is a major problem in the application of engineering metals in many types of service. This form of deterioration has been noted from antiquity, but the development of modern smelting and refining operations of steel has made the economic consequences of atmospheric corrosion very significant in modern times. As a result, there has been an ongoing effort to understand this phenomenon and to develop standards that can be used to predict the severity of the process in service [*1*].

These concerns caused the International Organization for Standardization (ISO), at the organization meeting of Technical Committee 156 in Riga, Latvia in 1976, to identify atmospheric corrosion as a priority area for standards development. At the next meeting

[1] President, Dean Corrosion Technology, 1316 Highland Court, Allentown, Pennsylvania, 18103.
[2] Lead Materials Engineer, Corporate Engineering Department, Air Products and Chemicals, Inc. 7201 Hamilton Boulevard, Allentown, Pennsylvania, 18195-1501.

of TC156 in Borås, Sweden in 1978, the committee decided to form a working group (TC156/WG4) to develop standards for the classification of corrosion under the leadership of members from the Czech Republic. As a result of this effort, four standards were promulgated: ISO 9223, 9224, 9225, and 9226. These standards were based on an extensive review of atmospheric corrosion results in Europe and North America [*2*].

The ISO CORRAG Collaborative Exposure Program was instituted in 1986 for the purpose of establishing a worldwide program through ISO/TC156 that would use consistent standards, uniform exposure times, and standard materials. In addition, data was to be obtained on temperature, humidity, sulfur dioxide concentrations, and chloride deposition at 51 sites. Mass loss data was to be obtained on four metals: carbon steel, zinc, copper, and aluminum using flat panels, 100 x 150 mm, and wire helices, 2-3 mm diameter and 1 m long. Specimen removals were planned with six removals after one-year exposures, one two-year, one four-year, and one eight-year. Three replicate specimens of each metal and specimen type were to be removed at the end of each interval. The one-year specimen exposures were to be spaced at six-month intervals [*3*].

The program has now been closed, and the results are being analyzed. Several studies have been published comparing the relative performance of metals at different sites [*4*]. The comparability of the panels and helices [*5*] and the predictability of corrosion rates based on atmospheric variables have been published [*6*]. However, these studies have focused on the one-year results and little attention has been given to the multi-year specimens. The purpose of this paper is to examine the multi-year exposure data to understand better the kinetics of the process and to determine to what degree the atmospheric variables of time of wetness, sulfation, and chloride deposition can be used to predict multi-year corrosion.

Procedures and Results

Input Data

The ISO CORRAG program has been described in detail earlier. The program consisted of six one-year exposures of flat panels (100x150x2 mm) and helix specimens beginning every six months for three years. Multi-year exposures of two, four, and eight years were initiated at the beginning of the exposure period. Triplicate specimens were used for each exposure. The metals selected were a low carbon steel from a single supplier and commercially pure zinc, copper, and aluminum. These nonferrous metals were obtained from local sources in each of the participating nations. There were 51 sites in 14 nations at the end of the program. The program was initiated in 1986 and officially closed in 1998. At the conclusion of each exposure, the specimens were retrieved and sent to the laboratory that had done the initial weighing for cleaning and evaluation. Mass loss values were obtained and converted to corrosion thickness loss values in µm units. The results from the various sites have been collected and tabulated by the Czech member, SVUOM, and reported previously [*6*].

Data Analysis

The mass loss values were averaged for each exposure. In the case of the one-year results, the averages of the data from all six exposures were used in this study. Average values were calculated for time of wetness (TOW), hrs./year, sulfur dioxide concentration (SO_2), mg/m^3, and chloride deposition rate (Cl), mg/m^2 day for the eight-year period. Only the sites with complete data on these variables were included in this study.

Regression analyses were carried out for the flat panel specimens at each site. The mass loss data was converted to logarithmic values (base 10) and regressed against the logarithmic exposure time in years as the independent variable. Previous studies have found that atmospheric corrosion kinetics follow a power law relationship

$$M = aT^b \tag{1}$$

where M = mass loss per unit area,
T = exposure time,
a = mass loss in the first year, and
b = mass loss time exponent (referred to as "slope").

This expression becomes as follows after the logarithmic conversion

$$\log M = a' + b \log T \tag{2}$$

where $a' = \log a$ (referred to as "intercept").

The Microsoft Excel 2000 spreadsheet program was used to carry out the regressions. The correlation coefficient, R, is a measure of the goodness of fit of a regression, and the value of R^2 represents the fraction of total variance of the data explained by the regression. For this study, there were only four exposure periods so that the degrees of freedom of the regression are two. The minimum value of R^2 for a 5% significance level (95% confidence level) is 0.83. The regressions with values below this level were excluded from the analysis. This left 22 sites for steel, 23 for zinc and copper, and 21 for aluminum. The results of these regressions are plotted in Figure 1.

The values of a′ and b from these regressions were then averaged and the standard deviations were calculated for each metal and are shown in Table 1. The values were plotted on probability paper to determine whether the values were normally distributed. Correlation analyses were performed to determine if there was any correlation between the a′ and b values. The results of these analyses indicated that the distribution was normal, and there was no significant correlation between a′ and b. The correlation coefficients are reported in Table 1.

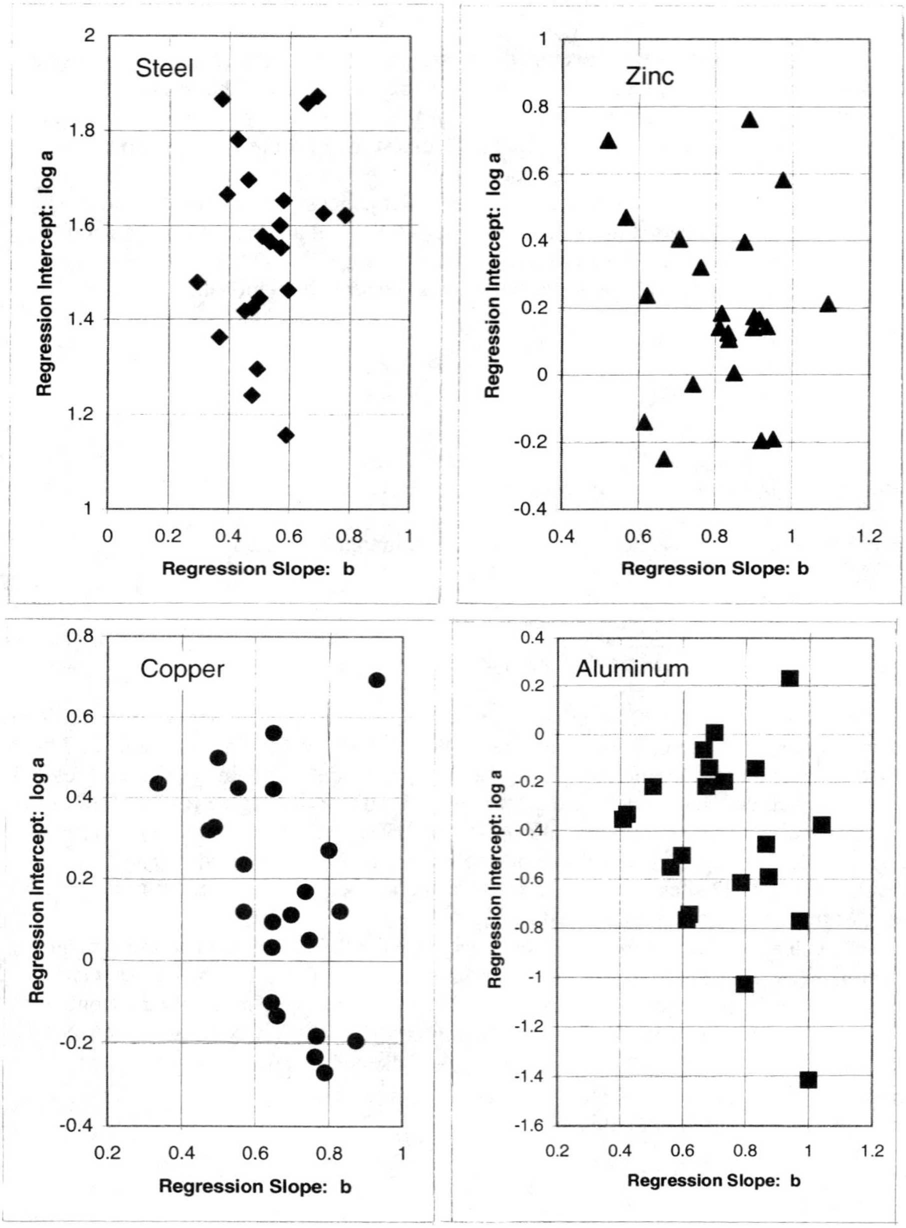

Figure 1 *Results of regression analyses on mass loss vs. exposure time using logarithmic conversion of the data.*

Table 1 - *Regression summary for log mass loss vs. log exposure time*

	Metal			
	Fe	Zn	Cu	Al
Slope Ave	0.523	0.813	0.667	0.728
SD	0.121	0.143	0.142	0.181
Log Int Ave	1.556	0.194	0.160	-.439 (1)
SD	0.198	0.272	0.264	0.376
N	22	23	23	21
R	.186	0.125	0.322	-0.222

(1) Plot of points on probability paper shows abnormality at ends
R=Correlation coefficient between slope and Log Int values
(>0.423 at 95% confidence level, DF=20)

Regression analyses were then carried out to determine to what degree the measured environmental variables affected the a′ and b values. Previous studies have shown that the environmental variables have a strong effect on the a′ value, but there have been no studies of environmental variables on the b value. The results of these regressions are shown in Tables 2A-D. In each case, regressions were made using all three variables and then two at a time and finally single variable regressions. The reason for this procedure was to attempt to eliminate variables that are not significant contributors to the relationship.

Finally, it was desired to determine how closely the best expression was to matching the measured eight-year corrosion losses. For each of the four metals, the regression analyses for slope and intercept that yielded the smallest standard error were chosen. The environmental variables were used in these expressions to predict the eight-year metal loss. These calculations were made in each case. The first was based on the regression expression used to fit the data from the exposures. The second was based on environmental measurements and used predicted values for slope and intercept based on Tables 2A-D. The third calculation used the measured one-year value and a predicted slope from the regression in Tables 2A-D and environmental data. These results are shown in Figure 2.

Discussion

Equation 1 has been widely used to describe the atmospheric corrosion kinetics [*7*]. The "a" term represents the corrosion loss in one year, while the "b" term represents the long-term performance with "b" values less than one in most cases. Previous studies [*6*] have focused on the effects environmental variables have on the one-year results but have not considered the longer-term performance. Townsend [*8*] has examined the performance of weathering steel and has discovered that alloying elements can, in some cases, change the "b" value significantly. The lower the "b" is, the more protective the corrosion product layer on the metal surface. The results in Figure 1 demonstrate clearly that the "b" values show a significant variation for all four alloys. It was of interest to try to understand how environmental variables affect the "b" value in this case of a single composition exposure.

Table 2A - *Summary of regression analyses on slope and intercept values with environmental variable steel - 22 data points.*

	Slope Regressions									
				SO_2 (1)		TOW (2)		Cl (3)		
Regression Variables	R^2	SE	F	Coef	t	Coef	t	Coef	t	Int(4)
SO_2,TOW, Cl	**0.236**	0.114	1.85	**2.16**	0.23	6.22	**2.33**	-1.37	-0.80	0.319
SO_2, TOW	**0.209**	0.113	2.51	**2.20**	0.24	5.24	**2.24**	--	--	0.341
TOW, Cl	**0.234**	0.111	2.90	--	--	6.19	**2.38**	-1.37	-0.82	0.327
SO_2, Cl	0.005	0.126	0.04	0.85	0.08	--	--	0.49	0.29	0.516
SO_2	0.000	0.124	0.01	0.74	0.07	--	--	--	--	0.520
TOW*	**0.207**	0.110	**5.21**	--	--	5.20	**2.28**	--	--	0.348
Cl	0.004	0.123	0.09	--	--	--	--	0.48	0.29	0.519

	Intercept Regressions									
				SO_2 (1)		TOW (2)		Cl (3)		
Regression Variables	R^2	SE	F	Coef	t	Coef	t	Coef	t	Int(4)
SO_2,TOW, Cl	**0.526**	0.147	**6.66**	41.9	**3.50**	2.37	0.69	4.92	**2.22**	1.316
SO_2, TOW	**0.395**	0.161	**6.22**	41.7	**3.17**	5.93	1.77	--	--	1.238
TOW, Cl	**0.204**	0.185	2.44	--	--	1.64	0.38	4.88	1.75	1.461
SO_2, Cl*	**0.514**	0.145	**10.03**	41.4	**3.51**	--	--	5.63	**2.92**	1.391
SO_2	**0.296**	0.170	**8.41**	40.1	**2.90**	--	--	--	--	1.441
TOW	0.076	0.195	1.65	--	--	5.17	1.28	--	--	1.383
Cl	**0.198**	0.181	**4.95**	--	--	--	--	5.37	**2.23**	1.512

(1) - SO_2 is average SO_2 concentration in 8 years mgSO_2/m^3; coefficient multiplied by 10^{-4}
(2) - TOW is average time of wetness, hrs. per year when t >0°C, RH >80%, 8 years, x 10^{-5}
(3) - Cl is average deposition rate, mg Cl/m^2day for 8 years, coefficients multiplied by 10^{-4}
(4) Int is the intercept value for regression. (log of metal loss in µm)
R^2 is the square of the multiple correlation coefficients.
SE is the standard error of the regression.
F is the ratio of regression variance to residual variance.
t is the ratio of coefficient value to its standard deviation..
Bold and underlined values are significant at the 95% CL.
* Regressions used for Figure 2 calculations

Table 2B - *Summary of regression analyses on slope and intercept values with environmental variable zinc - 23 data points*

	Slope Regressions									
				SO_2 (1)		TOW (2)		Cl (3)		
Regression Variables	R^2	SE	F	Coef.	t	Coef	t	Coef	t	Int (4)
SO_2,TOW, Cl	0.171	0.140	1.31	7.72	1.01	-1.48	-0.47	3.77	1.81	0.804
SO_2, TOW	0.029	0.147	0.30	5.15	0.65	1.08	0.36	--	--	0.758
TOW, Cl	0.127	0.140	1.46	--	--	-0.98	-0.31	3.38	1.65	0.817
SO_2, Cl	0.162	0.137	1.93	7.15	0.96	--	--	3.33	1.82	0.759
SO_2	0.022	0.144	0.48	5.40	0.70	--	--	--	--	0.794
TOW	0.009	0.145	0.18	--	--	1.25	0.43	--	--	0.769
Cl*	0.123	0.137	2.94	--	--	--	--	3.10	1.71	0.786

	Intercept Regressions									
				SO_2 (1)		TOW (2)		Cl (3)		
Regression Variables	R^2	SE	F	Coef	t	Coef	t	Coef	t	Int(4)
SO_2,TOW, Cl	**0.405**	0.226	**4.32**	42.6	**3.43**	0.64	0.12	3.68	1.09	0.007
SO_2, TOW	**0.368**	0.227	**5.82**	40.1	**3.26**	3.15	0.69	--	--	0.053
TOW, Cl	0.037	0.280	0.38	--	--	3.44	0.55	1.53	0.37	0.061
SO_2, Cl*	**0.405**	0.220	**6.80**	42.8	**3.59**	--	--	3.87	1.31	0.012
SO_2	0.353	0.224	**11.46**	40.8	**3.38**	--	--	--	--	0.053
TOW	0.030	0.274	0.65	--	--	4.46	0.80	--	--	0.040
Cl	0.022	0.275	0.48	--	--	--	--	2.52	0.69	0.172

(1) - SO_2 is average SO_2 concentration in 8 years mgSO_2/m^3; coefficient multiplied by 10^{-4}
(2) - TOW is average time of wetness, hrs. per year when t >0°C, RH >80%, 8 years, x 10^{-5}
(3) - Cl is average deposition rate, mg Cl/m^{2}day for 8 years, coefficients multiplied by 10^{-4}
(4) Int is the intercept value for regression.
R^2 is the square of the multiple correlation coefficients.
SE is the standard error of the regression.
F is the ratio of regression variance to residual variance.
t is the ratio of coefficient value to its standard deviation..
Bold and underlined values are significant at the 95% CL.
* Regressions used in Figure 2 calculations

Table 2C - *Summary of regression analyses on slope and intercept values with environmental variable copper - 23 data points*

	Slope Regressions									
				SO_2 [1]		TOW [2]		Cl [3]		
Regression Variables	R^2	SE	F	Coef.	t	Coef	t	Coef	t	Int [4]
SO_2,TOW, Cl	**0.204**	0.136	1.62	2.15	0.29	4.71	1.67	-3.95	1.96	0.531
SO_2, TOW	0.043	0.145	0.45	5.33	0.69	1.72	0.68	--	--	0.589
TOW, Cl*	**0.200**	0.133	2.50	--	--	4.78	1.74	-4.08	**2.12**	0.537
SO_2, Cl	0.087	0.142	0.95	3.26	0.42	--	--	-2.12	1.19	0.676
SO_2	0.021	0.143	0.46	5.14	0.68	--	--	--	--	0.650
TOW	0.021	0.144	0.44	--	--	1.65	0.66	--	--	0.609
Cl	0.078	0.139	1.78	--	--	--	--	-2.27	1.34	0.688

	Intercept Regressions									
				SO_2 [1]		TOW [2]		Cl [3]		
Regression Variables	R^2	SE	F	Coef.	t	Coef	t	Coef	t	Int [4]
SO_2,TOW, Cl*	**0.500**	0.205	**6.35**	19.25	1.72	4.85	1.14	8.58	**2.81**	-0.163
SO_2, TOW	**0.293**	0.238	4.14	12.36	0.98	11.35	**2.74**	--	--	-0.288
TOW, Cl	**0.422**	0.215	**7.31**	--	--	5.50	1.24	7.42	**2.38**	-0.111
SO_2, Cl	**0.466**	0.207	**8.74**	20.40	1.82	--	--	10.47	**4.06**	-0.014
SO_2	0.027	0.273	0.59	11.10	0.77	--	--	--	--	0.114
TOW	**0.259**	0.238	**7.33**	--	--	11.20	**2.71**	--	--	-0.242
Cl	**0.378**	0.218	**12.76**	--	--	--	--	9.51	**3.57**	0.062

(1) - SO_2 is average SO_2 concentration in 8 years $mgSO_2/m^3$; coefficient multiplied by 10^{-4}
(2) - TOW is average time of wetness, hrs. per year when $t > 0°C$, $RH > 80\%$, 8 years, x 10^{-5}
(3) - Cl is average deposition rate, mg Cl/m^2day for 8 years, coefficients multiplied by 10^{-4}
(4) Int is the intercept value for regression.
R^2 is the square of the multiple correlation coefficients.
SE is the standard error of the regression.
F is the ratio of regression variance to residual variance.
t is the ratio of coefficient value to its standard deviation..
Bold and underlined values are significant at the 95% CL.
* Regressions used in Figure 2 calculations

Table 2D - *Summary of regression analyses on slope and intercept values with environmental variable aluminum - 21 data points*

	Slope Regressions									
				SO_2 (1)		TOW (2)		Cl (3)		
Regression Variables	R^2	SE	F	Coef	t	Coef	t	Coef	t	Int (4)
SO_2,TOW, Cl	0.008	0.196	0.05	5.29	0.05	1.00	0.20	-1.80	-0.35	0.704
SO_2, TOW	0.001	0.191	0.01	1.15	0.11	-0.29	-0.09	--	--	0.734
TOW, Cl	0.008	0.190	0.07	--	--	1.03	0.21	-1.84	-0.38	0.705
SO_2, Cl	0.006	0.191	0.06	7.61	0.07	--	--	-1.05	-0.31	0.733
SO_2	0.001	0.186	0.01	1.17	0.11	--	--	--	--	0.724
TOW	0.000	0.186	0.01	--	--	-0.29	-0.10	--	--	0.738
Cl	0.006	0.185	0.11	--	--	--	--	-1.09	-0.33	0.735

	Intercept Regressions									
				SO_2 (1)		TOW (2)		Cl (3)		
Regression Variables	R^2	SE	F	Coef.	t	Coef	t	Coef	t	Int(4)
SO_2,TOW, Cl	**<u>0.430</u>**	0.307	4.28	56.16	**<u>3.33</u>**	7.41	-0.95	14.05	1.77	0.467
SO_2, TOW	**<u>0.326</u>**	0.324	4.35	51.29	**<u>2.92</u>**	2.65	0.47	--	--	0.701
TOW, Cl	0.058	0.383	0.55	--	--	4.62	-0.48	9.71	0.99	0.348
SO_2, Cl*	**<u>0.400</u>**	0.306	**<u>6.00</u>**	54.44	**<u>3.26</u>**	--	--	8.53	1.57	0.679
SO_2	0.318	0.317	**<u>8.84</u>**	51.14	**<u>2.97</u>**	--	--	--	--	0.609
TOW	0.007	0.383	0.12	--	--	2.35	0.35	--	--	0.521
Cl	0.046	0.375	0.91	--	--	--	--	6.31	0.96	0.483

(1) - SO_2 is average SO_2 concentration in 8 years mgSO_2/m^3; coefficient multiplied by 10^{-4}
(2) - TOW is average time of wetness, hrs. per year when t >0°C, RH >80%, 8 years, x 10^{-5}
(3) - Cl is average deposition rate, mg Cl/m^{2}day for 8 years, coefficients multiplied by 10^{-4}
(4) Int is the intercept value for regression.
R^2 is the square of the multiple correlation coefficients.
SE is the standard error of the regression.
F is the ratio of regression variance to residual variance.
t is the ratio of coefficient value to its standard deviation..
<u>Bold and underlined</u> values are significant at the 95% CL.
* Regression used in Figure 2 calculations. Slope value was average from Table 1.

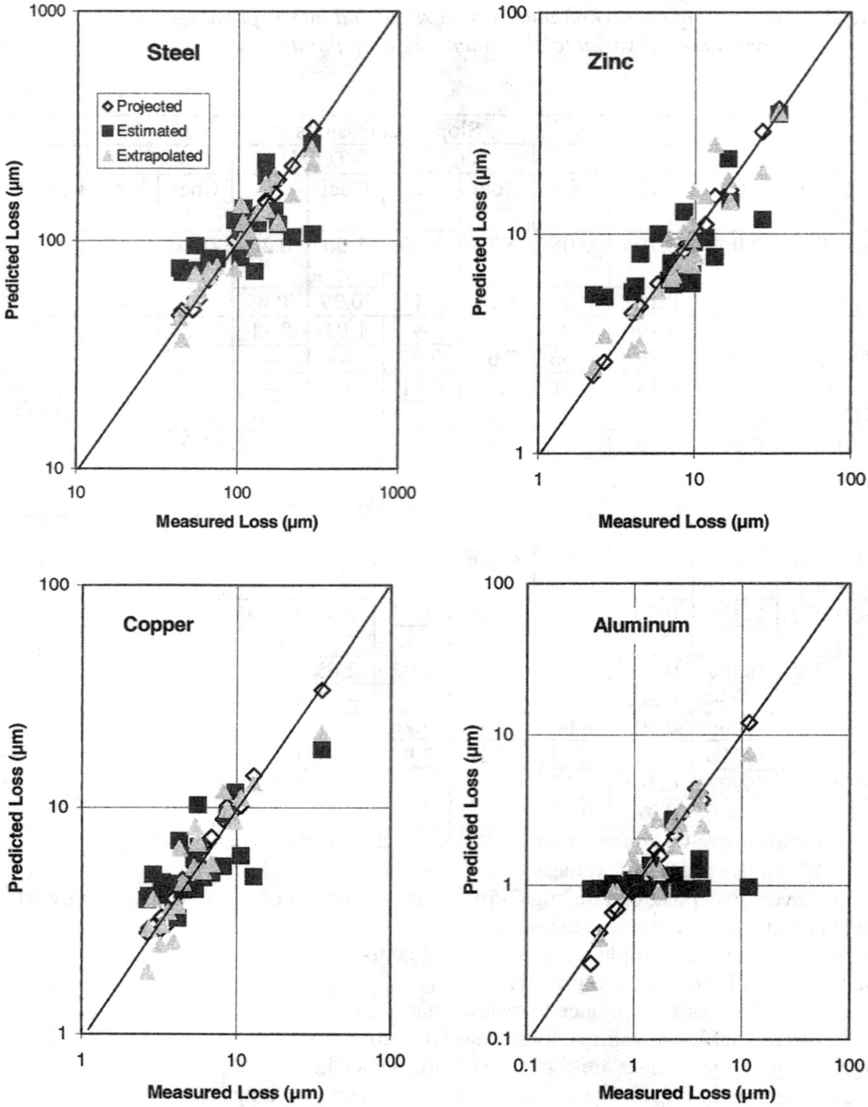

Figure 2 - *Comparison between actual eight-year loss and values calculated from regression analyses for steel.*

◇ Eight-year projected using regression results shown in Figure 1

■ Eight-year estimated using regression results in Tables 2A - 2D at lowest SE

▲ Eight-year extrapolated using slope regression from Tables 2A - 2D at lowest SE and 1-year measured rate

The results in Table 1 provide further insight into the scope and nature of this variation. In all cases, the variation in slope values showed normal behavior when probability paper plots were examined.

None of the correlations between the slope and intercept values were significant. However, it was of interest to note that the correlation coefficients in the cases of zinc, copper, and aluminum were negative, and this suggests a mechanism whereby an initial high corrosion rate contributes a more protective corrosion product layer, thus suppressing corrosion at a later time. The very weak correlation makes this concept very tentative, except that three of four data sets showed similar behavior.

The large variation in slope values has been observed before in the case of aluminum [*9*]. It is not clear why this parameter should exhibit such variability. In the case of weathering steels, the alloying elements do affect the slopes, but in the present study, there is little variation in alloy content to affect the performance. All the steel samples came from the same lot of metal, so no alloy variation is available to explain the variation in slope. The other metals were of commercial purity, and it is unlikely there were significant variations in alloy content.

It was of interest to examine the data set to determine if the measured environmental parameters caused significant variation in the slope values that were measured. These regressions are shown in Tables 2A-D. Six regressions were carried out looking at the three environmental parameters separately and in all combinations. This procedure has the advantage of revealing cases where nonrandom interactions between the environmental variables cause effects to look significant spuriously. When two effects have opposite signs in multiple regressions and look to be significant but become nonsignificant when examined separately, one should be suspicious of a false positive conclusion. The only case where this set of circumstances was seen was in the copper slope regression, Table 2C. Chloride and time of wetness showed this type of behavior with both effects becoming nonsignificant in single variable regressions. This makes any conclusion regarding these variables speculative on the basis of the data considered.

In reviewing the steel results in Table 2A, it is clear that the environmental data had a very small affect in reducing the variance of the slope variable. The R^2 values were barely significant at the 95% confidence level, and only the time of wetness showed a significant effect. The best regression in terms of producing the lowest standard error of the slope also gave the highest F value. This regression was the single variable time of wetness expression, but the R value was barely significant compared to random error at the 95% confidence level. It is of interest to note that the TOW effect is positive for steel, i.e. higher TOW causes the rust layer to be less protective. The other environmental effects do not appear to be significant in affecting the slope.

The intercept regressions showed much larger R^2 values and both sulfation and chloride deposition showed significant effects. The best regression in terms of minimizing the standard error was the two-variable regression with SO_2 and Cl. This also gave the highest F value. There is a comparison of the multi-year, three-variable intercept value to the previously determined single-year regressions in Table 3. In this case, none of the differences were significant, although the numbers may look somewhat different. The single-year regressions were based on 32 sites, while the multi-year regression was based on 22 sites. Therefore, the single-year values are probably more reliable.

Table 3 - *Comparison of intercept values from multi-year regression to one-year regression; three variable regressions.*

	Fe			Zn			Cu			Al		
Variable	MY	SY	δ/SE	MY	SY	δ/SE	MY	SY	δ/SE	MY	SY	δ/SE
SO_2	4.19	2.94	1.04	4.26	2.98	1.03	1.92	1.30	0.56	5.62	5.02	0.36
TOW	2.37	7.07	1.37	0.64	6.08	1.07	4.85	2.46	0.56	-7.41	3.26	1.37
Cl	4.92	8.34	1.55	3.68	7.18	1.04	8.58	8.82	0.08	14.05	6.71	0.92
int	1.32	1.17	1.25	0.01	0.11	0.58	0.163	0.076	0.60	0.467	0.468	0.00

SO_2 values x 10^{-3},
TOW values x 10^{-5},
Cl values x 10^{-4},
SY = Single year regression *[5]*,
MY = Multi-year regressions,
SE = Standard error of MY regression coefficient,
int = Log base 10 of intercept (corrosion loss in μm), and
δ = |MY-SY|.

In the case of zinc, none of the slope regressions were significant. This suggests that the environmental variables do not affect the protectiveness of the corrosion product layer to a significant degree. The intercept regressions also were not as strong as was seen with steel, but the sulfation effect showed consistently significant values. The chloride-sulfation regression gave the lowest standard error and highest R^2 value while the sulfation regression gave the largest F value. The comparison between the single-year and multi-year effects for zinc is shown in Table 3 and again, the differences are not significant. However, the single-year results are probably more reliable.

In the case of copper (Table 2C) the slope regression with time of wetness and chloride gave the lowest standard error and highest F value. The R^2 was significant at the 95% confidence level, but only chloride was significant. It is important to note that the TOW effect was positive as seen with steel, suggesting that the corrosion products were less protective at high TOW value. The chloride effect was negative for all the slope regressions suggesting that chloride somehow makes the corrosion products more protective. The copper intercept regressions, also shown in Table 2C, showed minimum standard error with all three environmental variables. However, only chloride appeared to be significantly greater than zero as seen by the relatively low "t" values for the other variables. Of the three variables, the SO_2 effects were the least significant in improving the data fit. Both chloride deposition and time of wetness were significant in most of the regressions. It should be noted that the sign of the intercept effect of chloride is positive while the slope effect is negative. This means that chloride initially accelerates the corrosion but ultimately reduces the rate. For example, at the Kure Beach 250m site, the time it would take for a copper panel to reach a rate equivalent to no chloride exposure would be 4.8 years. Sites with lower chloride levels would reach that point in a shorter time. It is of interest to note that the TOW effect was much smaller in the single-variable regression suggesting that this effect may be spuriously large in the smaller data set.

The aluminum slope regressions are shown in Table 2D. None of those regressions were significant suggesting that the slope values are not strongly affected by environmental variations. This may be a result of the inherently different corrosion process in the case of aluminum. Aluminum tends to corrode by a pitting mechanism rather than general corrosion that builds a corrosion product with increasing thickness. The exponent in Equation 1 reflects the pit geometry rather than the corrosion product protectiveness, and this may explain why the slope regressions show no significant environmental effects.

The intercept regressions for aluminum were not very effective in explaining the variance in this variable. The regression that produced the lowest standard error was the SO_2, Cl, two-variable regression. This regression showed a significant R^2 value and F value. There was close agreement between the SO_2 effects in the single-year and multi-year regressions, but the other two variables showed rather large discrepancies. This was not unexpected because of the rather unpredictable nature of aluminum atmospheric corrosion. Because of the random localized nature of the corrosion process, the measured rates are much more subject to random variations.

Although the behavior of these regressions analyses can be inferred from the calculated statistics of R^2, SE, and F, it is instructive to examine how these regressions would predict the mass loss values at the various sites, and compare these predictions to the measured results after eight years of exposure. These values are shown in Figure 2 for all the sites used in this study. The projected values were based on the best fit of the four exposures to Equation 1. The estimated values were based on the regressions for slope and intercept giving the smallest standard error as shown in Tables 2A-D. In the case of aluminum, the slope values used were the average slope from Table 1 since none of the regressions were significant and the standard error of the slope expression was greater than the standard deviation of the slope shown in Table 1.

The extrapolated values shown in Figure 2 were based on the measured one-year measured corrosion loss and the slope estimates from Tables 2A-D using the expressions giving the smallest standard error as with the estimated values. It was desired to show this comparison in order to evaluate the accuracy of the slope projection as a way of estimating corrosion losses when one-year exposure data is available. The ISO classification method recommends obtaining one-year exposure data as a preferred way to determine site corrosion class, so it was of interest to examine to what degree this extrapolation method would better approximate long-term results.

The results in Figure 2 clearly show that the projections were close to the measured values in most cases, but the estimated values showed dramatic variation from the measured values and, in many cases, deviated significantly for the measured values. In order to make this conclusion more quantitative, the ratios of projected-to-actual and estimated-to-actual values were calculated and the standard deviations (SD) of these ratios were then computed. These values are shown below in Table 4.

Table 4 - *Standard deviation of ratios of projected and estimated results to actual values.*

Metal	Projected SD	Estimated SD	Estimated SD/Mean	Extrapolated SD	Extrapolated SD/Mean
Steel	0.043	0.395	0.358	0.193	0.194
Zinc	0.052	0.496	0.455	0.297	0.284
Copper	0.074	0.404	0.372	0.266	0.258
Aluminum	0.112	0.614	0.701	0.398	0.367

These results demonstrate that the power law model used to fit the results is reasonably close to the measured values at eight years and should be within ±10% in 95% of the cases, except for aluminum.

On the other hand, the use of the environmental parameters that were chosen for this study only allows a much lower degree of accuracy in predicting long-term atmospheric corrosion behavior. In the cases of steel, zinc, and copper, the prediction using environmental variables is within a factor of two of the measured value in almost all cases. Aluminum is significantly worse with estimates more than a factor of two in many cases.

The extrapolated corrosion losses using the one-year loss and calculated "b" values are clearly more accurate than the procedure that estimated both the slope and intercept. However, it gives from three to five times the error seen in the projected values.

These facts suggest that other environmental variables may be necessary to describe atmospheric corrosion rates of engineering metals. An obvious deficiency in the results is the fact that temperature is not included. Because corrosion is a chemical process, there should be a temperature effect. The problem with the measurement of environmental temperature is that the specimens are often dry during the exposure. It would be closer to reality if dew point temperatures were measured and used as the environmental temperature variable. The presence of soluble salts on the metal surface may cause the surface to remain wet above the dew point temperature, but this is a relatively small effect, typically 2-3°C for dryness at 80% relative humidity. It would be interesting to see if a regression analysis using dew point temperature or the reciprocal of absolute dew point temperature would give better quality results.

Conclusions

- The atmospheric corrosion loss of all four metals investigated showed power law kinetics with the exponent varying about ±20% around the average. The one-year corrosion rate coefficient of the expression did not show a significant correlation with the exponent for any of the metals. However, there was an indication in the cases of zinc, copper, and aluminum that the correlation might have been negative in a larger study. Statistical analyses indicated that the distributions of both coefficient (intercept) and exponent (slope) were normal.

- Regression analyses of the slope and intercept values against the environmental factors of time of wetness, sulfation, and salinity indicated the following. For all four metals, the slope regressions were barely or not at all significant with environmental variables. On the other hand, the intercept regressions were very significant in most cases. For steel, only time of wetness seemed to affect the slope while sulfation and salinity strongly affected the intercept (coefficient) values. In the case of zinc, none of the environmental variables were significant in the slope regressions, but sulfation was a significant contributor in the intercept regressions. In the case of copper, the slope regressions showed significant effects from both chloride and time of wetness. The chloride effect was negative indicating that chloride made the corrosion products more protective. Both chloride deposition and time of wetness were significant in affecting the intercept value. None of the environmental variables affected the aluminum slope value significantly. Chloride and SO_2 affected the intercept regression but, in this case, the significance was less than the other three.

- Comparison of the intercept values from this study to previously calculated results from the one-year exposures in environmental factor regressions gave comparable results, but the earlier study is probably more reliable because it included a larger selection of sites.

- A comparison of the estimated eight-year results for all four metals based on environmental data to the actual results showed that for steel, zinc, and copper the results were within a factor of two. Aluminum showed greater deviation. This suggests that some other variable should be included. Dew point temperature is suggested as an additional variable that could be included in the regression analyses.

Acknowledgment

The authors gratefully acknowledge Air Products and Chemicals, Inc. for support and permission to publish this paper. In addition, the authors wish to acknowledge the work of Dr. Dagmar Knotkova and her colleagues at SVUOM, together with all the participants in the ISO CORRAG program for their efforts in generating this data compilation and for sharing it.

References

[1] Dean, S. W., "Atmospheric Corrosion - Lessons from the Past and Challenges for the Future," *Corrosion Prevention 94 - Proceedings*, Australasia Corrosion Association, Inc., Melbourne, 28-30 November 1994, Paper 63, pp 1-12.

[2] Dean, S. W., "Analysis of Four Years of Exposure Data from the USA Contribution to ISO CORRAG Program," *Atmospheric Corrosion, ASTM STP 1239*, W. W. Kirk and H. H. Lawson, Eds., American Society for Testing and Materials, West Conshohocken, PA, 1990, pp 163-176.

[3] Dean, S. W. "ISO CORRAG Collaborative Atmospheric Exposure Program: A Preliminary Report," *Degradation of Metals in the Atmosphere, ASTM STP 965*, S. W. Dean and T. S. Lee, Eds., American Society for Testing and Materials, West Conshohocken, PA 1988, pp. 385-431.

[4] Dean, S. W. and Reiser, D. B., "Analyses of Data from ISO CORRAG Program," Paper No. 340, *Corrosion 98,* NACE, Houston, TX, 1998.

[5] Dean, S. W. and Reiser, D.B. "Comparison of the Atmospheric Corrosion Rates of Wires and Flat Panels," Paper No. 455, *Corrosion 2000*, NACE, Houston, TX, 2000.

[6] Knotkova, D., "ISO CORRAG International Testing Program in the Frame of ISO/TC156/WG4 Classification of Corrosivity of Atmospheres," National Research Institute for the Protection of Materials, 170 04 Praha 7, Czech Republic, 1993.

[7] Dean, S. W. "Corrosion Tests and Metals Under Natural Atmospheric Conditions," *Corrosion Testing and Evaluation: Silver Anniversary Volume ASTM STP 1000*, R. Baboian and S. W. Dean, Eds., American Society for Testing and Materials, West Conshohocken, PA 1990, pp. 163-176.

[8] Townsend, H. E., "The Effects of Alloy Elements on the Corrosion of Steel in Industrial Atmospheres," *Proceedings of the 14th International Corrosion Congress*, Corrosion Institute of South Africa, Kelvin (1999).

[9] Dean, S. W. and Anthony W. H., "Atmospheric Corrosion of Wrought Aluminum Alloys During a Ten-Year Period," *Degradation of Metals in the Atmosphere, ASTM STP 965*, S. W. Dean and T. S. Lee, Eds., American Society of Testing and Materials, West Conshohocken, 1988, pp. 199-205.

Robert D. Klassen, [1] Pierre R. Roberge,[2] Derek R. Lenard,[3] and Geoffrey N. Blenkinsop[4]

Corrosivity Patterns Near Sources of Salt Aerosols

Reference: Klassen, R. D., Roberge, P. R., Lenard, D. R., and Blenkinsop, G. N., **"Corrosivity Patterns Near Sources of Salt Aerosols,"** *Outdoor Atmospheric Corrosion, ASTM STP 1421*, H. E. Townsend, Ed., American Society for Testing and Materials International, West Conshohocken, PA, 2002.

Abstract: Corrosivity patterns near a de-iced highway and a salt-water body were studied with wire-on-bolt coupons and simulations of the airflow. De-iced highways in the winter produce zones where the corrosivity is as high as that near a salt-water body and extend beyond 150 m from the road edge. In a study of the effects of wind sheltering, there was a 34-fold difference in corrosivity between the most wind-protected and the least wind-protected site even though each set was exposed to the same relative humidity. This is consistent with the concept that atmospheric corrosion rates in marine-equivalent environments depend primarily on salt aerosol deposition rates, which in turn depend on local wind velocity and turbulence patterns. At the salt-water body site, there was a seasonal trend to the corrosivity with a three-fold difference between the maximum in December and the minimum in July. The seasonal trend in corrosivity correlated with the seasonal trend in the monthly average relative humidity. The corrosivity pattern around two buildings near the salt-water body was quite non-uniform due to differences in wind speeds.

Keywords: salt aerosols, corrosivity, de-iced highway, wire-on-bolt coupons

---

Introduction

The corrosive effect of salt aerosols that are carried by the wind from sources such as salt-water bodies is well recognized. In the Pacer Lime algorithm, locations that are

---

[1] Research Associate, Dept. Chem./Chem. Eng., Royal Military College of Canada, PO Box 17000, Stn Forces, Kingston, Ontario, Canada K7K 7B4.

[2] Professor, Dept. Chem./Chem. Eng., Royal Military College of Canada, PO Box 17000, Stn Forces, Kingston, Ontario, Canada K7K 7B4.

[3] Scientist, DREA/Dockyard Laboratory (Pacific), Building 199, CFB Esquimalt, PO Box 17000,Stn Forces, Victoria, B.C. V9A 7N2.

[4] Technologist, DREA/Dockyard Laboratory (Pacific), Building 199, CFB Esquimalt, PO Box 17000, Stn Forces, Victoria, B.C. V9A 7N2.

within 4.5 km from a sea are given the highest of four corrosivity ratings without other considerations [1]. In another study, atmospheric damage functions for four metals were developed that each included a term for the chloride deposition rate [2]. The chloride deposition rate was either measured by a salt candle or calculated as wet deposition from rainfall rates and average chloride concentration in precipitation. Also, the ISO atmospheric classification algorithm requires a chloride deposition rate from salt candle measurements as well as time-of-wetness and sulfur dioxide measurements [3].

There are two modes of aerosol deposition that are relevant for particles in the size range of marine-type aerosols: (i) inertial impaction and (ii) turbulent diffusion. Inertial impaction is significant for objects that are small enough not to cause gross changes in the surrounding air flow pattern. Essentially, some of the upstream aerosol particles are not able to follow the flow lines around the object due to inertia. Examples include devices used for measuring atmospheric corrosivity such as CLIMAT coupons [4] and salt candles [3]. The deposition rate due to inertial impaction is

$$R_d = C_\infty U_\infty A_t \eta \tag{1}$$

where R_d is the aerosol deposition rate, C_∞ is the upstream aerosol concentration, U_∞ is the upstream wind velocity, A_t is the projected target area, and η is the capture efficiency of the target. The capture efficiency, η, depends on the upstream fluid velocity at an infinite distance, the target size, the aerosol size and density and the fluid density and viscosity [5]. For practical evaluations, the air velocity two diameters upstream from the surface of a cylinder, termed the approach velocity, may be considered to be an approximation of that at an infinite distance.

Aerosol particles are transported via winds from sources, such as salt-water bodies and deiced highways, by convection and turbulent diffusion. Turbulent diffusion can also deposit particles onto surfaces that are normal to the main airflow, although it is less effective than inertial deposition. The deposition rate by turbulent diffusion is highly dependent on the wind speed and the shape and size of the object as well as aerosol concentration in the bulk stream [6-10].

Experimental

Corrosivity measurements were made by exposing modified CLIMAT [4] or wire-on-bolt units according to the ASTM Standard Practice For Conducting Wire-On-Bolt Test For Atmospheric Galvanic Corrosion (G 116-93). Instead of employing copper, steel and nylon rods to support the aluminum wire, three copper rods were employed for each unit. The percent mass loss of aluminum wire after ninety days of exposure is considered an index of corrosivity. Corrosivity measurements were made near two de-iced highways in the winter, near a set of wind-sheltered microenvironments and also near a salt-water body.

In order to gain insight into the corrosivity patterns for certain geometries, the air flow pattern was modeled with commercial software (Fluent Inc., version 5.2). The air was modeled as a turbulent fluid using the "realizable" k-ε model and "standard" wall functions. The grid size for the control volumes nearest the surfaces was optimised such that the dimensionless distance factors, y+ and y*, were both within the range of 50-500.

The program calculated the air velocity, pressure, turbulent kinetic energy, rate of turbulent energy dissipation and velocity gradients for each control volume.

Corrosivity Near a De-Iced Highway

Seasonal Effects

CLIMAT units were placed at three sites on the campus of the Royal Military College of Canada (RMC) for two seasons, winter and summer. The units at the first site (Point Frederick) were placed 2 m above the ground and about 30 m from a fresh water body (St. Lawrence River). The units at the second site were placed near a highway (#2) 2 m above the ground and about 3 m from the edge of the road. The units at the third site were placed on the roof of a building (Sawyer mod 4). Fig. 1 shows the corrosivity in terms of the average mass loss of aluminum wire. By far the highest response was the site near the road during the winter. This highway is de-iced in the winter with a mixture of sand and salt. Since the CLIMAT unit was outside of the direct splash zone from traffic, the high corrosivity was caused by the generation of salt aerosols by traffic that were then transported by wind.

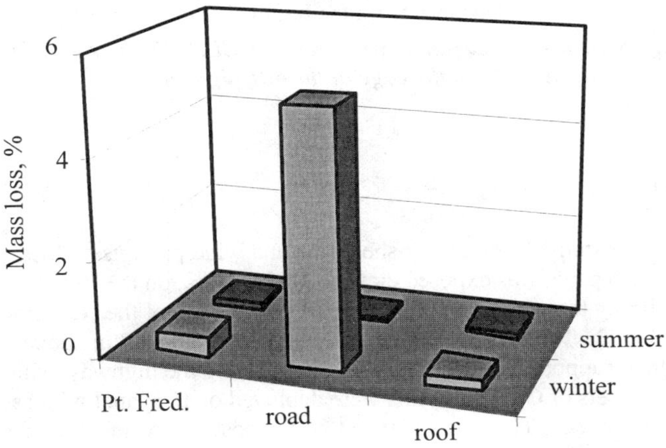

Figure 1. Average m*ass loss of aluminum wire from CLIMAT units at three locations on the RMC campus during two seasons.*

Effect of Distance

The effect of distance from a salted highway was measured by placing a set of CLIMAT units at differing distances in a perpendicular line from the road edge. The highway chosen was the 401, which is a well-traveled four-lane expressway. The CLIMAT units were placed on a fence line, about 1.5 m above the ground. The fence-line was isolated in terms of nearby roads, except for the highway. The unit closest to the highway unfortunately disappeared during the exposure period. The results in terms of average mass loss of aluminum wire for the remaining units are shown in Fig. 2. The level of corrosivity is within the range measured at sea-coast locations [4]. The drop in corrosivity is nearly linear between 20 to 140 m from the edge of the highway. Further measurements are necessary to determine how far from a salted road that corrosion rates are elevated.

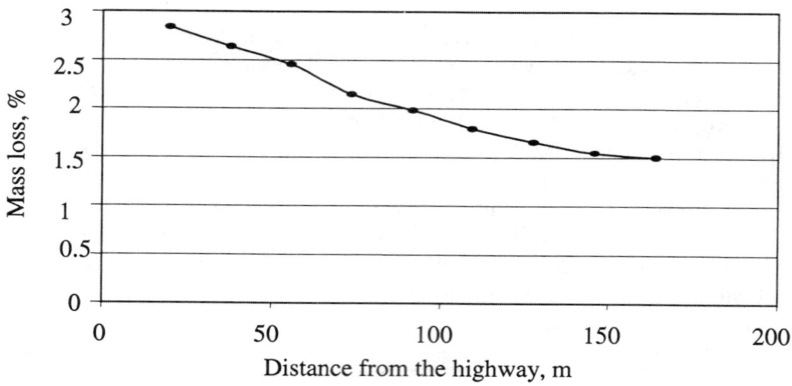

Figure 2-*Average mass loss of aluminum wire from the CLIMAT units as a function of distance from the edge of the 401 highway.*

Wind-Sheltered Microenvironments

In order to further study the effects of sheltering and wind patterns on local corrosivity a corrosion panel was exposed close to Highway #2, on the Royal Military College campus, during the winter of 1999/2000. A photograph of the corrosion panel in place is shown in Fig. 3. The bottom of the panel was placed about 2 m above the ground on the south concrete support for the pedestrian bridge facing the highway, which was about 4 m away. Six sets of CLIMAT units were deployed on the panel with each set in a different microenvironment. The bottom left set was almost completely boxed in. The top left set was oriented parallel to the panel and the middle right set was oriented perpendicular to the panel in order to compare differences in the wind structure very close to the panel. The top right set had baffles designed to partially shelter the CLIMATs from east-west winds and the middle left set had baffles designed to partially shelter winds approaching the CLIMAT set from an angle. The bottom right set had a roof designed to provide more shelter from the rain than what the concrete wall already provides. A seventh set was exposed on a signpost away from the panel in order to provide a reference point as a fully-exposed site. Therefore, the relative humidity was the same for each set, the differences being in the rate of aerosol deposition. The mass loss values are shown in Table 1.

Figure 3-*Panel of CLIMAT units in different microenvironments on the wall of a bridge near a salted highway.*

Table 1-*Percent mass loss of aluminium wire for the CLIMAT coupons on the corrosion panel in Fig. 3.*

Orientation/Shelter	East side	Middle	West side	Average
Box	0.34	0.13	0.32	0.26
Side baffles	3.80	3.62	3.04	3.49
Roof	4.31	1.44	6.56	4.10
Angle baffles	6.25	3.57	4.32	4.71
Flat	6.15	4.24	7.62	6.00
Perpendicular	7.40	6.93	7.50	7.28
Nearby sign post	9.43	8.59	8.85	8.96

Comparison between the flat (upper left) and perpendicular (middle right) CLIMATs reveals the effect of the boundary layer near the wall. The flat CLIMATs were about 2 cm from the wall whereas the ends of the perpendicular CLIMATs were about 11 cm from the wall. Typically, wind velocity near a wall increases logarithmically until it reaches the air velocity that is unaffected by the presence of the wall [11]. The higher mass loss of the perpendicular set (7.28%) than the flat set (6.00%) is consistent with a higher wind velocity further out from the wall.

The effect of baffling the wind parallel to the wall is evident by comparing the side-baffle CLIMATs with the flat CLIMATs. There was an average corrosivity reduction of 42% with this type of partial shelter. This effect can be entirely attributed to wind shielding because there was no vertical shielding from precipitation. The effect of the angle baffles was not as obvious because there was a higher degree of variability within the angle baffles CLIMATs. However, the average of the angle-baffle CLIMATs indicates a reduction of 21% in corrosivity. Again this is entirely attributable to wind sheltering and not precipitation sheltering.

Another trend was that the middle CLIMAT in each microenvironment, with the exception of the side baffles, exhibited lower corrosivity. A plausible explanation for this trend is that the outside CLIMATs provided a small, but significant, degree of wind shielding. The modeled airflow patterns for the flat and side-baffle cases were examined for this possibility.

One obvious conclusion from these measurements is that shielding, whether from wind or direct precipitation, can dramatically reduce the corrosion rate to samples that are exposed to the same time-of-wetness factor. In fact, there was a 34-fold difference between the average mass loss in the boxed-in CLIMATs and the signpost CLIMATs. This is consistent with the concept that atmospheric corrosion rates depend primarily on aerosol deposition rates, which in turn depend on wind velocity and turbulence patterns.

Terminology could be modified to better reflect the localized nature of atmospheric corrosivity. It is suggested that macrocorrosivity refer to the characterization of an area that is on the scale of kilometres such as a city or county; that microcorrosivity refer to the characterization of locations that are on the scale of meters such as different sites near a building or vehicle; and that nanocorrosivity refer to characterizing spots that are on the scale of centimetres.

Air Flow Patterns in Two Microenvironments

Inertial deposition is expected to be the dominant mode of aerosol deposition to the CLIMAT cylinders. Airflow contours around the flat and side-baffle CLIMAT sets are shown in Figs. 4 and 5 respectively. The air splits around the leading cylinder, going mostly over the top. The approach air velocity, two diameters upstream of the leading cylinder, is in the 4.1 to 4.9 m/s range. The side baffles significantly change the airflow pattern around the CLIMAT cylinders compared to the flat set and cause the highest air velocity to be well away from the cylinders. The side baffles essentially create a quiescent recirculating zone. According to the simulation, the upstream velocity coming underneath the cylinder is in the 2.0 to 3.0 m/s range. This drop in approach air velocity is consistent with the 42% reduction in the average corrosivity measured between the flat and side baffle sets.

The air velocity profiles also explain why the flat set experienced a micro-shielding effect whereas the side-baffle set did not. The drop in approach air velocity of the middle cylinder in the flat set was in the 2.4 to 3.3 m/s range is proportional to the drop in corrosivity of the middle cylinder (Table 1). There was little variation in the approach air velocity upstream of the three cylinders in the side-baffle set, which is consistent with their similar corrosivities (Table 1).

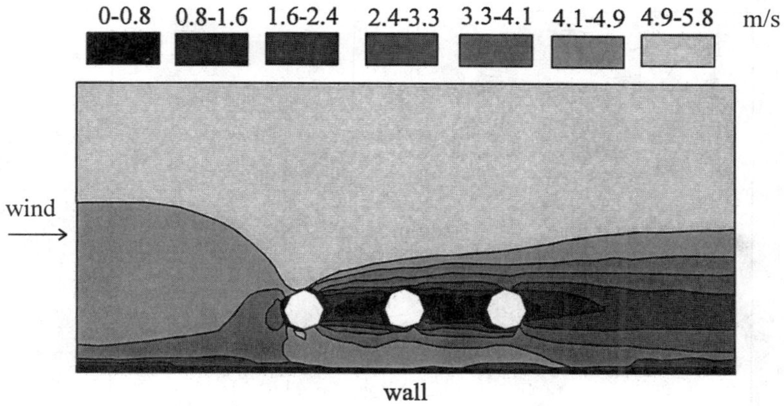

Figure 4-*Simulated air flow around the Flat CLIMAT set. The three white octagonal shapes are cross-sections of the copper bolts.*

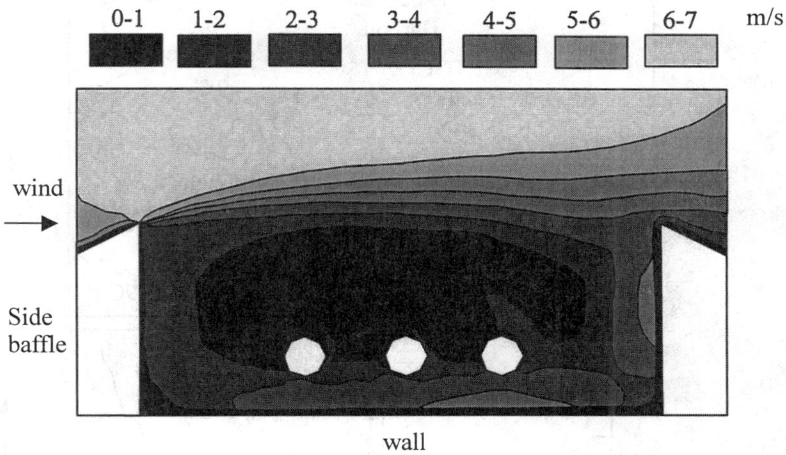

Figure 5-*Simulated air flow around the Side Baffle CLIMAT set.*

Corrosivity Near a Salt-Water Body

Monthly trends

Corrosivity near a salt-water body was also studied using CLIMAT coupons. Figure 6(a) shows the location of the nearest city to the test site (Victoria/Esquimalt) on Vancouver Island, British Columbia, Canada. Figure 6(b) shows the Meterological and Oceanographic Centre (METOC) and buildings 199A and 199C within Canadian Forces Base (CFB) Esquimalt. Figure 6(c) shows the locations where the CLIMAT coupons were exposed on and near buildings 199A and 199C as well as the materials test rack.

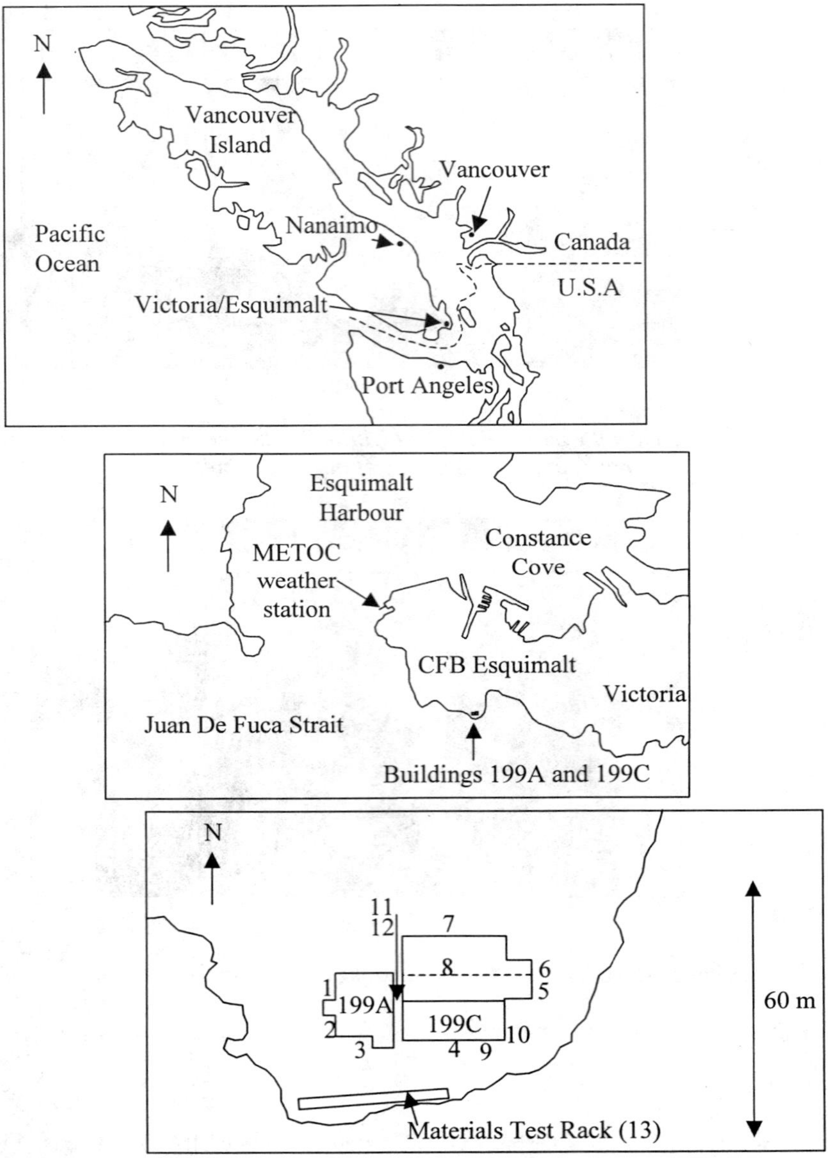

Figure 6-*Maps for the salt-water test site: (a) Vancouver Island, (b) Esquimalt area showing the METOC weather station and buildings 199A and 199C and (c) Buildings 199A and 199C with the numbers corresponding to the locations of CLIMAT units.*

CLIMAT units were exposed at the beginning of each month at the materials test rack for the recommended duration of ninety days. A monthly composite mass loss was computed based on the results from three overlapping exposure periods at the Materials Test Rack and these results are shown in Fig. 7. Also included are the monthly average wind speed and relative humidity (RH) for the same time frame as measured at the nearby METOC station.

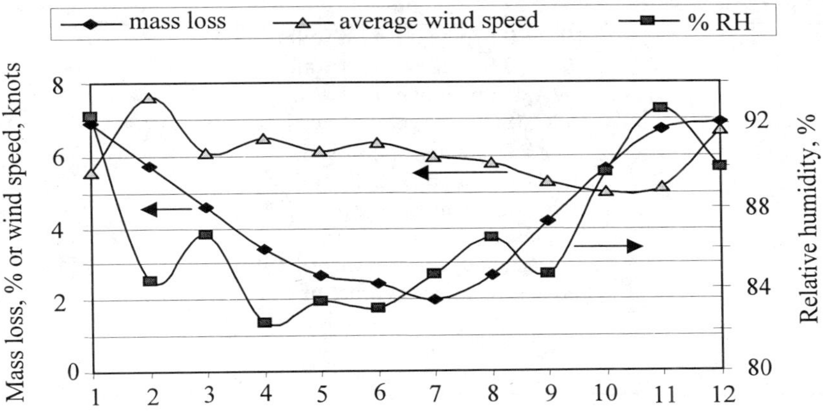

Figure 7-*Monthly CLIMAT mass loss, average wind speed and average relative humidity (month 1 corresponds to January). The monthly CLIMAT mass loss values were computed from overlapping three month exposures at the Materials Test Rack. The wind speed and relative humidity data were taken from the METOC weather station.*

The CLIMAT mass loss results show a seasonal trend, with the highest corrosivity in December and the lowest corrosivity in July. The average monthly wind speed does not have a seasonal trend but stays fairly constant from month to month. However, the average monthly RH has a seasonal trend that is fairly close to the corrosivity trend. Thus it appears that changes in the monthly relative humidity drives the seasonal changes in the overall level of corrosivity at this site.

Corrosivity Near Two Buildings

The corrosivity near two buildings (199A and 199C) as shown in Fig. 6(c) was measured with a set of CLIMAT units that were exposed over the months of November, December and January. These results, as well as a set exposed on the materials test rack, are listed in Table 2.

There are five microenvironments that can be identified by these measurements. In order of increasing corrosivity they are: (i) south-east corner of 199C, (ii) alley between 199A and 199C, (iii) materials test rack, (iv) roof perimeter and (v) roof peak. The highest corrosivity at 2 m above the roof peak was expected because the wind velocity is highest at this site due to two factors. As previously mentioned, wind velocity increases logarithmically with elevation over a flat surface and also accelerates as it passes over the top of obstacles on the ground such as buildings.

Table 2-*Average mass loss of aluminum wire for the CLIMAT units exposed for ninety days near the two buildings shown in Fig. 6(c).*

CLIMAT unit #	Location	Average mass loss, %
1	199A, 0.5 m above west wall	6.02
2	199A, 0.5 m above west wall	6.08
3	199A, 0.5 m above south wall	5.75
4	199C, 0.5 m above south wall	5.52
5	199C, 0.5 m above east wall	5.93
6	199C, 0.5 m above east wall	6.19
7	199C, 0.5 m above north wall	5.66
8	2 m above roof peak of 199C	6.54
9	2 m above ground, south corner of 199C	2.54
10	2 m above ground, east corner of 199C	1.75
11	1 m above ground, in alley	3.27
12	1 m above ground, in alley	3.59
13	Materials Test Rack	4.79

The corrosivity around the perimeter of the roof would experience, on average, lower wind velocities because they were at a lower elevation than above the roof peak. There is a slight difference in the corrosivities between the north/south walls (average 5.64%) and the east/west walls (average 6.05%). This may be related to the higher wind speeds coming westerly winds as shown in Fig. 12. The corrosivity at the materials test rack (4.79%) was lower than that on the roof perimeter sites and this is consistent with a lower elevation and therefore lower wind velocities. The corrosivity in the alley between the two buildings was lower than that on the roof. Modeling of the airflow pattern in this region, as described later, helps to explain qualitatively the lower corrosivity in the alley. The corrosivity near the corner of building 199C can be explained in terms of the nearness to the wall, compared to the other CLIMATs, and also the predominate wind directions patterns as described later.

The airflow patterns around buildings 199A and 199C in two dimensions were modeled in order to gain more insight into the corrosivity pattern. In Fig. 9, the wind velocity pattern is shown around building 199A with winds coming from due west at 8 m/s. The profile illustrates higher wind speeds at the top of the building and lower winds in the alley. Thus, if all the winds were from the west then the corrosivity in the alley should be much less than that on the perimeter of the roof. However, there was not such a drastic drop in corrosivity. The airflow pattern resulting from winds from due south towards the alley was modeled in three dimensions. The wind speed at the southern face was 8 m/s. A top view of the wind speed pattern is shown in Fig. 10. This is a contour plot of a plane 1m above the ground. The domain centered on the alley between 199C and 199A and included 10 m of space in front of and after the buildings. The flow pattern shows increased wind velocities in the alley at the same elevation as the CLIMATs were placed.

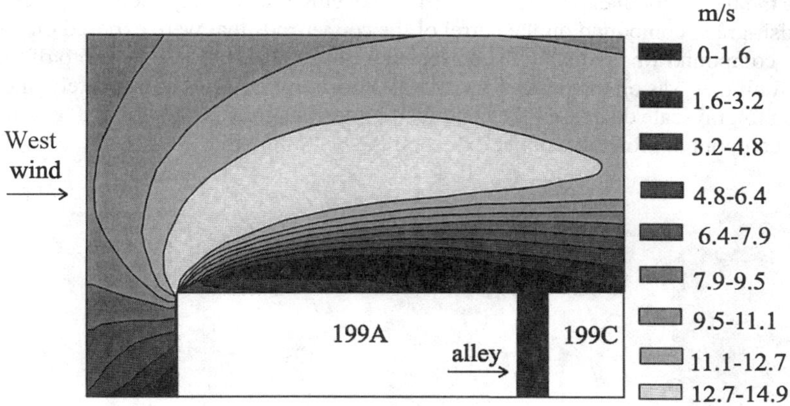

Figure 9-*Contour plot of the simulated wind velocity around building 199C with winds coming from the west at 8 m/s.*

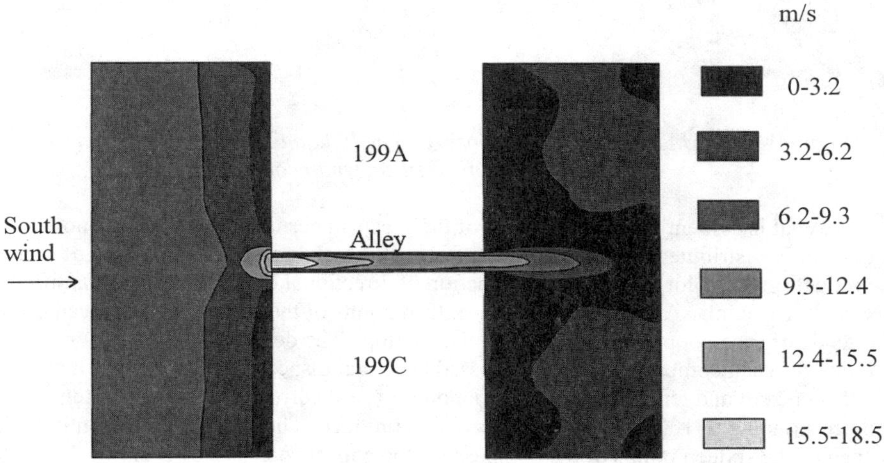

Figure 10-*Top view of the contour plot of wind velocity at 1m above the ground with wind coming from due south at 8 m/s.*

Corrosion Patterns on Copper Bolts

After the three month exposure period from November to January inclusive, there was a bluish-green compound on the barrel of the copper rods that were exposed on the roof. The compound was probably the corrosion product, $CuCl_2.2H_2O$. A photo of the CLIMAT unit that was on the peak of the roof is shown in Fig. 11. Unfortunately, the conversion to grayscale decreased the contrast between the greenish compound and the brownish compound, probably copper oxide, on the rest of the rod.

Figure 11-*CLIMAT unit #9 after exposure. The light material on the copper barrel is bluish-green when shown in color.*

What is evident is that the pattern of the greenish corrosion product was not uniformly distributed around the circumference of each barrel. The intensity of corrosion product (green color intensity) as a function of directional orientation was quantified by visual inspection. A template with the sixteen points of the compass was placed onto the outside of each copper rod that was on the rooftop. The degree of greenness or corrosion index for each compass point was assessed by visual inspection by assigning a number between zero and ten with zero corresponding to zero corrosion product and ten corresponding to 100% coverage of corrosion product. The average corrosion index for each of the sixteen points of the compass for the copper rods is shown in Fig. 12.

The fraction of time that winds came from the sixteen points of the compass during the three-month exposure period is shown in Fig. 13. The dominant direction was the north to north-east. However, the pattern of corrosion product did not correspond to the most frequent wind direction but to the direction with the highest wind speeds, which are in the west to south quadrant at this site (Fig. 12).

There are two possibilities to explain these observations. One is that the aerosol concentration in the winds coming from the north are much less than those coming from the other directions. This cannot be determined a priori by considering the location of

CFB Esquimalt in Fig. 6 but needs to be measured experimentally. The other possibility is that the wind speed must be above a certain minimum before retention of aerosol particles can occur on a smooth metallic surface. In order for the corrosion product to develop on the copper surface, the salt aerosol must impinge onto the surface and then be retained long enough to promote localized corrosion and react to form part of the corrosion product. The impingement step is well understood because it follows the principles of particle inertia [5]. However, the characteristics of aerosol retention on a smooth metallic surface are not as well understood. In either case, the phenomenon of aerosol retention after impingement onto a smooth metallic surface is worth further study.

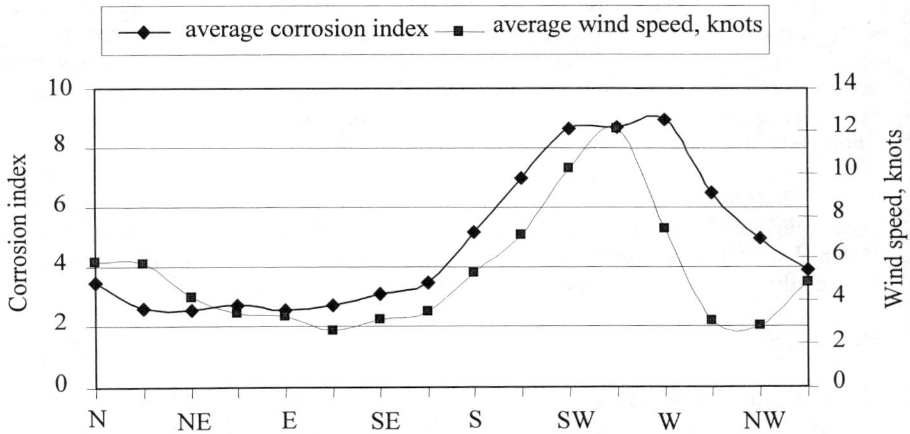

Figure 12-*Average corrosion index for the copper rods exposed on the rooftop and the average wind speed recorded at the METOC station as a function of the sixteen points of the compass.*

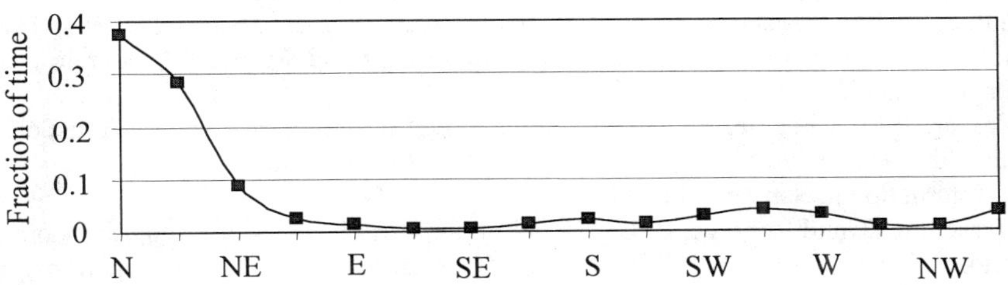

Figure 13-*Fraction of time that winds came from the sixteen points of the compass during the CLIMAT exposure period (Nov, Dec., and Jan.).*

Conclusions

Measurements near a de-iced highway indicated that corrosivity is as high as that near a salt-water body. They also indicted that the zone of corrosivity extends beyond 150 m from the highway edge.

Wind shielding can dramatically reduce the corrosion rate to samples that are exposed to salt aerosols. The latter observation is consistent with the concept that atmospheric corrosion rates near sources of salt aerosols depend primarily on aerosol deposition rates, which in turn depend on wind velocity and turbulence patterns. Relative humidity is a secondary factor in these situations. Thus modeling of airflow patterns is helpful in interpreting corrosivity data where salt deposition is a predominate factor.

The corrosivity measurements near a salt-water body (CFB Esquimalt) indicated that there is a seasonal trend to the corrosivity with a three-fold difference between the maximum in December and the minimum in July. This seasonal trend correlated with the seasonal trend in the monthly average relative humidity. These measurements also indicate that the phenomenon of aerosol retention on smooth metallic surfaces, after impingement, is worth further study.

Acknowledgments: The authors thank Mrs. Hong Nguyen for her work in preparing and processing the CLIMAT coupons and Mr. Ken McRae (Air Vehicles Research Section, Defence and Civil Institute of Environmental Medicine, Ottawa, Ontario, Canada) for financially supporting this work within the context of aircraft corrosion.

References

[1] Summitt, R. and Fink, F.T., "The USAF corrosion testing program and a corrosivity severity index algorithm, Chap. 17, *Atmospheric Corrosion*, Ed. Ailor, W.H., 1982, pp. 245-263.

[2] Lipfert, F.W., Benarie, M., Daum, M.L., "Metallic corrosion damage functions, *Proceedings Electrochemistry Society*, Vol. 86, No.6, 1986, pp.108-154.

[3] ISO 9223, "Corrosion of metals and alloys-Classification of corrosivity of atmospheres. International Organization for Standardization.

[4] Doyle, D.P., Wright, T.E., "Rapid methods for determining atmospheric corrosivity and corrosivity resistance, Chap. 16, *Atmospheric Corrosion*, Ed. W.H. Ailor, 1982

[5] Hidy, G.M., "*Aerosols—An Industrial and Environmental Science*," Academic Press, Inc, 1984.

[6] Klassen, R.D., Roberge, P.R. and Tullmin, M.A., "Modeling of Aerosol Transport as an Aid to Corrosivity Assessment," Paper No. 489, Corrosion 1999, NACE International, San Antonio, TX.

[7] Klassen, R.D. and Roberge, P.R., "Modeling the Effects of Local Topography on Atmospheric Corrosivity," Paper No. 272, Corrosion 2000, NACE International, Orlando, FL.

[8] Klassen, R.D. and Roberge, P.R., "Aerosol Transport Modeling as an Aid to Understanding Atmospheric Corrosivity Patterns," *Materials and Design*, Vol. 20, 1999, pp. 159-168.

[9] Klassen, R.D. and Roberge, P.R., "The Effects of Wind on Local Atmospheric Corrosivity," Paper No. 1544, Corrosion 2001, NACE International, Houston, TX.

[10] Klassen, R.D., Hinton, B. and Roberge, P.R., "Aerosol Model Aids Interpretation of Corrosivity Measurements in a Tropical Region of Australia, *Marine Corrosion in Tropical Environments, ASTM STP 1399*, S.W. Dean, H.D. Delgadillo and J.B. Bushman, Eds., West Conshohocken, PA, 2000.

[11] Bird, R.B., Stewart, W.E. and Lightfoot, E.N., "*Transport Phenomenon*," John Wiley & Sons, Inc, 1960.

Johan Tidblad,[1] Vladimir Kucera,[1] Alexandre A. Mikhailov,[2] Jan Henriksen,[3] Katerina Kreislova,[4] Tim Yates,[5] and Birgit Singer[6]

Field Exposure Results on Trends in Atmospheric Corrosion and Pollution

---

Reference: Tidblad, J., Kucera, V., Mikhailov, A. A., Henriksen, J., Kreislova, K., Yates, T., and Singer, B., "**Field Exposure Results on Trends in Atmospheric Corrosion and Pollution**," *Outdoor Atmospheric Corrosion, ASTM STP 1421*, H. E. Townsend, Ed., American Society for Testing and Materials International, West Conshohocken, PA, 2002.

Abstract: The International Co-operative Programme on Effects on Materials including Historic and Cultural Monuments (ICP Materials) is an extensive field exposure programme within the United Nations Economic Commission for Europe (UN ECE). In its network of test sites, which presently includes 30 test sites in 14 European countries and in Israel, the United States and Canada, several one-year exposures of unalloyed carbon steel, zinc, copper, cast bronze, limestone and steel panel with alkyd paint have been performed during the period 1987-97. The present work summarizes and analyses the one-year exposures for trend effects in Europe and, in particular, quantifies the part of the trend attributable to changes in sulfur dioxide concentration. SO_2 is the largest single contributing factor to the decreasing corrosion trends. The decreasing H^+ in precipitation is also a contributing factor, its effect is, however, much smaller than that of dry deposition. In addition to the ICP Materials results, long term trend examples of zinc and carbon steel corrosion and SO_2 concentration for the period 1946-1997 are shown for Stockholm, Moscow, Prague and Kopisty.

Keywords: atmospheric corrosion, acid deposition effects, materials degradation, time dependence

---

[1]Doctors of Sciences, Swedish Corrosion Institute, Kräftriket 23, SE-104 05 Stockholm, Sweden.

[2]Doctor of Sciences, Institute of Physical Chemistry, Russian Academy of Sciences, Leninskij Prospekt 31, 117915 Moscow, Russian Federation.

[3]Doctor of Sciences, Norwegian Institute for Air Research, P.O. Box 100, N-2007 Kjeller, Norway.

[4]Doctor of Sciences, SVUOM, U Mestanského Pivovaru 4, CZ-17000 Praha, Czech Republic.

[5]Doctor of Sciences, Building Research Establishment Ltd., Garston, Watford WD2 7JR, U. K.

[6]Doctor of Sciences, Bavarian State Department for Historical Monuments, Hofgraben 4, D-80539 Münich, Germany.

Introduction

Air pollution in Europe has been documented with written accounts that span more than two thousand years. In medieval times the primary concern was about health while complaints on enhanced corrosion of building materials were frequent by the beginning of the seventeenth Century in London. Serious attempts to monitor concentrations of air pollutants begun as late as through the 1920s and 1930s with the designing of a range of air pollution monitoring equipment including for example the lead candle for determination of SO_2 deposition and filter paper black smoke monitors [*1*].

Thus, it has been known for several centuries that air pollutants emitted by burning of fossil fuels have a serious impact on buildings. The effects include loss of mechanical strength, leakage and failure of protective coatings due to degradation of materials. Also, the disagreeable appearance in all larger towns of soiled but otherwise beautiful buildings is caused by deposition of particulate matter arising from atmospheric pollution. On one hand, there are many parameters that can influence the damage to materials, it is an interplay between chemical, physical and biological parameters. On the other hand, systematic laboratory exposures in the 1930s demonstrated the corrosive effect of SO_2 on metals, which was later also proved by field exposures, and SO_2 has for a long time been regarded as the main corrosive pollutant. Thanks to powerful international efforts within the scope of the UN ECE Convention on Long-range Transboundary Air Pollution (CLRTAP) SO_2 has decreased substantially and is no longer the only important corrosion stimulator. Now, the effect of SO_2 needs to be considered in a multi-pollutant situation in combination with other gaseous pollutants such as NO_2, O_3, and their reaction products, including particles.

The decrease of especially SO_2 emissions during the last three centuries has significantly improved the situation and has resulted in substantial reductions of corrosion rates for many materials. This paper will show evidence of this success story by summarising the main results of the trend exposures performed within the UN ECE ICP Materials for the period 1987 to 1997 and also by giving examples of trend results in individual cities located in Europe where long-term data are available.

General Trends in UN ECE Countries of Zinc and Carbon Steel Corrosion and Pollution for the Period 1987-1994

The results presented in this section are based on the UN ECE International Co-operative Programme on Effects on Materials including Historic and Cultural Monuments (ICP Materials). A short summary of the programme is given here but the interested reader should consult the results presented at the workshop on "Quantification of Effects of Air Pollutants on Materials", held in Berlin, Germany in 1998, which also includes an overview of the programme [*2*]. Updated information about the programme is also available at the ICP Materials' web page [*3*].

ICP Materials started in 1985 and the first field exposure was started in 1987. It was initiated in order to provide a scientific basis for new protocols and regulations developed within the Convention on Long-Range Transboundary Air Pollution. The main aim is to perform a quantitative evaluation of the effects of multi-pollutants such as S and N compounds, O_3 and particles as well as climate parameters on the atmospheric corrosion of important materials, including materials used in objects of cultural heritage. The primary objective is to collect information on corrosion and environmental data in order to evaluate dose/response functions and trend effects. This is achieved by exposing material specimens in a network of test sites, by measuring gaseous pollutants, precipitation and climate parameters at or near each test site and by evaluating the corrosion effects on the materials. A Task Force is organising the programme with Sweden as lead country and the Swedish Corrosion Institute serving as the main research centre. Sub-centres in different countries have been appointed, each responsible for their own group of materials. All environmental measurements are reported and compiled by the environmental sub-centre, the Norwegian Institute for Air Research (NILU). In each country a national contact person has been appointed responsible for the sub-centre and/or test sites. The original network, used during the period 1987-1995, consisted of 39 exposure sites in 12 European countries and in the United States and Canada [*3*].

The aim of the trend exposure, which constitutes a part of the programme, is to elucidate the environmental effects of pollutant reductions achieved under the Convention on Long-Range Transboundary Air Pollution (CLRTAP) and identify extraordinary environmental changes that result in unpredicted materials damage. The trend exposure consisted of repeated one-year exposures of steel and zinc on the 39 test sites included in the programme during the period 1987 to 1995. These have been performed on all test sites for unsheltered and sheltered carbon steel and zinc. In general, the exposure periods have been 1987-88, 1992-93 and 1994-95 with the exception of unsheltered zinc where the first exposure period was replaced with a repeated exposure 1990-91. The reason for using steel and zinc is that ISO 9223 "*Corrosion of metals and alloys – Corrosivity of atmospheres – Classification*" recommends these metals for classification purposes and consequently corrosion data from several exposures have been published and may be used for comparison.

The trends in corrosion of zinc and carbon steel and pollution have been treated in several papers and here only a summary is given. The environmental sub-centre, the Norwegian Institute for Air Research (NILU), is responsible for the evaluation of trends of environmental parameters [*4*]. SVÚOM, Czech Republic is the sub-centre responsible for the trend analysis of carbon steel and zinc [*5*]. Analyses of simultaneous trends in environmental and corrosion data using statistical methods have been carried out by this sub-centre [*5*] and by the main research centre [*6-7*], who has also contributed to the official UN/ECE report on trends in impacts of long-range transboundary air pollution [*8*].

Overall trends of corrosion and pollution are shown in Figure 1. In order to present the results on a common scale for all parameters the values are compared to the initial value, corresponding to the 1987/88 period. The corrosivity change is substantial, resulting in a reduction in corrosion rate of carbon steel of about 40 % which is, however, relatively smaller than the reduction in zinc corrosion (about 50 to 60 % reduction). Of the environmental parameters SO_2 shows the largest decreasing trend (about 50 to 60 % reduction). The O_3 level has only been measured at 22 of the test sites in the programme. Not all of these registered O_3 values during the entire programme and gaps were quite frequent. When analysing the O_3 data no specific trends were observed, either increasing or decreasing.

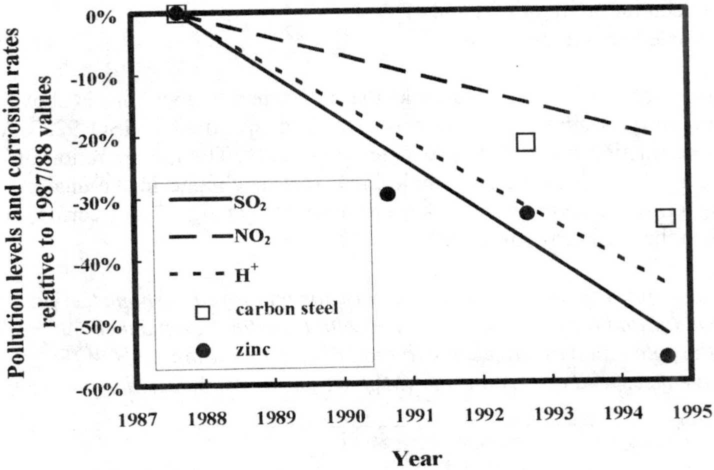

Figure 1 – *Trends of SO_2, NO_2, acidity (H^+), and corrosion of unsheltered unalloyed carbon steel and zinc. All values are expressed relative to the initial (1987/88) value. For environmental data the average trend during the 8-year period is indicated instead of the individual annual averages.*

The results presented clearly show a trend where the environmental parameters SO_2, NO_2 and H^+ as well as the corrosion attack on steel and zinc are decreasing. Dry deposition of SO_2 is the most important parameter for atmospheric corrosion of steel and zinc and is responsible for a large part of the decrease in corrosion attack, as will be shown in the following. Wet deposition refers to deposition by precipitation while dry deposition refers to deposition by any other process.

Effect of Dry Deposition of SO_2

In a first attempt using correlation analysis and analysis of variance it was proven that [SO_2] or TOW[SO_2] contributed significantly to the trend in corrosion. It was also shown that other sources might be important [*5*].

This is illustrated in Table 1, which shows result from a regression analysis using the model

$$ML = b_0 + b_1 TOW[SO_2] \qquad (1)$$

where

ML = mass loss in g m^{-2},
TOW = time of wetness in time fraction h h^{-1},

$[SO_2]$ = SO_2 concentration in μg m^{-3}, and
b_0 and b_1 = estimated parameters.

The time of wetness (TOW) is calculated as the time when temperature is above 0 °C and relative humidity is simultaneously above 80% as specified in ISO 9223. When measured as time fraction it ranges between zero and unity. The incorporation of TOW in the equation makes it possible to include the effect of climate in the analysis. The TOW and SO_2 terms are combined to take into account that significant corrosion only takes place when the surface is sufficiently wet.

Table 1 – *Regression results based on Equation 1 for unalloyed carbon steel and zinc. The coefficients (b_0 and b_1) are shown together with their 95% confidence intervals(±). In addition, the calculated average contribution of dry deposition ($b_1TOW[SO_2]$) and the average measured corrosion rate (ML) are shown for selected exposures.*

	Exposure period	Steel, unsheltered	Steel, Sheltered	Zinc, unsheltered	Zinc, Sheltered
b_0	1987-88	166±32	48±23		
	1990-91			6.7±1.2	
	1992-93	114±32	26±25	5.9±1.4	2.2±1.2
	1994-95	96±29	21±18	3.8±0.9	1.8±0.7
b_1	1987-88	10.2±2.6	4.3±1.9		
	1990-91			0.24±0.12	
	1992-93	14.1±4.2	5.5±3.3	0.40±0.18	0.42±0.16
	1994-95	13.0±4.2	7.3±2.6	0.32±0.13	0.23±0.09
$b_1TOW[SO_2]$	1987-88	94	40	1.9[1]	2.5[2]
	1994-95	65	36	1.6	1.1
ML	1987-88	258	87	8.7[1]	4.7[2]
	1994-95	164	55	5.5	2.8

[1]1990-91
[2]1992-93

By comparing b_0-values from equations obtained for different exposure periods, conclusions can be made regarding contributions to corrosion trends other than those from SO_2. The parameter b_0 is the part of the mass loss that is SO_2-independent and should be constant with time if other effects besides SO_2 are negligible. Looking at Table 1, which shows the b_0-values, a decreasing trend with time is observed. For example, the SO_2-independent corrosion attack on steel has decreased from 166 to 96 g m^{-2} during the period 1987/88 to 1994/95. This trend is observed for both steel and zinc and for both unsheltered and sheltered samples, however, only trends for unsheltered samples are significant. For unsheltered materials the analysis will be extended further when discussing wet deposition. To summarise, other parameters besides SO_2 are contributing to the decreasing corrosion rates, definitely for the unsheltered samples but possibly also for the sheltered steel and zinc samples.

Further analysis of NO_2 as an additional source for the decreasing corrosion trends were made. However, the NO_2 effect could not be proven, possibly due to correlations with other parameters like SO_2 and O_3.

Effect of Wet Deposition of H^+

The effect of wet deposition on unsheltered samples was investigated in the form mm[H^+], i.e., the total acidity load of precipitation. The effect could be quantified and the results are shown in Table 2. For both metals a dose-response relation has been used including the term TOW[SO_2] for dry deposition and the term mm[H^+] for wet deposition, in addition to other terms. Figure 2 is based on data presented in Tables 1-2 and shows a summary of the analysis of unsheltered and sheltered carbon steel and zinc.

Table 2 – *Calculated contributions to mass loss in g m^{-2} averaged over all test sites of unsheltered unalloyed carbon steel and zinc for two trend exposure periods, and based on a dose-response relation including a dry deposition term (TOW[SO_2]) and a wet deposition term (mm[H^+]).*

Material	Exposure period	Dry deposition	Wet deposition	Other	Total
Steel	1987-88	120	28	110	258
Steel	1994-95	56	17	90	163
Zinc	1990-91	2.2	1.6	4.6	8.4
Zinc	1994-95	1.4	1.4	2.7	5.5

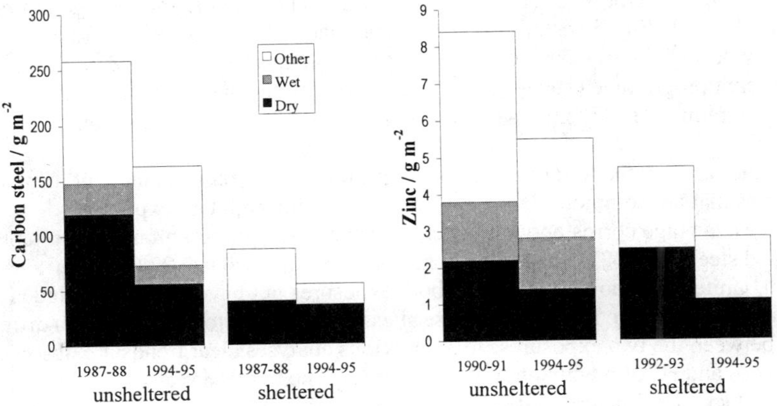

Figure 2 – *Mass loss of unalloyed carbon steel (left) and zinc (right), averaged over all ICP Materials test sites, and for unsheltered and sheltered positions. Each bar consists of the parts corresponding to dry and wet deposition using data from Tables 1-2.*

For steel, two periods have been compared, the 1987/88 and 1994/95 exposure. All test sites have been used to calculate and average contribution of dry and wet deposition. The total average, calculated from the original mass loss values, are also shown together with the remaining part due to effects from other sources than SO_2 and rain acidity, e.g., particulate matters. A smaller part may also be due to NO_2, however, this part can not be quantified. The dry deposition is responsible for the largest decrease during this period (67%) but the wet deposition part is also substantial (12%). The remaining decrease is due to unknown sources. The zinc result is similar to the steel results in that dry deposition dominates over wet deposition regarding the corrosion trend. However, a much larger part remains to be allocated. It should be noted that the results to a certain extent depend on the assumed model. The difference in importance of the dry and wet deposition effects is, however, high and the assumption of another model should not affect their relative ranking.

General Trends in UN ECE Countries of Copper, Bronze and Limestone Corrosion and Pollution for the Years 1987 and 1997

The original field exposure programme, which was described in the previous section and lasted between 1987 and 1995, was designed primarily for the evaluation of the effect of sulphur pollutants in combination with NO_x on important materials and also for the evaluation of trend effects of carbon steel and zinc. A new four-year multipollutant exposure programme started in the fall of 1997. Compared with the original programme changes have been made to the network of test sites, the exposed materials and the characterisation of environment. The network of test sites has been improved by excluding redundant sites from the original network and by incorporating new test sites having new combinations of climate and pollution. The present network consists of 30 exposure sites in 15 European countries and in Israel, the United States and Canada. Fewer materials are exposed, but the increased importance of cultural heritage has been recognised by including limestone as a new trend material. Even if results from specifically designed trend exposures are not yet available, the one-year data of the new multipollutant programme (1997) can be compared with the one-year data of the original programme (1987) since several materials and test sites are common in the two exposures.

Using the same methodology as was applied for carbon steel and zinc, but using only the test sites that are common to the original and the multipollutant exposure programmes, average corrosion values were calculated for copper, bronze, limestone and painted steel (Table 3). Site 36 (Lisbon, Portugal) and also site 27 (Lincoln Cathedral, United Kingdom) included in both exposures but have not been included in the calculation due to the fact that for these sites the racks had to be moved to a different location between the two exposures. As in previous analyses clear trends are observed for SO_2, NO_2 and H^+ of precipitation but not for O_3. The trends are also clear for the corrosion attack of all materials at almost all test sites.

Figure 3 shows the relative decrease in the same manner as was presented in Figure 1 and based on the values given in Table 3. Note that the resulting values are not necessarily directly comparable to those obtained in the previous trend analysis due to the different selection of test sites. However, the trends in SO_2, NO_2 and H^+ seem to fit relatively well with Figure 4, which makes a comparison possible. The average reduction is for most materials between 30% and 50% for the ten-year period with the lowest value for unsheltered bronze (compare Figure 1).

Table 3 – *Comparison of average values from 1987/88 and 1997/98 of SO_2, NO_2, acidity, corrosion of unsheltered and sheltered copper and bronze, and corrosion of unsheltered limestone and degradation of paint on steel expressed as spread of damage from scribe. The values are based on data from the sites common to the original and multipollutant exposure programs.*

Parameter / Material	unit	1987/88	1997/98	Sites with decreasing values
SO_2	µg/m^3	16	5.3	17 of 18
NO_2	µg/m^3	29	22	14 of 16
H^+	mg/l	0.043	0.015	10 of 10
Copper, unsheltered	g/m^2	9.9	6.8	16 of 19
Copper, sheltered	g/m^2	6.3	4.0	16 of 19
Bronze, unsheltered	g/m^2	6.9	3.5	19 of 19
Bronze, sheltered	g/m^2	2.3	1.5	17 of 19
Limestone, unsheltered	µm	12	6.4	17 of 19
Paint on steel, unsheltered	mm	2.1	1.0	17 of 19

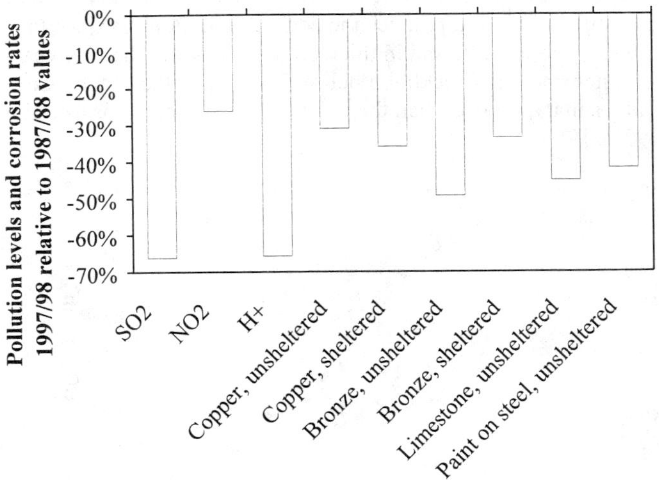

Figure 3 – *Relative trends during the period 1987/88 to 1997/98 of SO_2, NO_2, acidity (H^+), corrosion of unsheltered and sheltered copper and bronze, and corrosion of unsheltered limestone and painted steel.*

Trend Examples of Zinc and Carbon Steel Corrosion and SO_2 Concentration for the Period 1946-1997

The results presented so far clearly shows significant trends in both corrosion and pollution all over Europe during the period 1987-1997. This is part of a longer trend with values of corrosion and pollution peaking between 1950-1970 depending on the location. In the following a few examples of long-term trends are shown, in order to put the trends of the ICP Materials programme into a context. Data have been gathered from Stockholm (Sweden), Moscow (Russian Federation), Prague and Kopisty (Czech Republic), where long-term data on both corrosion and SO_2 concentration exist. The materials and exposure conditions follow the recommendations specified in ISO 9223.

Corrosion of Zinc in Stockholm, Sweden

The trends of zinc corrosion and SO_2 pollution gathered at the site 'Vanadis' in Stockholm are shown in Figure 4. A series of events has contributed to the changes in the corrosion rate. Before 1945 individual domestic heating with coal, coke, wood, gas was prevailing. In the beginning of the 1950s oil with high S content replaced gradually the earlier used fuels. As a consequence both the SO_2 concentration and the corrosion rate raised dramatically. In 1959 the first district heating work in 'Värtan' was started and the district heating has continued its extension since then. In 1966, when the SO_2 concentration peaked, the S content of oil was on average 1.6 %. It has then in pursuance of legislation been lowered. Light oil was lowered to 0.3 % in 1980 and to 0.2 % in 1987. Thick oil was in 1970 lowered from 2.5 % to 1.0 % and in 1989 to 0.8 %. These values are still valid but in practice the content in light oil is since the beginning of the 1990's about 0.1 % and in thick oil about 0.4 %. The first year corrosion attack of zinc is now in Stockholm below 5 g m^{-2} and the SO_2 concentration is below 5 µg m^{-3}, values that are lower than the most optimistic scenarios as estimated by scientists in the 1970's [9].

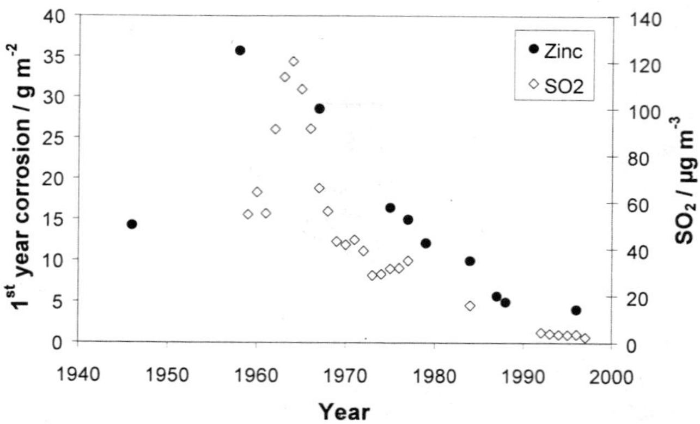

Figure 4 – *Trends in first year corrosion attack of zinc and SO_2 concentration in Stockholm, Sweden.*

Corrosion of Zinc in Prague, Czech Republic

The trends of zinc corrosion and SO_2 pollution gathered in Prague are shown in Figure 5. The decrease in SO_2 pollution in Prague is largely due to changes in the use of fuels, and especially the increased use of gas (Table 4). Another reason is that many local boiler houses were closed and they are now connected to the central heating with the heat delivered from one large heating plant 20 km outside Prague. Also, several industrial companies have moved the production to newly built plants outside the central parts of Prague.

Table 4 – *Distribution between use of gas, liquid, coke and other solid fuels in Prague.*

Year	Gas	Oil	Coke	Other solid
1985	25 %	27 %	15 %	33 %
1992	56 %	10 %	11 %	23 %
1994	58 %	10 %	9 %	23 %
1996	64 %	12 %	6 %	18 %

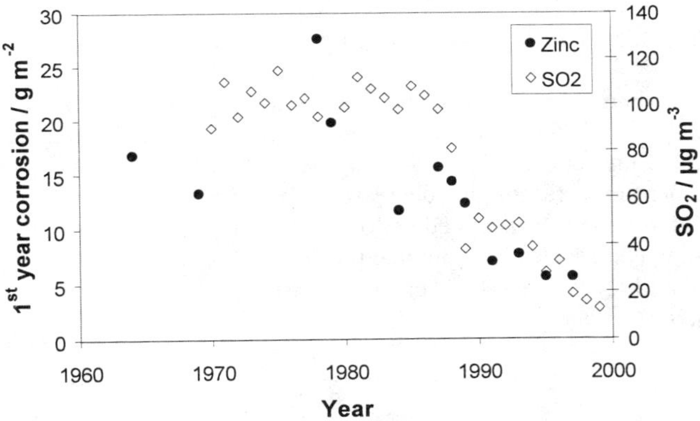

Figure 5 – *Trends in first year corrosion attack of zinc and SO_2 concentration in Prague, Czech Republic.*

Corrosion of Carbon Steel in Kopisty, Czech Republic

The trends of carbon steel corrosion and SO_2 pollution gathered at Kopisty are shown in Figure 6. This industrial site is located about 90 km northwest of Prague. The surroundings consist mainly of industrial and coal mine areas and communications with medium traffic intensity. There are also several power plants using brown coal. In 1994, CEZ a.s. (Czech Electricity Plants) started a program of desulphurization in all power plants and the total invested sum has been 46 billion Czech crowns. Also, the changes in industrial structure and technologies have resulted in a decrease of the total power consumption of about 20%. The decreasing trend of SO_2 concentration has resulted in a decrease of the corrosion rate of steel of about 70%.

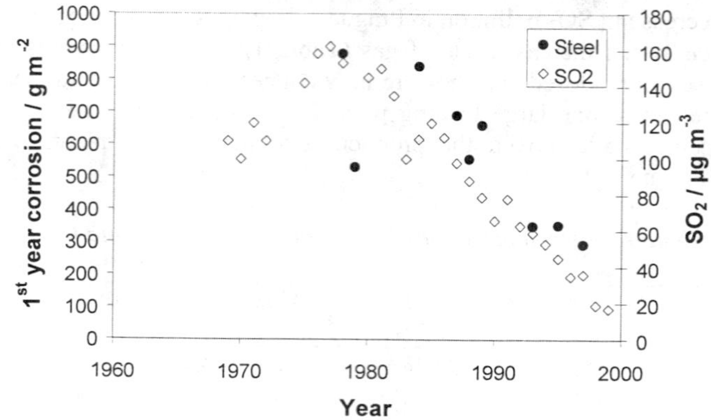

Figure 6 – *Trends in first year corrosion attack of carbon steel and SO_2 concentration in Kopisty, Czech Republic.*

Corrosion of Carbon Steel in Moscow, Russian Federation

The trends of carbon steel corrosion and SO_2 pollution gathered in Moscow are shown in figure 7. Since 1953 a gradual transition from coal to gas heating has reduced the corrosion rate of unsheltered carbon steel in the open atmosphere more than 75 %. For zinc and copper the corresponding decrease is about 60 to 70 %. The decreasing trend is evident all over the Russian Federation during the 1980's for carbon steel, as well as for zinc and copper [*10*]. It should be noted that the increasing trend of the last point (1997), most likely is due to local changes nearby the test site rather than overall changes in Moscow City.

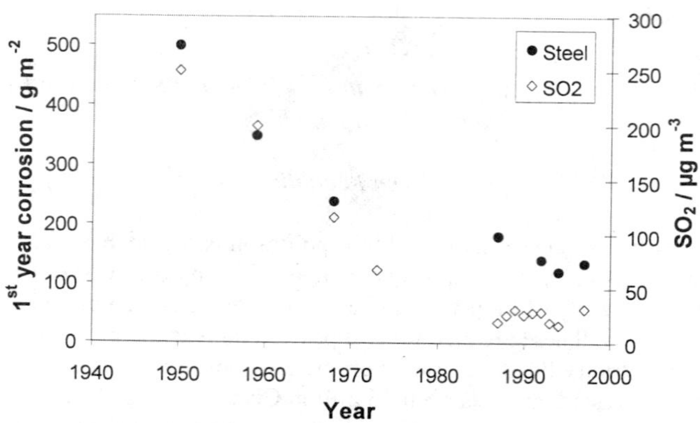

Figure 7 – *Trends in first year corrosion attack of carbon steel and SO_2 concentration in Moscow, Russian Federation.*

Conclusions

- The analyses show that the purposeful actions for reduction of emission of S pollutants undertaken in order to fulfill the demands of the protocols under the LRTAP Convention has resulted in very substantial reductions in corrosion rates of materials, including objects of cultural heritage all over Europe.

- Decreasing corrosion trends in the network of urban and rural sites of ICP Materials during the period 1987 to 1997, between 30 % to 50 % depending on the material, have been observed for carbon steel, zinc, copper and bronze, limestone and paint coated steel. The decrease in corrosivity occurred first in Scandinavia and later in the Western and Central Europe.

- Decreasing pollution trends have been observed for SO_2 (about 70%), NO_2 (about 30%) and H^+ concentration in precipitation (about 70%), while no trends were found for O_3 concentration.

- SO_2 is the largest single contributing factor to the decreasing corrosion trends. For unsheltered materials the decrease in corrosivity also relates partly to decreasing H^+ in precipitation. The decrease in corrosivity is generally larger than what would be expected from the decreases in SO_2 and H^+ concentration. Likely contributors to this unexplained part are particles or other parameters that are not measured in the programme.

- Some sites where long time series of corrosion and pollution are available give evidence of the reasons for the increase and subsequent decrease of the corrosivity of atmospheres.

Acknowledgments

ICP Materials is the result of co-operation between the following organisations. Each was responsible for gathering meteorological and pollution data, for providing sites for the exposure of materials, and/or by acting as a sub-centre for analysis and exposure of materials:

- Institute of Chemistry and Academy of Fine Arts, Austria;
- Department of Chemistry, University of Antwerp, Belgium;
- National Research Council of Canada and the Ministries of the Environment of Canada and of Ontario, Canada;
- SVÚOM, Czech Republic;
- Ministry of the Environment, Environmental Information Centre , Estonia;
- Technical Research Centre of Finland (VTT), Manufacturing Technology, Finland;
- LISA - Université Paris XII, France;
- Bavarian State Department of Historical Monuments, Germany;
- Israel Antiquities Authority, Conservation Department, Israel;

- Agency for Energy Sources (ENEA), Environmental Department, Italy;
- Institute of Environmental Sciences (TNO-MEP), Netherlands;
- Norwegian Institute for Air Research (NILU), Norway;
- Technical University of Lisbon, Laboratory of Mineralogy, Portugal;
- Academy of Sciences, Institute of Physical Chemistry, Russian Federation;
- Ministerio de Fomento, Spain;
- Swedish Corrosion Institute, Sweden;
- EMPA, Corrosion/Surface Protection, Switzerland;
- Building Research Establishment (BRE), United Kingdom;
- Environmental Protection Agency, USA; and
- Getty Conservation Institute, Museum services, USA.

References

[1] Brimblecombe, P., "History of Urban Air Pollution," *Urban Air Pollution - European Aspects*, J. Fenger, O. Hertel and F. Palmgren, Eds., Kluwer Academic Publishers, Dordrecht, Netherlands, 1998, pp. 7-20.

[2] Fitz, S. and Kucera, V., "The UN/ECE Material Exposure Programme. Aim and Scope," *Proceedings of the UN/ECE Workshop on Quantification of Effects of Air Pollutants on Materials,* Berlin 1998, Germany, Umweltbundesamt, 1999, pp. 1-10.

[3] UN ECE Convention on Long-range Transboundary Air Pollution, International Co-operative Programme on Effects on Materials, including Historic and Cultural Monuments, URL: http://www.corr-institute.se/ICP-Materials/, Swedish Corrosion Institute, Stockholm, Sweden, 2000.

[4] Henriksen J., and Bartonova, A.,"Trends in the environmental data during 8 years of the UN/ECE materials programme." *Proceedings of the UN/ECE Workshop on Quantification of Effects of Air Pollutants on Materials,* Berlin 1998, Germany, Umweltbundesamt, 1999, pp. 139-148.

[5] Kreislova, K., Knotkova, K., and Boschek, P.,"Trends of corrosivity based on corrosion rates. Period 1986-1995." *Proceedings of the UN/ECE Workshop on Quantification of Effects of Air Pollutants on Materials,* Berlin 1998, Germany, Umweltbundesamt, 1999, pp. 149-156.

[6] Tidblad, J., Mikhailov, A. A., and Kucera, V.,"Allocation of corrosion trends to dry and wet deposition based on ICP Materials data" *Proceedings of the UN/ECE Workshop on Quantification of Effects of Air Pollutants on Materials,* Berlin 1998, Germany, Umweltbundesamt, 1999, pp. 157-164.

[7] Tidblad J., Kucera V., Mikhailov A. A., Henriksen J., Kreislova K., Yates T., Stöckle B., and Schreiner M., "Final Dose-Response Function and Trend Analysis from the UN ECE Project on the Effects of Acid Deposition," *Co- operation in Corrosion Control, 14th International Corrosion Congress,* Document Transformation Technologies, Cape Town, South Africa, 1999, Paper 337,1, 12 pp.

[8] "Trends in Impacts of Long-range Transboundary air pollution", Technical report prepared by the Bureau, the International Co-operative Programmes (ICP) of the Working Group on Effects of the UN/ECE Convention on Long-range Transboundary Air Pollution. ISBN 1 870393 52 X, 1999.

[9] Sweden's Case Study for the UN Conference on Human Environment, "Air Pollution Across National Boundaries. The Impact on the Environment of Sulfur in Air and Precipitation," Royal Ministry for Foreign Affairs, Stockholm, 1972.

[10] Mikhailov, A. A., Suloeva, M. N., and Vasileva E. G., "Environmental Aspects of Atmospheric Corrosion," *Water, Air and Soil Pollution*, Vol. 85, 1995, pp. 2673-2678.

Lucien Veleva[1] and Mario A. Alpuche-Aviles[2]

Time of Wetness (TOW) and Surface Temperature Characteristics of Corroded Metals in Humid Tropical Climate

---

Reference: Veleva. L., and Alpuche-Aviles, M. A., "**Time of Wetness (TOW) and Surface Temperature Characteristics of Corroded Metals in Humid Tropical Climate**," *Outdoor Atmospheric Corrosion, ASTM STP 1421,* H. E. Townsend, Ed., American Society for Testing and Materials International, West Conshohocken, PA, 2002.

Abstract: In the description of an environment and explanation of corrosion phenomena the most important data are represented by the combination of relative humidity (RH) and temperature (T), often described as temperature-humidity complex (T-RH). It is well known that in open air the metallic corrosion correlates with the time during which a thin film of electrolyte covers the metal surface (i.e., TOW), which determines the duration of the electrochemical corrosion process. According to ISO 9223 TOW can be calculated from the T-RH complex or measured with appropriated sensors. The metal surface real temperature, which effect is not included in ISO 9223:92, can provide very important information about the kinetics of the corrosion process.

During five years, metal samples (low carbon steel, copper and zinc) were exposed to tropical humid climate of the Peninsula of Yucatan, Mexico, in two typical environments, marine-coastal (MC) and rural-urban (RU). The registered metal temperature (Tm) and TOW (by copper/gold sensors, reported by ASTM Practice G84-89 (1999)) strongly depend on the kind of environment. Our registers show that each of the studied metals could be at different temperatures, even though they are exposed in the same location. It was also found that the distribution of TOW in different temperature ranges was needed in order to explain metal corrosion process and rates. SEM images reveal that the Tm and TOW influence the morphology of the corrosion products formed.

Key Words: atmospheric corrosion, tropical climate, time of wetness, metal temperature, corrosion product morphology

---

[1] Research Professor, Applied Physics Department, Centro de Investigación y Estudios Avanzados (CINVESTAV-IPN), Unidad Mérida, C. P. 97310, Mérida, Yucatán, México. Email: lveleva@hotmail.com.

[2] Graduate Student, Chemistry Department, Box 9573, Misssissippi State University, Mississippi 39762. Email: ama8@msstate.edu.

Introduction

Climatic characteristics play a key role on the atmospheric electrochemical corrosion process and to be able to fully understand this corrosion phenomenon, it is very important to properly describe the environment that causes metal degradation [1-7]. In recent years the parameter time of metal wetness (TOW) has deserved special attention, since it is the time during which the metal surface is covered by a thin electrolyte layer, containing different air contaminants, and the corrosion cell can operate [8]. There is a limiting value of relative air humidity (RH) at which condensation starts on the metals [9]. The TOW can be used to classify the corrosivity of an atmosphere [10], and when it is used, instead of exposure time to correlate with the corrosion metal rate, greatly improves the correlation factor [11].

The TOW is usually calculated, according to the International Standard ISO 9223:92 "Corrosion of metals and alloys. Corrosivity of atmospheres classification", by the daily temperature-relative humidity complex (T-RH), using as a critical RH value the minimum of 80% ($t \geq 0°C$), when the condensation starts on the metal. Then the environmental corrosion aggressiveness category can be assigned based on the annual TOW value. However, this procedure is adequate to use only in atmospheres free from chloride, because if the deposition of hygroscopic contaminants occurs on the metal surface (for example, in MC) this lowers the RH_{cr} value and the corrosion can start at RH ~40-50% [9, 12]. This fact implies that in a marine-coastal (MC) environment the higher concentration of chlorides increases the real TOW with respect to a rural-urban (RU) zone, even when the climate is the same for both of these environments [13].

Based on statistical analysis of the T-RH complex, it was previously shown [13] that in tropical humid climate the annual and monthly average values of T and RH (suggested by ISO 9223) could be similar. However during the day two different T-RH complexes are characteristic for RH and MC environments. For these environments the calculated annual TOW value was practically the same, however the experimental corrosion rates of standards metals (low carbon steel, copper, zinc) were very different. In the MC test site, the metals had much higher corrosion rates [11]. This fact is due to some typical characteristics of the humid tropical climate and especially at the coastal area, where the metal degradation occurs in the presence of hygroscopic surface pollutants, such as chlorides. Their presence decreases the RH_{cr} value for wetness formation on the metal and the corrosion could start at RH =50%-60%. Besides, a combination of constant high daily and monthly RH and T, and measured annual TOW (up to 8,500 h) accelerates the electrochemical corrosion of metals exposed to MC atmosphere. Due to this fact, the tropical climate (Southeastern Mexico, at the Caribbean area) falls into the highest corrosivity category of ISO 9223.

It is also important to notice that a statistical analysis [13] shows very different distributions of calculated TOW in temperature regions for the RU and MC zones (both part of the tropical climate), as shown on Figure 1. (TOW was calculated for RH values higher than 60% in the MC site and 80% for the R-U environment). It can be seen that in the RU atmosphere a larger percentage of TOW (54-66%) occurs in the 20-25°C T range, but in the MC site 64-73% of TOW is in the 25-30°C. These characteristics determine different rates of metal degradation for each site.

Another important characteristic of the MC zone atmosphere is that it presents an almost constant range of values of T and RH, because the sea, the biggest thermodynamic natural source (with a large volume of water), has a permanent buffering effect on these environmental parameters. On RU zones, this effect is not present and the T and RH show much larger fluctuation during the night and day. Our statistical analysis of T-RH complex of tropical humid climate showed that the previously mentioned changes (in RU and MC environments) can be predicted [13]. However, the environment and the metal surface do not necessarily have the same temperature, and this fact can make a difference between the calculated and the real TOW value (i.e., corrosion time) and also affect the kinetics of the metal degradation process. To know the real metal T, TOW period and its distribution in temperature ranges, will improve the understanding of the corrosion process.

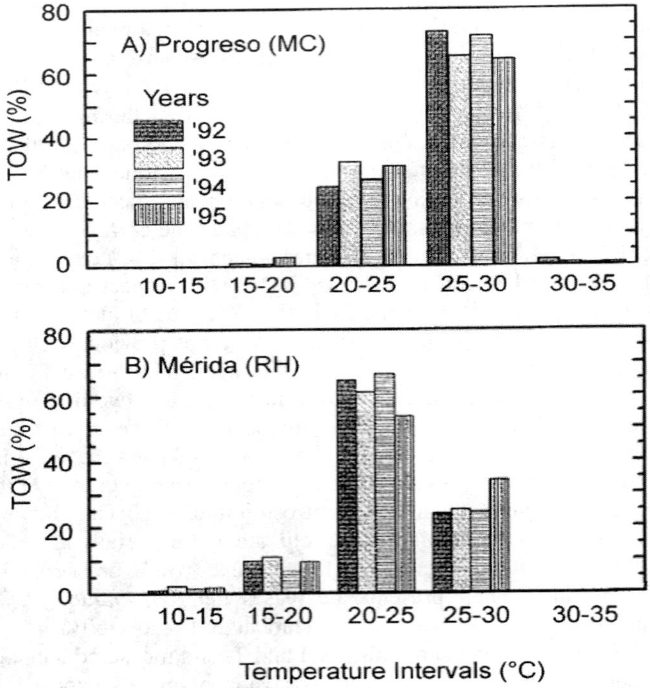

Figure 1. *Distribution of calculated TOW for temperature ranges for a MC Progreso) and a RH (Mérida) zone. The TOW for Progreso was calculated with a RH_{cr} of 60%*

The aim of this study is to monitor and record TOW and T values for different metal samples, exposed in two typical atmospheres: RU and MC, both part of the tropical humid climate. Images of examination by SEM show variations in the morphology of the corrosion products due to differences of the TOW values found on metal surfaces.

Experimental

Following the procedures of ISO 9226:92, "Corrosion of metals and alloys. Determination of corrosion rate of standard specimens for evaluation of corrosivity", flat specimens (100 x 150 x 10 mm) of low carbon steel (by mass percentage, %: C = 0.40, Mn = 0.97, S = 0.035, P = 0.014, Si = 0.15, Cr = 0.10, Ni = 0.08, Cu= 0.037, Sn = 0.03, Mo = 0.03), electrolytic copper and zinc (> 99.99 %) were exposed since 1995 to tropical humid climate, on racks at an angle of 45° to the horizontal. Their exposure were carried out at two test sites: Progreso (MC), 50 m from the seashore and Merida (RU), at a distance 30 km from the sea, both located on the Yucatan Peninsula (Southeastern Mexico, in the Caribbean area). Despite the close distance between both places, the differences in their daily T-RH complex make them interesting to study, because they show how variable the tropical climate can be.

Values of T and TOW of the metals samples (on their skyward and groundward faces) were recorded every hour by appropriated sensors (Fig.2a), and stored in electronic device, designed and built in CINVESTAV-IPN [14]. The copper/gold TOW sensor, proposed by the ASTM Practice G84-89(1999)e1, "Standard Practice for Measurement of Time-of-Wetness on Surfaces Exposed to Wetting Conditions as in Atmospheric Corrosion Testing", is shown in Figure 2b. When condensation occurs, a small potential develops across the sensor, which is amplified by an electronic circuit and reported as TOW. There are only two possible readings from the TOW sensor that indicate whether the surface is wet (V > 0) or not (no potential signal). Data for TOW and T were periodically transferred to a personal computer for their analysis.

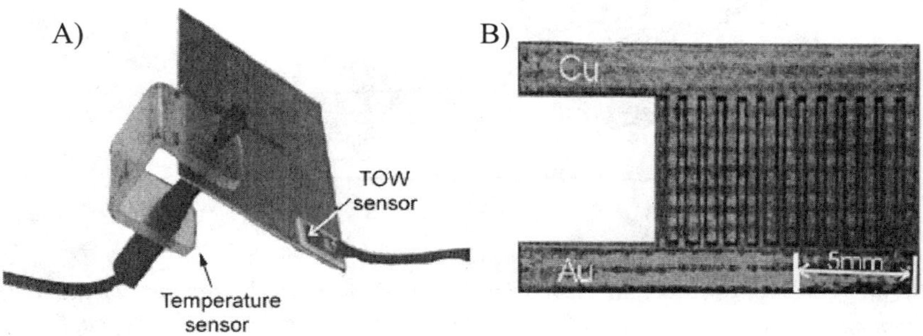

Figure 2. *Sensors used for the study. (A) metal sample with the temperature and TOW sensors. (B) The detector used for TOW measurements*

Results and discussion.

The analysis of the registered data reveals that the daily T values of the metal samples, exposed in the RU site, are very different from the atmosphere temperatures. Figure 3 shows the environmental and metal samples T fluctuations for a dry day (in March). It can be seen, that during the daytime period (from 8 a.m. to 6 p.m.) the metal T are higher that the air ones, and at night the metals are colder. For example, the copper samples are 44°-45°C (7°C hotter at 1 p.m., than the environment) and after the sunset they are 3-5°C cooler. On summer days the metal T can reach 56°C which is usually 15-16°C higher than the environmental. The nature and orientation of the metal samples also influence the T and TOW values, due to the different metal solar absorbance, emissivity and conductivity (including the specifics and color of the corrosion products) and in fact the exposed metals present a little T and TOW differences between them.

In a non-rainy day (Figure 3), in a RU zone the condensation (TOW) happens only at night and in the early hours of the day, when the t<25°C. The annual effective corrosion period (TOW) in RU atmosphere is usually 4,800 h (≈ 60% of the year). It can be seen that on the zinc samples there is more TOW, because they maintain lower temperature than copper and the low carbon steel specimens. The TOW record shows also that the skyward (sw) copper sensor got wet before the groundward (gw) one, because they are exposed to different air pressures. Dissimilarities in TOW cause individual corrosion periods for each of the metal faces.

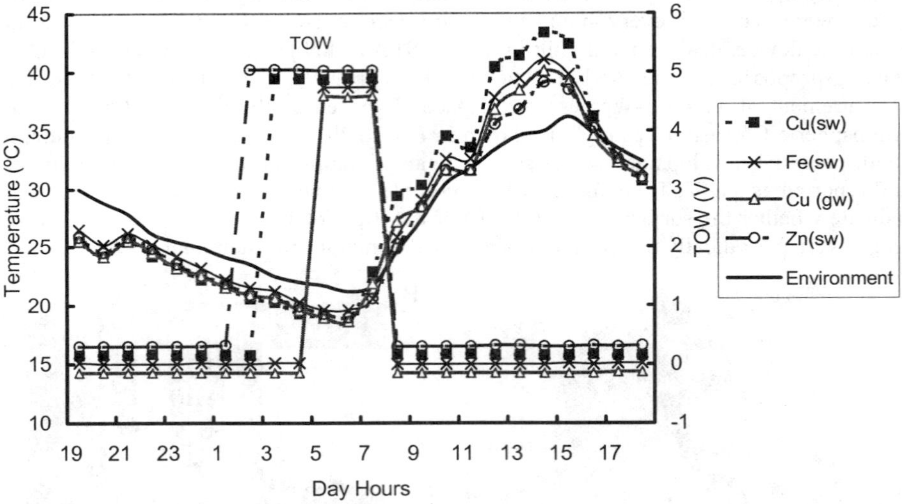

Figure 3. *Temperatures and TOW (amplified potential of the sensor) for the metal specimens registered on March 3 to 4, 1997, a dry day. Fe refers to low carbon steel*

Figure 4 presents data for a rainy day also in March. Again, the zinc and the copper have the lowest (at night) and highest temperatures (at noon) of all the samples. Due to the long rainy period on this day, there is a 15° C difference between the night/early morning T values and the registered ones at noon. The TOW starts at the early night and goes into the next day morning. The skyward face of the copper samples gets wet before the groundward face, as it did on the dry day, and also the copper groundward side got dry before the skyward one. Again zinc is the sample with the highest TOW value, it gets wet first and gets dry last, as expected from the lower temperatures that this material presents, but its TOW value is different from that of the low carbon steel and of the copper samples (both skyward and groundward sides). Different metals can therefore have different TOW (corrosion period) in the same environment.

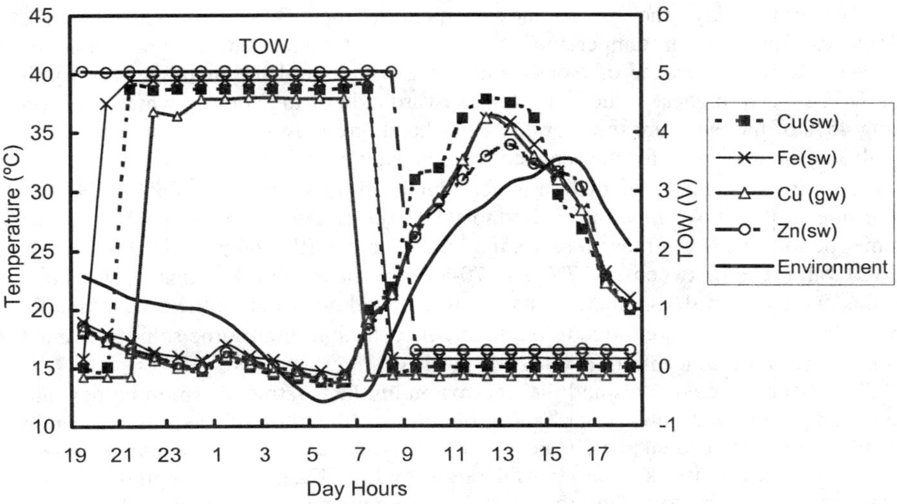

Figure 4. *Temperatures and TOW (amplified potential of the sensor) for the metal specimens registered on January 31 to February 1, 1997, a rainy day. Fe refers to low carbon steel.*

The data recorded in the RU test site during December, usually clouded and windy period (Figure 5), show an interesting behavior: a very stable night T values for environment and metals, accompanied by long TOW (excellent corrosion time).

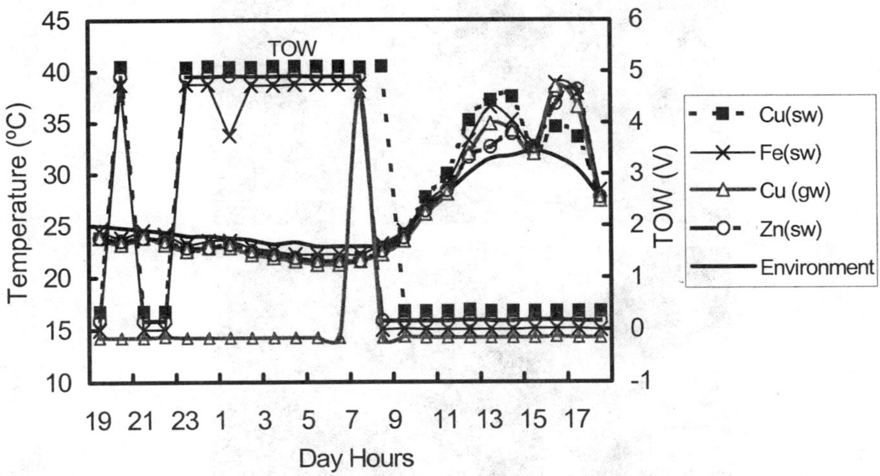

Figure 5 *Temperatures and TOW (amplified potential of the sensor) for the metal specimens registered on December 27 to 28 1997. Fe refers to low carbon steel.*

During the day, the metals slowly increase their T reaching again 7-8°C differences from the air temperature. Some drops of metal temperatures could be observed due to appearance of clouds and wind, which cool the metal surface to air values. The TOW highest value is for the skyward side of the copper, which was wet during 45% of the exposure time; on the other hand, groundward face remains wet for only 4% of the total time for the same conditions. However, in the MC environment, in the same period the air T fluctuation is very small, in the range of 21-25°C (RH ≈70-90%), due to the above mentioned thermostat effect of the sea. Even in July, in the summer period, the MC atmosphere has the same stable T-RH complex: insignificant T and RH variations (between 25-27°C and 70-85%, respectively). Practically, the metal samples maintain T values similar to the environmental ones and all the time (annually ≈8500 h) TOW is registered, due to the high concentration of airborne salinity (annual average of chloride concentration ≈ 362 mg $m^{-2}d^{-1}$).

The nature of the metals and their corrosion products (structure, morphology and color), sample orientation (skyward and groundward), direction of the winds and rain precipitation could account for different T and TOW values for the metals. However, none of these factors is taken into consideration by ISO 9223, even though they cause various corrosion periods in the same environmental conditions. The direct measurement and record of TOW and T values could provide important information about the duration of the corrosion process and the range of temperatures when it occurs. It has been previously shown [15] that the morphology of the corrosion product can vary with the environmental conditions. Figure 6 presents SEM images of copper samples corroded during one month in the MC.

A)

B)

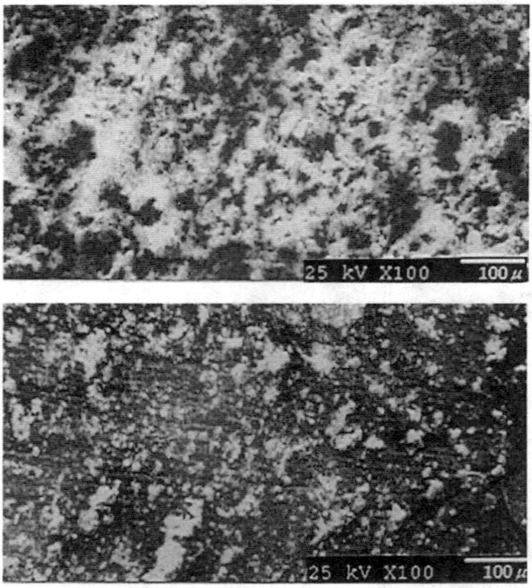

Figure 6. *SEM images of patina formed on the skyward (A) and groundward (B) faces of a copper specimen exposed to the RH zone for one month.*

A)

B)

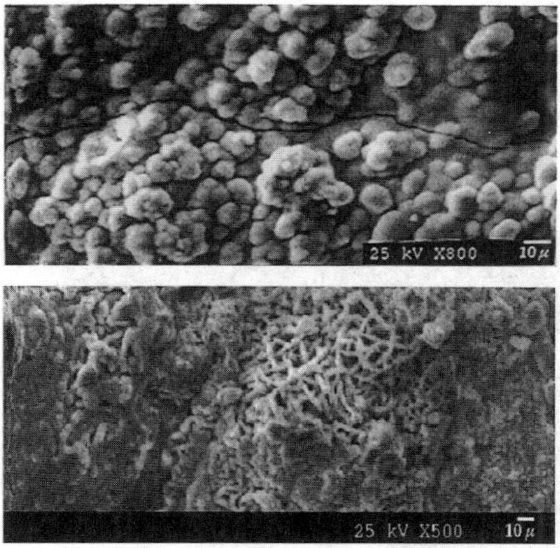

Figure 7. *SEM images of a low carbon steel sample exposed to the MC zone for one month, skyward (A) and the groundward (B) face.*

It can be seen that there is a difference in the density of the corrosion products formed on the skyward (a) and the groundward (b) faces of the specimen. The corrosion products of the skyward face are more abundant than on the groundward side due to the higher TOW registered on this face than that on the groundward face.

Figures 7 and 8 show SEM images of iron (low carbon steel) samples exposed to the MC and RU zones, respectively. It can be seen that the morphology of the corrosion products formed in the MC atmosphere, after one month on the iron skyward face (Figure 7a), is different from the image of the products on the iron groundward face (Figure 7b). In our previous study [17] was detected by X-ray diffraction that the first steel corrosion product correspond to γ-FeOOH, which morphology is presented in Figure 7a. In later stages of corrosion (longer TOW), γ-FeOOH undergoes transformation to α-FeOOH and its typical morphology is shown in Figure 7b. Figure 8a and 8b show the evolution of the corrosion products on iron skyward sample, exposed to the RU environment for one and two months, respectively. It can be noted the difference in the morphology of the corrosion products formed on the skyward side of iron samples in both test sites (Fig.7a and 8a). It was mentioned above that samples exposed in RU environment experience two times lower TOW than the registered values in MC. Besides metal surfaces in the RU site suffer significant T and TOW changes during the day-night cycle. This fact could provide internal stress in the layer of corrosion products, and this effect is not present for the samples exposed in MC environment. Also, after two months of exposure, the corrosion products formed in the RU zone become amorphous [17]. These variations on corrosion product morphology are due to the differences on the principal contaminant and their concentrations in the atmosphere. The pollutant classification, according to ISO 9223, makes the RU test site in the lowest categories of SO_2 (category P_O) and salinity (sodium chloride) concentration (category S_1). In contrast, the MC site presents concentration categories P_1 (low) for SO_2 and S_2 (the highest) for the salinity [11].

A)

B)

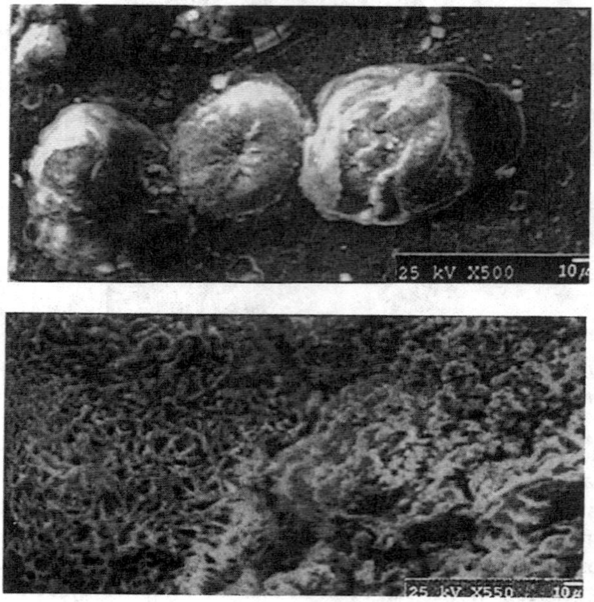

Figure 8. *SEM images of the skyward side of a low carbon steel sample exposed to the RH zone, (A) after one month of exposition, and (B) after two months.*

Conclusions

The recorded temperature values of different metals samples (electrolytic copper, zinc and low carbon steel) exposed to humid tropical climate, show how the metal temperature can be quite different from the environmental one. The metal temperature and time of wetness (TOW) greatly depend on the daily T-RH complex of the zones, even when they take part of the same climate. Some specific characteristics of the metal (e.g. solar absorbance, emissivity, and thermal conductivity) and its face orientation during exposure can also contribute to the surface temperature and TOW changes. The SEM reveals that those differences on TOW and T, registered on the metallic samples, may have caused the distinct morphology of the corrosion products formed on their surfaces.

We find that is highly recommended to measure directly the metal T and TOW of samples exposed for testing, to improve the understanding of corrosion processes, specially in tropical humid climate.

Acknowledgments.
The authors wish to acknowledge Mexican CONACYT (grant PR 2667-PA) and CINVESTAV-IPN, Unidad Merida, for the financial support of this study. We also thank Chem. Blanca Cauich, Eng. Felipe Kantun and Eng. Jose Bante for their technical assistance.

References

[1] Knotkova, D., Boschek, P., and Kreislova, K., "Results of ISO CORRAG program: Processing of one-year data in respect to corrosivity classification," *Atmospheric Corrosion, ASTM STP 1239*, W. W. Kirk and H. H. Lawson, Ed., American Society for Testing and Materials, West Conshohocken. PA, 1995, pp. 38-55.

[2] Dean, S. W., and Reiser, D. B. "Time of wetness and dew formation: A model of atmospheric heat transfer," *Atmospheric Corrosion, ASTM STP 1239*, W. W. Kirk and H. H. Lawson, Ed., American Society for Testing and Materials, West Conshohocken. PA, 1995, pp. 3-10.

[3] Feliu, S., Morcillo, M., and Chico, B., 2001, "Effect of state of sea on atmospheric corrosion in coastal zones," *British Corrosion Journal*, Vol. 32, No. 2, pp. 157-160.

[4] Feliu, S., Morcillo, M., and Chico, B., 1999, "Effect of distance from sea on atmospheric corrosion rate," *Corrosion* (Houston), Vol. 55, No. 6, pp 883-891.

[5] Morcillo, M., Simancas, J., and Feliu, S., "Long-term atmospheric corrosion in Spain: Results after 13-16 years of exposure and comparison with worldwide data," *Atmospheric Corrosion, ASTM STP 1239*, W. W. Kirk and H. H. Lawson, Eds., American Society for Testing and Materials, West Conshohocken. PA, 1995, pp. 195-214.

[6] Tidblad, J., Mikhailov, A. A., and Kucera, V., 2000, "Model for the prediction of the time of wetness from average annual data on relative air humidity and air temperature," *Protection of Metals*, Vol. 36, No. 6, pp. 533-540.

[7] Tidblad, J., Mikhailov, A. A., Kucera, V., "Application of a model for prediction of atmospheric corrosion in tropical environments," *Marine Corrosion in Tropical Environments, ASTM STP 1399*, S. W. Dean, G. Hernandez-Duque Delgadillo, and J. B. Bushman, Eds., American Society for Testing and Materials, West Conshohocken. PA, 1995, pp. 250-285.

[8] Lee, T. S., and Baker, E. A., "Calibration of Atmospheric Corrosion Test Sites," *Atmospheric Corrosion of Metals, ASTM STP 767*, S. W. Dean Jr., and E. C. Rhea,

Eds., American Society for Testing and Materials, West Conshohocken. PA, 1982, pp. 250-266.

[9] Shreir, L. L., Jarman, R. A., and Burstein, G. T., Eds., *Corrosion, Vol.1, Metal/environmental reactions*, Butterworth-Heinemann,1994, Oxford, U. K.

[10] Evans, U., 1965, "Electrochemical mechanism for atmospheric rusting," *Nature*, Vol. 206, pp. 980-982.

[11] Veleva L. and Maldonado L., 1998, "Classification of the atmosphere corrosivity in the humid tropical climate", *British Corrosion Journal*, Vol. 33, No. 1, pp.53-57.

[12] Roy, S. K. and Ho, K. H., 1994, "Corrosion of steel in tropical marine atmospheres", *British Corrosion Journal*, Vol. 24, No. 4, pp. 287-291.

[13] Veleva L., Pérez G. and Acosta M., 1997, "Statistical analysis of the temperature-humidity complex and time of wetness of a tropical climate in the Yucatan Peninsula in Mexico," *Atmospheric Environment*, Vol. 31, No.5, pp. 773-776.

[14] Veleva, L., Bante, F., Kantun, F., *Mexican Patent*, Number 18033, 1998.

[15] Veleva L., and Luja, M., 1999, "SEM characterization of copper corrosion products (patina) formed in a tropical humid climate," *British Corrosion Journal*, Vol. 34, No.1, pp.34-36.

[16] Vu, D. H., *Atmospheric corrosion of metals in tropics*, Nauka 1994, Moscow, Russia

[17] Veleva L., Castro P., Hernandez Duque G. and Schorr M., 1998, "The corrosion performance of steel and reinforced concrete in a tropical humid climate. A review", *Corrosion Reviews*, Vol.16, No.3, pp.235-284.

M. Morcillo,[1] E. Almeida,[2] B. Chico,[1] and D. de la Fuente[1]

Analysis of ISO Standard 9223 (Classification of Corrosivity of Atmospheres) in the Light of Information Obtained in the Ibero-American Micat Project

Reference: Morcillo, M., Almeida, E., Chico, B., and de la Fuente, D., "**Analysis of ISO Standard 9223 (Classification of Corrosivity of Atmospheres) in the Light of Information Obtained in the Ibero-American Micat Project**," *Outdoor Atmospheric Corrosion, ASTM STP 1421*, H. E. Townsend, Ed., American Society for Testing and Materials International, West Conshohocken, PA, 2002.

Abstract: The publication in 1992, by the International Standards Organization, of ISO standard 9223 on the corrosivity of atmospheres represented an extraordinary advance in classification of the aggressivity of atmospheres from the viewpoint of metallic corrosion in outdoor exposure. In 1988, basically following the experimental methodology of the ISOCORRAG programme of ISO/TC156/WG4, the MICAT project began. This was a four-year programme of field research on atmospheric corrosion, conducted both in laboratories and in a network of 75 atmospheric exposure test sites throughout the Ibero-American region, thus considering a broad spectrum of climatological and pollution conditions. This work compares, for the four reference materials (carbon steel, zinc, copper and aluminium), the corrosion data obtained in the ISOCORRAG and MICAT programmes, as a function of the environmental variables of the different atmospheres (time of wetness and deposition rates of SO_2 and chlorides), analysing the corrosivity categories obtained and those estimated on the basis of ISO standard 9223. In a good number of cases deviations were shown in comparison with ISO standard 9223. The paper makes recommendations for a possible future revision of ISO 9223, with the aim of making a better estimation of atmospheric corrosivity categories on the basis of environmental variables.

Keywords: atmospheric corrosivity, classification, carbon steel, zinc, copper, aluminium

Introduction

The MICAT project was launched in 1988, sponsored by the Ibero-American Programme "Science and Technology for Development" (CYTED). Fourteen countries were involved in the project: Argentina, Brazil, Chile, Colombia, Costa Rica, Cuba, Ecuador, Mexico, Panama, Peru, Portugal, Spain, Uruguay and Venezuela. Research was conducted both at laboratories and in a network of 75

[1] Dept. of Materials Engineering, Degradation and Durability. National Centre for Metallurgical Research (CENIM-CSIC). Gregorio del Amo, 8, 28040-Madrid (Spain).

[2] LTR/IMP. Instituto Nacional de Engenharia e Tecnologia Industrial. Estrada do Paço do Lumiar 1649 – 038-Lisbon (Portugal).

atmospheric exposure test sites throughout the Ibero-American region, considering a broad spectrum of climatological and pollution conditions.

The organizational structure, methodology and some results have been given elsewhere [*1*]. The MICAT project has basically followed the experimental methodology of the ISOCORRAG collaborative programme [*2*]. A final report on the MICAT project has been published [*3*].

The materials investigated were reference metals, in the form of flat plate specimens, with the following characteristics: carbon steel (unalloyed), zinc (98.5% min.), copper (99.5% min.), and aluminium (99.5% min.).

Specimens were withdrawn from the test sites after one year (for three consecutive years), two, three and four years of atmospheric exposure. On each occasion, four specimens of each material were withdrawn, three of which were used to determine mass losses, according to ISO Determination of the Corrosion Rate of Standard Specimens for the Evaluation of Corrosivity (ISO 9226). The fourth specimen was used for laboratory studies: analysis of corrosion products, microscopic examination of the morphology of the corrosion products layer and attack of the base metal, etc.

The data obtained in this project has been reported in a number of publications. The following are of special importance, since they present the final refined data for each of the materials considered in the project: carbon steel [*4,5*], zinc [*6,7*], copper [*8*] and aluminium [*9-11*].

Databases

This work presents, for the four reference materials (carbon steel, zinc, copper and aluminium), information on corrosion data obtained in the MICAT project, as well as the environmental variables (time of wetness and deposition rates of SO_2 and chlorides) of the different test sites participating, comparing the corrosivity categories obtained with those estimated in Table 6 of ISO Classification of Corrosivity of Atmospheres (ISO 9223). With a view to the comparison of results, information obtained in the ISOCORRAG project is also included.

Before proceeding to make a combined analysis of the information obtained in the two projects, it is first important to refine the information contained in the databases and to consider the degree of comparability of the databases.

Refining of Databases

For different reasons, this study has not taken into account the information obtained at several of the test sites. The reasons that justify this selection are as follows:

- As is known, atmospheric corrosion in extremely cold climatic zones (polar, etc.) follows mechanisms which are notably different to those presented in other climatic regions [*12*]. For this reason, it has not been considered appropriate to include in this study the testing stations of Jubany, Artigas and Marsh (MICAT project) or Oymiakon, Jubany and Svanvik (ISOCORRAG project).
- The MICAT project considered 18 rural atmospheres practically without sulphur dioxide pollution (<10 mg $SO_2 \cdot m^{-2} \cdot d^{-1}$) or chloride pollution (<3 mg $Cl^- \cdot m^{-2} \cdot d^{-1}$) [*9*]. In these atmospheres aluminium does not present significant attack (corrosivity category C_1 according to ISO 9223). The corrosion rate values obtained in these atmospheres may simply be due to metal attack by the chemical reagent used in gravimetric mass loss determination, rather than corresponding to

the action of the atmosphere on aluminium. The corrosion values shown by the blank test were in the interval 0.0-0.3 g· m^{-2}. For this reason, it is surprizing to see the not inconsiderable aluminium corrosion rates found in the testing stations of Stockholm, Point Reyes and Los Angeles (ISOCORRAG project) and Paulo Alonso (MICAT project), which in view of the SO_2 and chloride deposition rates reported, are atmospheres that can be considered to correspond to category S_0P_0 as defined by ISO 9223. Aluminium corrosion rate data from Stockholm, Point Reyes and Los Angeles were excluded from the analysis. The rates measured can be real and repeatable but the corrosion values were higher than expected taken into account the extreme low values of SO_2 and chloride contamination of the atmosphere.

- The testing stations situated in certain industrial areas have also been excluded. The presence in the atmosphere of certain specific pollutants of industrial origin, other than SO_2 or chlorides are not considered in ISO 9223, having in certain cases a decisive effect on metallic corrosion rates. This is the case of the testing stations of Rio de Janeiro (high concentration of H_2S in the atmosphere), Esmeraldas, Idiem and Petrox (MICAT project).
- Due to the lack of sufficient environmental information or corrosion data, the following testing stations have also been excluded: Puerto Cabello, Fortaleza, Piura, Punta del Este, San Cristobal and Porto Velho (MICAT project).
- Due to great contradictions between the environmental information and the corrosion data supplied, the Sao Paulo (MICAT project) and Kvarvik (ISOCORRAG project) testing stations have been excluded. The relatively low corrosion rates of the four metals in the Sao Paulo atmosphere do not correspond with the high SO_2 contents reported for this testing station. The same occurs with the Kvarvik testing station, in this case in relation with the high chloride deposition rates found in this atmosphere.

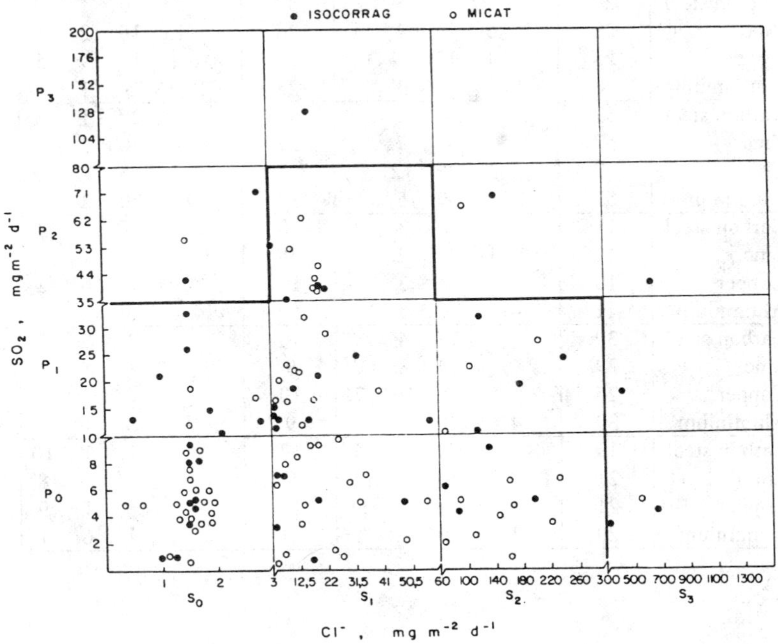

Figure 1 - *ISO 9223 grid, with the situation of the testing stations participating in the ISOCORRAG and MICAT projects. The bold outlining indicates the pollution conditions for which there is a greater volume of information.*

Comparability of databases

Figure 1 presents a grid prepared on the basis of the levels and categories of SO_2 and chloride pollution defined by ISO 9223. All the test sites participating in the ISOCORRAG and MICAT projects have been situated within this grid. The bold outlining in the figure highlights the environmental pollution conditions for which there is a sufficient number (>6) of testing stations. The outlined area contains the categories S_0P_0, S_0P_1, S_1P_0, S_1P_1, S_1P_2, S_2P_0 and S_2P_1.

Table 1 - *Corrosivity category distribution of the information supplied by the ISOCORRAG and MICAT databases, according to the environmental conditions of exposure of each of the reference metals.*

ISO Category	Material	ISOCORRAG database						MICAT database					
		Sample population	C_1	C_2	C_3	C_4	C_5	Sample population	C_1	C_2	C_3	C_4	C_5
S_0P_0	Carbon steel	28		26	2			36		32	4		
	Zinc	30		14	14	2		31	2	22	7		
	Copper	29		10	17	2		45		13	27	5	
	Aluminium	30	30					42	42				
S_0P_1	Carbon steel	24		7	17			5		3	2		
	Zinc	27		4	20	3		5		1	3	1	
	Copper	27		6	10	11		5		4	1		
	Aluminium	25	16		9			5	4		1		
S_1P_0	Carbon steel	29		13	13	3		32		24	7	1	
	Zinc	29		5	19	4	1	26		13	10	2	1
	Copper	28		1	13	14		29		1	14	9	5
	Aluminium	28	17		8	3		25	19		5	1	
S_1P_1	Carbon steel	57		5	47	5		37		7	27	1	2
	Zinc	57			48	7	2	27		4	20	3	
	Copper	57		3	42	12		33			15	17	1
	Aluminium	57	15		37	5		33	10		20	2	1
S_1P_2	Carbon steel	11		1	7	3		10			6	1	3
	Zinc	11		1	7	2	1	9		3	5	1	
	Copper	10			7	3		10			2	6	2
	Aluminium	11	4		2	5		10	4		5	1	
S_2P_0	Carbon steel	30			17	4	9	10			2	3	5
	Zinc	29			14	8	7	9			1	5	3
	Copper	25			2	16	7	10				4	6
	Aluminium	30	4		18	8		9			8	1	
S_2P_1	Carbon steel	19			8	8	3	12			1	1	10
	Zinc	21			9	5	7	9				1	8
	Copper	21				10	11	10				1	9
	Aluminium	20	2		10	7	1	11			4	6	1

Table 1 summarizes for each of the reference materials and environmental conditions the corrosivity categories defined by ISO 9223 found in the different testing stations and Table 2 shows the average values of all series of one year of exposure in each study, ISOCORRAG and MICAT. As can be seen, both databases are comparable in most of the different situations presented. The slight differences found between the two databases relate above all to high pollution conditions (S_1P_2, S_2P_0, S_2P_1).

Table 2 – *Average values of series of one year of exposure in each study, ISOCORRAG and MICAT. Corrosion values of carbon steel, zinc and copper are expressed in μm but the corresponding to aluminium in g· m^{-2}*

ISO Category	Material	ISOCORRAG database				MICAT database			
		Sample population	Values			Sample population	Values		
			Max.	Min.	Ave.		Max.	Min.	Ave.
S_0P_0	Carbon steel	5	19.7	4.6	12.2	20	28.1	1.4	13.2
	Zinc	5	2.29	0.19	1.09	19	3.31	0.11	1.12
	Copper	5	1.25	0.18	0.82	20	1.94	0.09	0.72
	Aluminium	5	0.27	0.08	0.17	18	0.27	0.01	0.12
S_0P_1	Carbon steel	6	36.2	25.2	28.6	2	31.1	9.7	20.4
	Zinc	6	1.96	0.66	1.33	2	1.77	0.82	1.30
	Copper	6	1.39	0.53	0.89	2	0.64	0.53	0.59
	Aluminium	6	0.76	0.16	0.41	2	0.54	0.12	0.33
S_1P_0	Carbon steel	5	45.2	12.8	26.8	17	69.3	11.1	28.6
	Zinc	5	2.11	0.70	1.37	11	2.53	0.19	1.08
	Copper	5	2.24	0.76	1.42	16	3.19	0.35	1.66
	Aluminium	5	2.19	0.24	0.83	15	1.93	0.06	0.54
S_1P_1	Carbon steel	9	43.8	24.0	34.2	11	49.4	17.4	32.0
	Zinc	9	2.30	1.10	1.52	9	2.46	0.56	1.35
	Copper	9	2.04	0.66	1.14	10	1.98	0.71	1.30
	Aluminium	9	1.46	0.62	0.97	10	3.56	0.27	1.11
S_1P_2	Carbon steel	4	61.7	27.9	43.5	3	158.9	28.2	98.4
	Zinc	4	3.80	0.94	2.26	2	1.25	0.45	0.85
	Copper	4	1.38	0.93	1.20	2	2.86	1.23	2.05
	Aluminium	4	3.16	0.46	1.74	2	0.78	0.33	0.56
S_2P_0	Carbon steel	4	87.4	35.2	60.3	5	371.5	29.3	184.2
	Zinc	4	4.29	1.40	2.65	3	7.07	2.89	4.94
	Copper	4	3.66	1.35	2.20	5	5.80	2.51	3.88
	Aluminium	4	3.24	0.70	1.47	4	1.68	0.86	1.35
S_2P_1	Carbon steel	4	99.4	37.9	63.4	3	365.0	33.5	197.6
	Zinc	4	5.13	1.52	3.30	3	7.47	4.03	5.80
	Copper	4	3.19	1.80	2.73	3	4.85	3.61	4.39
	Aluminium	4	4.13	0.78	2.23	3	3.79	1.67	2.80

Towards a Better Classification of Corrosivity Based on Environmental Data

The publication in 1992, by the International Standard Organization, of ISO standard 9223 on the corrosivity of atmospheres represented an extraordinary advance in classification of the aggressivity of atmospheres from the viewpoint of metallic corrosion in outdoor exposure.

However, despite the relatively good fit that the standard generally provides between environmental factors and corrosion rates, some deviations are found in the information obtained in the MICAT project. Other researchers have also noted discrepancies between the data they have obtained in different parts of the world [*13,14*] and ISO 9223.

In the pursuit of a better fit between environmental information and estimations of atmospheric corrosion rates, the information obtained in the ISOCORRAG and MICAT projects constitutes a highly valuable contribution with a view to future revisions of ISO 9223.

Two approaches have been followed to achieve a better fit: one based on the revision of Table 6 of ISO 9223; and another relating to the establishment of damage functions that permit an estimation of corrosion rates (for the first year of exposure) on the basis of environmental information (time of wetness (TOW) and deposition rates of sulphur dioxide (SO_2) and chloride (Cl^-)).

In order to consider a greater volume of information, it has been preferred to include in the two approaches the information supplied by both the MICAT and ISOCORRAG databases.

Revision of Table 6 of ISO 9223

Table 6 of ISO 9223 (Table 3) makes an estimation, for each of the reference metals, of the corrosivity categories of the atmosphere as a function of the time of wetness and pollution categories.

Table 3 - *Estimated corrosivity categories of the atmosphere according to ISO 9223.*

Unalloyed carbon steel															
	τ_1			τ_2			τ_3			τ_4			τ_5		
	S_0-S_1	S_2	S_3	S_0-S_1	S_2	S_3	S_0-S_1	S_2	S_3	S_0-S_1	S_2	S_3	S_0-S_1	S_2	S_3
P_0-P_1	1	1	1 or 2	1	2	3 or 4	2 or 3	3 or 4	4	3	4	5	3 or 4	5	5
P_2	1	1	1 or 2	1 or 2	2 or 3	3 or 4	3 or 4	3 or 4	4 or 5	4	4	5	4 or 5	5	5
P_3	1 or 2	1 or 2	2	2	3	4	4	4 or 5	5	5	5	5	5	5	5
Zinc and copper															
	S_0-S_1	S_2	S_3	S_0-S_1	S_2	S_3	S_0-S_1	S_2	S_3	S_0-S_1	S_2	S_3	S_0-S_1	S_2	S_3
P_0-P_1	1	1	1	1	1 or 2	3	3	3	3 or 4	3	4	5	3 or 4	5	5
P_2	1	1	1 or 2	1 or 2	2	3	3	3 or 4	4	3 or 4	4	5	4 or 5	5	5
P_3	1	1 or 2	2	2	3	3 or 4	3	3 or 4	4	4 or 5	5	5	5	5	5
Aluminium															
	S_0-S_1	S_2	S_3	S_0-S_1	S_2	S_3	S_0-S_1	S_2	S_3	S_0-S_1	S_2	S_3	S_0-S_1	S_2	S_3
P_0-P_1	1	2	2	1	2 or 3	4	3	3 or 4	4	3	3 or 4	5	4	5	5
P_2	1	2	2 or 3	1 or 2	3 or 4	4	3	4	4 or 5	3 or 4	4	5	4 or 5	5	5
P_3	1	2 or 3	3	3 or 4	4	4	3 or 4	4 or 5	5	4 or 5	5	5	5	5	5
N.B. Corrosivity is expressed as the numerical part of the corrosivity category code (e.g. 1 instead of C_1)															

In view of the information generated in both the ISOCORRAG and MICAT projects, it may be of interest to revise this table, in order to improve its correspondence with the corrosivity categories really found in practice. This revision will necessary have to be limited to the environmental conditions (TOW, τ; SO_2, P; and chlorides, S) for which sufficient information is available in the two projects and which are highlighted in Table 3.

In this task it has been considered appropriate to distinguish between the two categories of pollution by sulphur-containing substances (SO_2), P_0 and P_1, and the two categories of pollution by airborne salinity (Cl^-), S_0 and S_1, which ISO 9223 considers in an undifferentiated way. A differentiation is also made between zinc and copper, which ISO standard also considers jointly.

Tables 4 to 7 below present, for the four reference metals and the different environmental conditions of TOW (τ) and pollution (S_xP_y), the distribution in corrosivity categories (C_1 to C_5) of the information obtained in the two projects. As can be seen, there is a considerable dispersion of the information obtained. In the light of this information, Tables 4 to 7 indicate the recommended corrosivity intervals. For comparative purposes, the final column of Tables 4 to 7 presents the corrosivity categories specified by ISO 9223 (Table 3).

Table 4 - *Corrosivity categories recommended for carbon steel in the light of information obtained in the ISOCORRAG and MICAT projects, as a function of environmental conditions of the atmosphere.*

TOW	Pollution	Number of experimental data					Corrosivity categories	
τ	S_xP_y	C_1	C_2	C_3	C_4	C_5	Recommended	ISO 9223
τ_3	S_0P_0		20				C_2	C_2-C_3
	S_0P_1		3	6			C_2-C_3	C_2-C_3
	S_0P_2		1	2	4	1	C_2-C_4	C_3-C_4
	S_1P_1		3	5	1		C_2-C_3	C_2-C_3
τ_4	S_0P_0		37	5			C_2-C_3	C_3
	S_0P_1		7	13			C_2-C_3	C_3
	S_0P_2			6			C_3	C_4
	S_1P_0		32	19	1		C_2-C_3	C_3
	S_1P_1		9	54	5		C_2-C_4	C_3
	S_1P_2		1	13	4	3	C_3-C_5	C_4
	S_1P_3					7	C_5	C_5
	S_2P_0			13	4	13	C_3-C_5	C_4
	S_2P_1			3	8	12	C_3-C_5	C_4
	S_2P_2				6	1	C_4-C_5	C_4
τ_5	S_0P_0		13	1			C_2-C_3	C_3-C_4
	S_1P_0		5	1	6		C_2-C_4	C_3-C_4
	S_1P_1			15		1	C_3-C_4	C_3-C_4
	S_2P_0			6	3	1	C_3-C_5	C_5
	S_2P_1			6	1	3	C_3-C_5	C_5

Due to the great dispersion of corrosivity categories found experimentally, it has almost always been preferred to recommend corrosivity range, rather than one single corrosivity category as is frequently done by ISO 9223. On the other hand, care has been taken to ensure a logical relationship between the recommended corrosivity categories and the influence of a change in environmental conditions on the corrosivity of the atmosphere, in accordance with the abundant literature existing on the influence of environmental parameters on atmospheric corrosion, and which has been clearly shown by the MICAT project [*4-11*]. The recommended range of corrosivity categories includes roughly the 90% of the experimental data.

Table 5 - *Corrosivity categories recommended for* <u>*zinc*</u> *in the light of information obtained in the ISOCORRAG and MICAT projects, as a function of environmental conditions of the atmosphere.*

TOW	Pollution	Number of experimental data					Corrosivity categories	
τ	S_xP_y	C_1	C_2	C_3	C_4	C_5	Recommended	ISO 9223
τ_3	S_0P_0	2	16	2			C_2-C_3	C_3
	S_0P_1		3	7	2		C_2-C_4	C_3
	S_0P_2		1	1	6		C_2-C_4	C_3
	S_1P_1		3	6			C_2-C_4	C_3
τ_4	S_0P_0		18	20	3		C_2-C_4	C_3
	S_0P_1		2	16	2		C_2-C_4	C_3
	S_0P_2			4	1	1	C_2-C_4	C_3-C_4
	S_1P_0		16	25	5		C_2-C_4	C_3
	S_1P_1		1	49	9	2	C_2-C_4	C_3
	S_1P_2		4	12	3	1	C_2-C_4	C_3-C_4
	S_1P_3					7	C_5	C_4-C_5
	S_2P_0			9	11	8	C_3-C_5	C_4
	S_2P_1			4	6	10	C_3-C_5	C_4
	S_2P_2				6	1	C_4-C_5	C_4
τ_5	S_0P_0		2	9	4		C_2-C_4	C_3-C_4
	S_1P_0		2	4	1	2	C_2-C_4	C_3-C_4
	S_1P_1			13	1		C_3-C_4	C_3-C_4
	S_2P_0			6	2	2	C_3-C_5	C_5
	S_2P_1			5		5	C_3-C_5	C_5

Table 6- *Corrosivity categories recommended for* <u>*copper*</u> *in the light of information obtained in the ISOCORRAG and MICAT projects, as a function of environmental conditions of the atmosphere.*

TOW	Pollution	Number of experimental data					Corrosivity categories	
τ	S_xP_y	C_1	C_2	C_3	C_4	C_5	Recommended	ISO 9223
τ_3	S_0P_0		15	3	2		C_2-C_4	C_3
	S_0P_1		5	3	4		C_2-C_4	C_3
	S_0P_2		1	1	2	4	C_3-C_5	C_3
	S_1P_1		1	8			C_3-C_5	C_3
τ_4	S_0P_0	1	8	29	3		C_2-C_4	C_3
	S_0P_1		5	8	7		C_2-C_4	C_3
	S_0P_2			2	4		C_3-C_4	C_3-C_4
	S_1P_0		2	24	16	3	C_3-C_4	C_3
	S_1P_1		2	37	26	1	C_3-C_4	C_3
	S_1P_2			9	9	2	C_3-C_4	C_3-C_4
	S_1P_3				6	1	C_4-C_5	C_4-C_5
	S_2P_0				19	11	C_4-C_5	C_4
	S_2P_1				5	16	C_4-C_5	C_4
	S_2P_2				4	3	C_4-C_5	C_4
τ_5	S_0P_0			12	2		C_3-C_4	C_3-C_4
	S_1P_0			3	7	2	C_3-C_5	C_3-C_4
	S_1P_1			12	3		C_3-C_5	C_3-C_4
	S_2P_0			2	6	2	C_3-C_5	C_5
	S_2P_1				6	4	C_4-C_5	C_5

Table 7 - *Corrosivity categories recommended for aluminium in the light of information obtained in the ISOCORRAG and MICAT projects, as a function of environmental conditions of the atmosphere.*

TOW	Pollution	Number of experimental data				Corrosivity categories	
τ	S_xP_y	C_1-C_2	C_3	C_4	C_5	Recommended	ISO 9223
τ_3	S_0P_0	20				C_1-C_2	C_3
	S_0P_1	5	7			C_1-C_3	C_3
	S_0P_2	1	4	3		C_1-C_4	C_3
	S_1P_1		9			C_3	C_3
τ_4	S_0P_0	36				C_1-C_2	C_3
	S_0P_1	15	3			C_1-C_3	C_3
	S_0P_2	1		5		C_1-C_4	C_3-C_4
	S_1P_0	29	11	4		C_1-C_4	C_3
	S_1P_1	20	41	4	1	C_1-C_4	C_3
	S_1P_2	8	7	6		C_1-C_4	C_3-C_4
	S_1P_3			5	2	C_4-C_5	C_4-C_5
	S_2P_0	2	19	8		C_3-C_4	C_3-C_4
	S_2P_1	2	11	8	1	C_3-C_4	C_3-C_4
	S_2P_2		1	6		C_3-C_4	C_4
τ_5	S_0P_0	14				C_1-C_2	C_4
	S_1P_0	7	2			C_1-C_3	C_4
	S_1P_1	5	7	3		C_1-C_4	C_4
	S_2P_0	2	7	1		C_1-C_4	C_5
	S_2P_1		3	5	1	C_3-C_4	C_5

In the case of carbon steel, though the discrepancies are not very great, 15 of the 19 environmental situations considered have been revised (Table 4), recommending broader corrosivity category intervals than those indicated by ISO 9223. In general terms it is observed that ISO 9223 makes an overestimation of corrosivity.

The reverse is true in the cases of zinc and copper, where ISO 9223 in general makes estimations of corrosivity by default. Tables 5 and 6 present the corrosivity category intervals recommended in the light of the information obtained in the ISOCORRAG and MICAT projects, having revised 18 of 19 (in the case of zinc) and 15 of 19 (in the case of copper) of the environmental situations considered in this analysis.

As for aluminium (Table 7), the estimations on corrosivity categories according to ISO 9223 are clearly excessive in relation with the information obtained in the ISOCORRAG and MICAT projects. In the revision, 15 of the 19 situations considered have been modified.

From the point of view of users of the standard would be helpful to know what the highest corrosion rate that can be anticipated instead of a range of corrosivity categories as is shown in Tables 4 to 7. A compromise should be to accept tha highest corrosivity category of the range shown in the tables unless other data is provided to show a lower corrosivity.

ISO 9223 does not permit for aluminium a differentiation between corrosivity categories C_1 and C_2, since the term "negligible" is not quantified with regard to corrosion rates. As has already been mentioned, in the MICAT project, which considered a large number of atmospheres without SO_2 or Cl^- pollution (S_0P_0), a critical threshold of 0.0-0.3g· m^{-2}· y^{-1} was observed, which would correspond to metal attack by the chemical reagent used in gravimetric mass loss determinations (blank test) [9]. The aforementioned critical value of 0.3 g· m^{-2}· y^{-1} could be used in ISO 9223 to differentiate between corrosivity categories C_1 and C_2, establishing the corrosion rates of ≤0.3 g· m^{-2}· y^{-1} for C_1 and between 0.3 g· m^{-2}· y^{-1} and 0.6 g· m^{-2}· y^{-1} for C_2.

Damage Functions

A second alternative for estimating the corrosivity of atmospheres from environmental information may consist of establishing damage functions between both. To this end, a statistical analysis was made with the information available in the ISOCORRAG and MICAT databases.

Data was fitted to a linear Equation 1

$$Corrosion\ (C) = a_1 + a_2\ TOW + a_3\ S + a_4\ Cl \tag{1}$$

where the a_i coefficients (i = 1 to 4) are constants, C (μm in the case of carbon steel, zinc and copper and g· m^{-2} in the case of aluminium) is the first year corrosion rate, TOW is the time of wetness (annual fraction), S (mg SO_2· m^{-2}· d^{-1}) is the annual average of SO_2 pollution and Cl (mg Cl^-· m^{-2}· d^{-1}) that of chlorides.

Table 8 presents, for each reference metal, the damage functions obtained and their corresponding resulting correlation coefficients, which consider the two databases separately and in conjunction.

The correlation coefficients obtained are not very high, from which it is deduced that the environmental parameters considered in this work only partly explain the corrosion data. Such a high unexplained variance is attributable not only to the oversimplification of the mathematical model used, but also to the uncertainties inherent in the data and the probable occurrence of other variables with marked effects on corrosion that were not considered in the statistical treatment.

With the aim of improving the correlation coefficient, an attempt has also been made to include the variable of atmospheric air temperature in the statistical treatment. Considering the ISOCORRAG and MICAT databases together in conjunction, the correlation coefficients obtained for the four reference metals were 0.54, 0.68, 0.66 and 0.59, respectively, which hardly improves the explained percentage of the variance.

Table 8- *Damage functions obtained for the different reference metals in the ISOCORRAG and MICAT databases.*

Reference metal	N	Damage function	R	95% Confidence interval
MICAT				
Carbon steel	177	C = -0.742 + 8.527 TOW + 1.114 S + 1.239 Cl	0.76	a_1=[-23.333 , 21.748] a_2=[-31.982 , 49.037] a_3=[0.568 , 1.660] a_4=[1.057 , 1.421]
Zinc	137	C = -0.165 + 2.530 TOW + 0.003 S + 0.025 Cl	0.68	a_1=[-0.858 , 0.529] a_2=[1.264 , 3.796] a_3=[-0.015 , 0.022] a_4=[0.020 , 0.031]
Copper	150	C = -0.079 + 1.930 TOW + 0.022 S + 0.020 Cl	0.80	a_1=[-0.461 , 0.304] a_2=[1.241 , 2.618] a_3=[0.012 , 0.032] a_4=[0.017 , 0.023]
Aluminium	141	C = -0.276 + 0.709 TOW + 0.030 S + 0.012 Cl	0.75	a_1=[-0.625 , 0.073] a_2=[0.072 , 1.346] a_3=[0.021 , 0.039] a_4=[0.010 , 0.015]
ISOCORRAG				
Carbon steel	242	C = 5.213 + 37.692 TOW + 0.597 S + 0.144 Cl	0.68	a_1=[-5.310 , 15.736] a_2=[15.160 , 60.223] a_3=[0.473 , 0.721] a_4=[0.113 , 0.176]
Zinc	249	C = -0.082 + 2.126 TOW + 0.034 S + 0.011 Cl	0.69	a_1=[-0.830 , 0.665] a_2=[0.515 , 3.736] a_3=[0.025 , 0.043] a_4=[0.009 , 0.013]
Copper	248	C = 0.829 + 0.557 TOW + 0.010 S + 0.005 Cl	0.69	a_1=[0.507 , 1.152] a_2=[-0.146 , 1.259] a_3=[0.006 , 0.014] a_4=[0.004 , 0.006]
Aluminium	248	C = 0.299 + 0.761 TOW + 0.024 S + 0.001 Cl	0.56	a_1=[-0.107 , 0.706] a_2=[-0.126 , 1.647] a_3=[0.019 , 0.029] a_4=[0.000, 0.002]
MICAT + ISOCORRAG				
Carbon steel	409	C = 1.051 + 45.185 TOW + 0.727 S + 0.242 Cl	0.51	a_1=[-13.599 , 15.701] a_2=[16.676 , 73.694] a_3=[0.507 , 0.947] a_4=[0.185 , 0.298]
Zinc	386	C = -0.224 + 2.554 TOW + 0.030 S + 0.011 Cl	0.67	a_1=[-0.747 , 0.300] a_2=[1.507 , 3.602] a_3=[0.022 , 0.038] a_4=[0.010 , 0.013]
Copper	398	C = 0.376 + 1.698 TOW + 0.012 S + 0.005 Cl	0.63	a_1=[0.097 , 0.655] a_2=[1.145 , 2.251] a_3=[0.008 , 0.016] a_4=[0.004, 0.006]
Aluminium	389	C = 0.136 + 0.745 TOW + 0.027 S + 0.002 Cl	0.58	a_1=[-0.153 , 0.425] a_2=[0.165 , 1.325] a_3=[0.022 , 0.031] a_4=[0.001 , 0.003]

N = Number of data
R = Multiple correlation coefficient

Other Observations Regarding ISO 9223

Time of Wetness (TOW)

During the course of the MICAT project great difficulties were encountered when determining the TOW, estimated according to ISO 9223 as the number of hours/year (or annual fraction) during which relative humidity (RH) ≥80% and the air temperature (T) is simultaneously above 0°C.

The uncertainty in the determination of TOW derived from problems posed by the use of time of wetness monitors and the lack of precision of indirect measurements (e.g. in calculating the time at which RH exceeds a certain level) should be noted here. It is doubtful whether reported TOW values can be compared to one another as they were obtained by different procedures. With a view to the future revision of ISO 9223, it is believed that meteorological parameters (RH, number of rainy days and temperature) are more reliable, in view of the fact that they are easily obtainable from meteorological data banks.

Uncovered Corrosion Rates

In some of the testing stations in the MICAT project, first-year corrosion rates were found to be greater than the upper limit of the corrosion rate contemplated for the highest corrosivity category (C_5) defined by ISO 9223. This situation, though not very frequent, arose with all the reference metals, exposed in general in marine atmospheres close to the coast and with high airborne salinity values. This fact has also been reported in other studies [*13,14*].

The establishment of an upper limit for the corrosion rate in the highest corrosivity category (C_5) makes it impossible to classify the corrosivity of some atmospheres, which represent environments that are not beyond the scope of the international standard ISO 9223. Would it not be recommendable to avoid placing an upper limit on the corrosion rate in category C_5?

Conclusions

With a view to a future revision of ISO 9223, the following recomendations are noted:

- To distinguish between the two categories of pollution by sulphur-containing substances (SO_2), P_0 and P_1, and the two categories of pollution by airborne salinity (Cl^-), S_0 and S_1, which ISO 9223 considers in an undifferentiated way. A differentiation must be also made between zinc and copper, which ISO standard also considers jointly.

- Adjustment of the corrosivity category intervals of ISO 9223 on the basis of the environmental characteristics of the atmosphere (τ, S and P). For certain environmental conditions, this paper recommends corrosivity category intervals that are more in concordance with the experimental results obtained in the ISOCORRAG and MICAT projects.

- Differentiation between corrosivity categories C_1 and C_2 for the case of aluminium. The following limits are recommended with regard to corrosion rates

	Lower limit	Upper limit
C_1	$0.0\ g \cdot m^{-2} \cdot y^{-1}$	$0.3\ g \cdot m^{-2} \cdot y^{-1}$
C_2	$0.3\ g \cdot m^{-2} \cdot y^{-1}$	$0.6\ g \cdot m^{-2} \cdot y^{-1}$

- Replacement of the parameter time of wetness (TOW) with other related meteorological parameters (e.g. temperature, relative humidity, etc.) which are more reliable and easier to obtain from meteorological data banks.

- Not to establish an upper limit for the corrosion rate in the highest corrosivity category (C_5).

Acknowledgments

The authors would like to thank the CYTED Programme for providing the framework and financial support for coordination of the MICAT project, and in particular the International Coordinator of the Subprogramme "Corrosion: Environmental impact on materials" Dr. Leonardo Uller (Brazil). Many individuals and public and private bodies in Ibero-America have lent their support to the development of this project, and without their assistance it would not have been possible. But above all, the authors would like to acknowledge the work of the national coordinators of the remaining 12 participating countries: Argentina (B. Rosales), Brazil (M. Marrocos), Chile (G. Joseph), Colombia (A. Valencia), Costa Rica (J.F. Alvarez), Cuba (A. Cabezas), Ecuador (J. Peña), Mexico (J. Uruchurtu and L. Mariaca), Panama (A. Bósquez), Peru (G. Salas), Uruguay (S. Rivero) and Venezuela (M.R. Prato and O.T. de Rincón).

The authors would also like to thank to Dagmar Knotkova for supplying ISOCORRAG database.

References

[1] Morcillo, M., "Atmospheric Corrosion in Ibero-America: The Micat Project, " *Atmospheric Corrosion, ASTM STP 1239*, W. W. Kirk and H. H. Lawson, Eds., American Society for Testing and Materials, Philadelphia, 1995, pp. 257-275.

[2] Knotkova, D. and Vrobel, L., "ISOCORRAG: The International Testing Program with ISO/TC156/WG 4. " *Proc. 11th International Corrosion Congress*, Vol. 5, AIM, Milano, 1990, p. 5.581.

[3] Several authors, *Corrosión y Protección de Metales en las Atmósferas de Iberoamérica. Parte I – Mapas de Iberoamérica de Corrosividad Atmosférica*, M. Morcillo, E. Almeida, B. Rosales, J. Uruchurtu and M. Marrocos, Eds., CYTED, Madrid, 1999.

[4] Almeida, E., Morcillo, M., Rosales, B., Marrocos, M., *Materials and Corrosion* 51, 2000, pp.859-864.

[5] Almeida, E, Morcillo, M., Rosales, B., Marrocos, M., *Materials and Corrosion* 51, 2000, pp.865-874.

[6] Almeida, E, Morcillo, M., Rosales, B., *British Corrosion Journal*, 35, 2000, pp.284-288.

[7] Almeida, E., Morcillo, M., Rosales, B., *British Corrosion Journal*, 35, 2000, pp.289-296.

[8] Morcillo, M., Almeida, E., Marrocos, M., Rosales, B., *Corrosion (NACE)*, 57, 11, 2001, pp. 967-980.

[9] Morcillo, M., Almeida, E., Rosales, B., *Aluminium*, Vol. 76, 2000, pp. 316-319.

[10] Morcillo, M., Almeida, E., Rosales, B., *Aluminium*, Vol. 76 ,2000, pp. 610-615.

[11] Morcillo, M., Almeida, E., Rosales, B., *Aluminium*, Vol. 76, 2000, pp. 1066-1070.

[12] Morcillo, M., de la Fuente, D., Chico, B., Almeida, E., Rivero, S., Rosales, B. and Joseph, G., "Atmospheric Corrosion of Reference Metals in Antarctic Sites," *Symposium on cold climate corrosion. Northern Area NACE Conference*, 2001.

[13] King, G.A. and Duncan, J.R., *Corrosion and Materials (Australia)*, Vol. 23, Nº1, 1998.

[14] Corvo, F., in *"Corrosion metálica"*, L. Maldonado, P. Castro and L. Díaz, Eds., Consejo Nacional de Ciencia y Tecnología, Mérida (Yucatán), México.

Johan Tidblad[1], Vladimir Kucera[1], Alexandre. A. Mikhailov[2], and Dagmar Knotkova[3]

Improvement of the ISO classification system based on dose-response functions describing the corrosivity of outdoor atmospheres

---

Reference: Tidblad, J., Kucera, V., Mikhailov, A. A., and Knotkova, D., **"Improvement of the ISO classification system based on dose-response functions describing the corrosivity of outdoor atmospheres,"** *Outdoor Atmospheric Corrosion, ASTM STP 1421*, H. E. Townsend, Ed., American Society for Testing and Materials International, West Conshohocken, PA, 2002.

Abstract: The investigation on the effects of environmental parameters on the atmospheric corrosion of technical metals performed in several field and laboratory exposures starting from the 1930's have created a basis for classification of the corrosivity of atmospheres. The present ISO 9223 "*Corrosion of metals and alloys – Corrosivity of atmospheres – Classification*" standard was at the time of its development in the 1970-80's a big step forward as it was the first classification system describing the corrosivity categories in a quantitative way. The standard has found a broad application and has been adopted in other national and international standards. It was mainly based on data from the temperate regions of Europe and Northern America, which were available at that time. Since the publication of the standard certain limitations have become obvious and many remarks concern limitations with respect to covering climatic influences. This paper will focus on the development of dose-response functions describing the corrosivity of outdoor atmospheres and the possible improvements that can be made of the standard using the developed functions.

Keywords: atmospheric corrosion, dry deposition, modelling, SO_2, chloride, time of wetness, temperature

Introduction

Earlier standards for classification of corrosivity of atmospheric environments used qualitative verbal description like 'urban', 'rural', 'industrial' and 'marine' to describe the environment. On one hand these descriptions were user friendly since anyone can relate to their meaning but on the other hand the descriptions were subjective and could easily be misleading. The present ISO 9223 "*Corrosion of metals and alloys – Corrosivity of atmospheres – Classification*" standard was at the time of its development in the 1970-

---

[1]Doctors of Sciences, Swedish Corrosion Institute, Kräftriket 23, SE-104 05 Stockholm, Sweden.

[2]Doctor of Sciences, Institute of Physical Chemistry, Russian Academy of Sciences, Leninskij Prospekt 31, 117915 Moscow, Russian Federation.

[3]Doctor of Sciences, SVUOM Praha a. s., U Mestanského Pivovaru 4, CZ-17004.

80's a big step forward as it was the first classification system describing the corrosivity categories in a quantitative way. The standard uses measurements of corrosion and environmental parameters and is based on the data that was available at that time.

In 1986 the ISOCORRAG programme was initiated in the framework of ISO / TC 156 / WG 4 - *Corrosion of metals and alloys - Classification of corrosivity of atmospheres*. It is an international exposure programme with more than 50 sites located in Europe, Argentina, Canada, Japan, New Zealand and USA [*1,2*]. The programme was initiated to get a systematic database including different types of climate and environment since the present standard is based mostly on data in a restricted interval from temperate sites in the temperate zone and it was already then an awareness that the system could be improved especially for very cold and warm regions.

The purpose of ISO 9223 is to describe how to obtain a corrosivity category for carbon steel, zinc, copper or aluminium. The corrosivity category, named from C1 to C5 where C1 corresponds to a very low and C5 corresponds to a very high corrosivity, is a technical characteristic, which provides a basis for the selection of materials and protective measures. It should be especially noted that the corrosivity category is not a general characteristic of the environment and, consequently, the category may for a particular environment be different depending on which metal (carbon steel, zinc, copper or aluminium) is used in the assessment of the corrosivity category.

ISO 9223 provides two ways of obtaining the corrosivity category, by measurement of the corrosion rate of standard specimens and by measurement of environmental parameters together with a lookup table. The first method is more reliable and is included in the normative part of the standard as determination of the category while the second, which is included in the informative part of the standard, is intended for assessment/derivation of the corrosivity category. This paper will focus on the second way of assessing the corrosivity category, in particular the development of dose-response functions describing the corrosivity of outdoor atmospheres and the possible improvements that can be made using the developed functions. There are other issues that needs to be decided before releasing a new version of the standard, for example the use of flat and/or helix specimens and specifications of standard specimens used for evaluation of corrosion attack but these are briefly discussed in another paper of the present symposium [*3*].

Classification of corrosivity based on environmental data in the present ISO 9223 standard

This section is only a summary of the relevant parts of the standard and is included for reference purposes. Orders for ISO International Standards and other ISO publications should be addressed to the ISO members (http://www.iso.ch/iso/en/aboutiso/isomembers/ MemberCountryList.MemberCountryList), which are the primary ISO sales agents in their country. For example, in the United States requests should be directed to the American National Standards Institute (ANSI). For those who would like to provide TC 156 / WG 4 with feedback the convenor of WG4 can be reached through the Czech Standards Institute (CSNI), Biskupsky dvur 5, CZ-110 02 Praha 1. (phone: +420 2 21 80 21 11; fax: +420 2 21 80 23 11; e-mail: internat.dept@csni.cz).

Table 1 shows the corrosivity categories for the different metals. It should be especially noted that the highest category (C_5) has specified upper limits. Measured values above these limits have been detected in the field but they are outside the scope of the present standard.

Three environmental parameters are used for the assessment of corrosivity categories: Time of wetness (TOW), sulphur compounds based on sulphur dioxide (SO_2) and airborne salinity contamination (Cl). For these parameters classification categories are defined as τ (TOW), P (SO_2) and S (Cl), based on measurements of the parameters (Table 2).

Table 1 – *ISO 9223 Corrosivity categories for carbon steel zinc, copper and aluminium based on corrosion rates.*

Corrosivity	Category	C. steel $\mu m\ yr^{-1}$	Zinc $\mu m\ yr^{-1}$	Copper $\mu m\ yr^{-1}$	Aluminium $g\ m^{-2}\ yr^{-1}$
Very low	C_1	≤ 1.3	≤ 0.1	≤ 0.1	negligible
Low	C_2	1.3 - 25	0.1 - 0.7	0.1 - 0.6	≤ 0.6
Medium	C_3	25 - 50	0.7 - 2.1	0.6 - 1.3	0.6 - 2
High	C_4	50 - 80	2.1 - 4.2	1.3 - 2.8	2 - 5
Very high	C_5	80 - 200	4.2 - 8.4	2.8 - 5.6	5 - 10

Table 2 – *Classification of time of wetness (TOW), sulphur compounds based on sulphur dioxide (SO_2) expressed as concentration, $\mu g\ m^{-3}$, or deposition rate, $mg\ m^{-2}\ d^{-1}$ and airborne salinity contamination (Cl) expressed as deposition on the "wet candle" according to ISO 9225 "Corrosion of metals and alloys – Corrosivity of atmospheres – Measurement of pollution".*

TOW	$h\ yr^{-1}$
τ_1	≤ 10
τ_2	10 - 250
τ_3	250 - 2500
τ_4	2500 - 5500
τ_5	> 5500

SO_2	$\mu g\ m^{-3}$	$mg\ m^{-2}\ d^{-1}$
P_0	≤ 12	≤ 10
P_1	12 - 40	10 - 35
P_2	40 - 90	35 - 80
P_3	90 - 250	80 - 200

Cl	$mg\ m^{-2}\ d^{-1}$
S_0	≤ 3
S_1	3 - 60
S_2	60 - 300
S_3	300 - 1500

TOW is estimated from the temperature-humidity (T-RH) complex as the length of time when RH is greater than 80 % at a T value greater than 0 °C. TOW calculated by this method does not necessarily correspond with the actual time of exposure to wetness because wetness is influenced by many other factors. However, for classification purposes the calculation procedure is usually sufficiently accurate.

Given the categories for TOW, SO_2 and Cl as defined in Table 2, the standard provides different lookup tables for carbon steel, zinc, copper and aluminium, which enables the estimation of the corrosivity categories as specified in Table 1. Part of the lookup table for carbon steel is given in Table 3 as an example.

Table 3 – *Estimated corrosivity category for unalloyed carbon steel for TOW category τ_3, τ_4 and τ_5 and different SO_2 (P_0-P_3) and Cl (S_0-S_3) categories.*

	τ_3			τ_4			τ_5		
	S_0-S_1	S_2	S_3	S_0-S_1	S_2	S_3	S_0-S_1	S_2	S_3
P_0-P_1	2-3	3-4	4	3	4	5	3-4	5	5
P_2	3-4	3-4	4-5	4	4	5	4-5	5	5
P_3	4	4	4-5	5	5	5	5	5	5

Possibilities of revision of ISO 9223 classification of corrosivity based on environmental data

Within ISO / TC 156 / WG 4 there is an on-going discussion regarding possible ways of improving the present standard. The first would be to verify the lookup tables using the recent data obtained from ISOCORRAG and other exposure programmes. However, there is a principal obstacle to overcome, which is illustrated in Table 4. Even after assembling data from many sites there are still significant gaps in combinations of climate and pollution that makes it impossible to update the tables simply based on corrosivity measurements. This is especially true if more environmental parameters are added, as will be discussed in the following, in order to increase the applicability of the standard. Therefore it is important to develop dose-response functions describing the corrosion attack in terms of the environmental parameters. These can then either be directly incorporated in the standard or used to improve the lookup tables where no experimental data are available.

Table 4 – *Number of sites with measured values of unalloyed carbon steel in the joint ISOCORRAG-MICAT and Russian database used in this paper for TOW category τ_3, τ_4 and τ_5 and different SO_2 (P_0-P_3) and Cl (S_0-S_3) categories.*

	τ_3			τ_4			τ_5		
	S_0-S_1	S_2	S_3	S_0-S_1	S_2	S_3	S_0-S_1	S_2	S_3
P_0-P_1	28			55	11	2	11	4	1
P_2	2			7	2		3		1
P_3				1					

There are many parameters that can influence the damage to materials, it is an interplay between chemical, physical and biological parameters. The standard only includes three of the most important and a natural way to improve the standard would be to include more parameters. Two key questions have been discussed: extension of the characterisation of climate by inclusion of T or by replacing TOW with T and RH, and inclusion of the effect of wet deposition. In the end, and as will be elaborated in the following, it was decided by ISO / TC 156 / WG 4 to complement the standard by the inclusion of T but not to include any other parameters. Thus the model development and development of dose-response functions have been restricted to functions involving the parameters TOW, T, SO_2 and Cl.

The temperature - relative humidity - time of wetness complex

Many remarks to the present standard concern limitations with respect to climate, for example, TOW seems to be inadequate to describe the situation in tropical marine environments [*4*]. This indicates that TOW as defined in the standard does not capture all aspects of the climate. In particular, very cold and very warm sites tend to result in corrosion rates that are either outside the scope of the present standard or do not agree with the lookup tables. Therefore the T-RH-TOW complex was investigated in more detail. This resulted in a model that can be used for calculation of TOW based on annual averages of T and RH [*5*]. Compared to the procedure of estimating TOW from almost continuous measurements of T and RH the procedure is very simple and timesaving. The explained variability of the model is relatively high (R^2=0.73), which is sufficiently good considering that both procedures are calculated from the T-RH complex and are thus estimates of the 'true' time of wetness. The model is illustrated in Figure 1 together with experimental annual averages of T and RH included in the database used for estimation of dose-response functions.

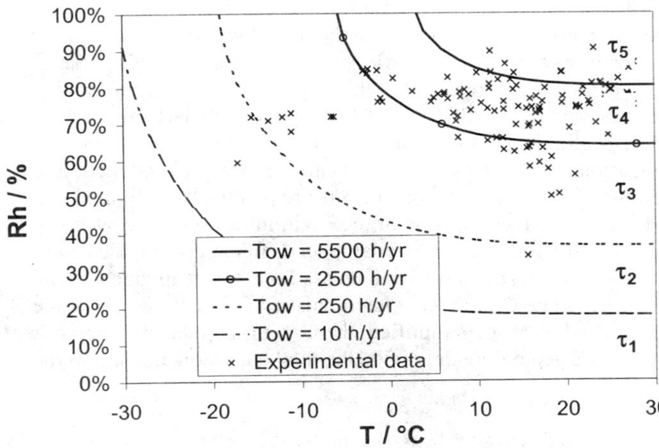

Figure 1 – *Calculated time of wetness category divisions based on a relative humidity and temperature model for calculation of time of wetness using annual data [5] and experimental annual averages of temperature and relative humidity from the joint ISOCORRAG-MICAT and Russian database.*

Since it is not sufficient with only TOW as a description of climate a natural step would be to replace TOW with T and RH. This was done in the analysis of data from the International Co-operative Programme on Effects on Materials including Historic and Cultural Monuments (ICP Materials) [*6*]. However, after discussions within ISO / TC 156 / WG 4 it was decided to instead add T to the present set of parameters. The decision was a compromise between what was considered to be best from a scientific point of view and what is more practical considering that the TOW concept in the present standard is widely used and accepted. The drawback is that the correlation between TOW and T is higher than the correlation between T and RH, which effects the certainty of the estimated coefficients.

Wet deposition

In cold and temperate climates the effect of dry deposition dominates over wet deposition. There are some indications, however, that the effect of wet acid deposition in warmer areas, *e.g.* southern China, could be higher than what could be expected when extrapolating data from Europe and comparable in magnitude to that of dry deposition [*7*]. It is also possible to have a beneficial effect of rain due to washing of previously deposited pollutants. Considering the availability of data, however, it is not possible at the present stage to include the effect of wet deposition in the developed dose-response functions.

General model description

This section is a summary of a paper in which the model was originally described by using unalloyed carbon steel as an example [*8*]. It should be noted that the dose-response function for carbon steel presented in this paper differs slightly from the previous due to the extension of the ISOCORRAG database with data from MICAT and Russian sites as described in the next section.

Atmospheric corrosion is a complex process. It can therefore be modelled in several ways and the selections of environmental parameters are practically unlimited. The present model is based on a characterisation of the environment by using, as much as possible, environmental parameters easily available on different geographical scales. The advantage of this is obvious, the model can be used by almost anyone without collection of sophisticated data. On the other hand, this requirement may for some locations enforce the model to be too simplified, thereby missing the incorporation of important phenomena. The general model divide the total corrosion attack into three dominating parts

$$K = f_{dry}(SO_2) + f_{dry}(Cl) + f_{wet}(H^+) \qquad (1)$$

where

K = total corrosion effect
$f_{dry}(SO_2)$ = effect of dry deposition of SO_2
$f_{dry}(Cl)$ = effect of dry deposition of chlorides
$f_{wet}(H^+)$ = effect of wet deposition of H^+ (acid rain)

At present stage it has not been possible to develop a model that includes all three contributions described in equation 1 due to lack of sufficient amount of data from field exposure programs. ICP Materials [*6*] do not measure dry deposition of chlorides while ISOCORRAG [*1*] do not include wet deposition of H^+. For the purposes of ISO 9223 only the dry deposition terms of SO_2 and Cl are needed and they will be elaborated individually in the following.

SO_2 polluted atmospheres

SO_2 is the main anthropogenic contributor to atmospheric corrosion. Chamber experiments and dry deposition tests have shown that the uptake of SO_2 by a given surface is significantly accelerated when the surface is wet. Therefore it is natural to express the SO_2 term in equation 1 as follows [8]

$$f_{dry}(SO_2) = A\ SO_2^{B}\ TOW^{C} \tag{2}$$

where

$f_{dry}(SO_2)$ = effect of dry deposition of SO_2
A, B and C = constants
SO_2 = SO_2 concentration
TOW = Time of wetness

This form was used in the analysis of data from ICP Materials [6], except that the expression for TOW was replaced with an expression involving the T-RH complex. The temperature part of this complex, which was valid for almost all metals, is schematically illustrated in Figure 2. According to the ICP Materials results, the effect of temperature has for many materials a maximum at about 9-11°C annual mean temperature. The increasing part in Figure 2 (a) can, in agreement with the T-RH-TOW model [5], be related to the increase of the time when temperature is above 0°C since this quantity is correlated with the annual average temperature. The decreasing part in Figure 2 (b), is interpreted as due to a negative correlation of ambient relative humidity and temperature values and also due to periods with a surface temperature above the ambient temperature, partly related to sun radiation. An elevated surface temperature leads to a faster evaporation of moisture after rain or condensation periods and to a decrease of the

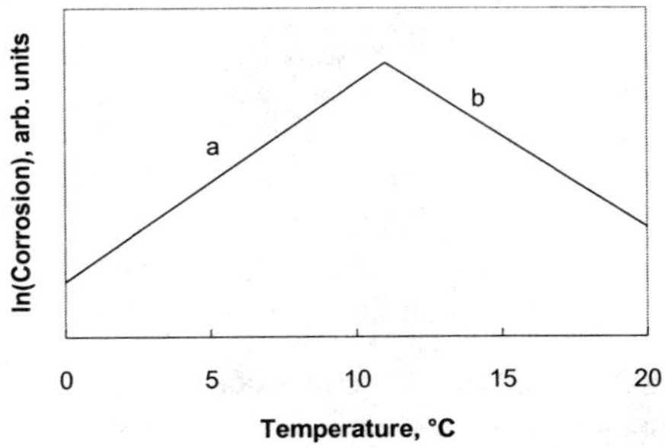

Figure 2 – *Schematic representation of the observed temperature dependence for many materials: a) increase of corrosion with temperature in the low temperature range and b) decrease of corrosion with temperature in the high temperature range*

thickness of the adsorbed water layer and, consequently, to a decrease of the time when the metal surface is wet. Thus, the intense solar radiation in inland tropics causes the time when surfaces are wet during unsheltered exposure to become relatively short.

In the present model, however, we are restricted to use the parameters TOW and T and the following expression has been used

$$f_{dry}(SO_2) = A \cdot SO_2^{B} \cdot TOW^{C} \cdot \exp\{g(T)\} \quad (3)$$

where

$f_{dry}(SO_2)$ = effect of dry deposition of SO_2
A, B and C = constants
SO_2 = SO_2 concentration
TOW = Time of wetness
g(T) = a temperature (T) function

In the low temperature range TOW increases with T while at about 10 °C. This means that g(T) may be independent of temperature below 10 °C, due to the correlation with TOW. As will be shown later, this behaviour is observed for zinc and aluminium but not for copper and steel, where an additional temperature effect besides that included in TOW is needed for low temperatures. For higher temperatures TOW is relatively independent of T, which means that g(T) will be similar to the temperature dependence given in Figure 2 (b).

Cl^- containing atmospheres

It is known that chlorides are a main accelerating factor of atmospheric corrosion in coastal regions. Moreover, which is important to stress, atmospheric corrosion in a chloride-containing atmosphere increases with temperature. The following equation was successfully used to describe ISOCORRAG data for carbon steel [*8*]

$$f_{dry}(Cl) = D \cdot Cl^{E} \cdot TOW^{F} \cdot \exp\{kT\} \quad (4)$$

where

$f_{dry}(Cl)$ = effect of dry deposition of chlorides
D, E, F and k = constants
Cl = dry deposition of chlorides
TOW = Time of wetness
T = temperature

In contrast to the expression for SO_2, Equation 4 describes corrosion in marine atmospheres that increases with temperature exponentially even above 10°C annual temperature. As will be shown below this is also valid for the other metals. The corrosion rate of most metals is strongly affected by the concentration of chlorides on the surface. One reason is that the chlorides have hygroscopic properties and thus contribute to the creation of an electrolyte layer. This leads to a prolongation of the periods when the surface is wet even at higher temperatures.

Extension of the database with data from MICAT and Russian sites

Before showing the main results an important change of the database, compared to the previous paper [*8*], needs to be mentioned. The present database has been extended to include also MICAT [*9*] and Russian sites from cold regions (Atka, Aldan, Bilibino, Komsomol'sk-na-Amure, Klyuchi, Oymyakon, Pobedino, Susuman, Tynda, Ust'-Omchug) in order to increase the number of sites with high and low temperature (Figure 3). Table 5 shows the range of the environmental parameters.

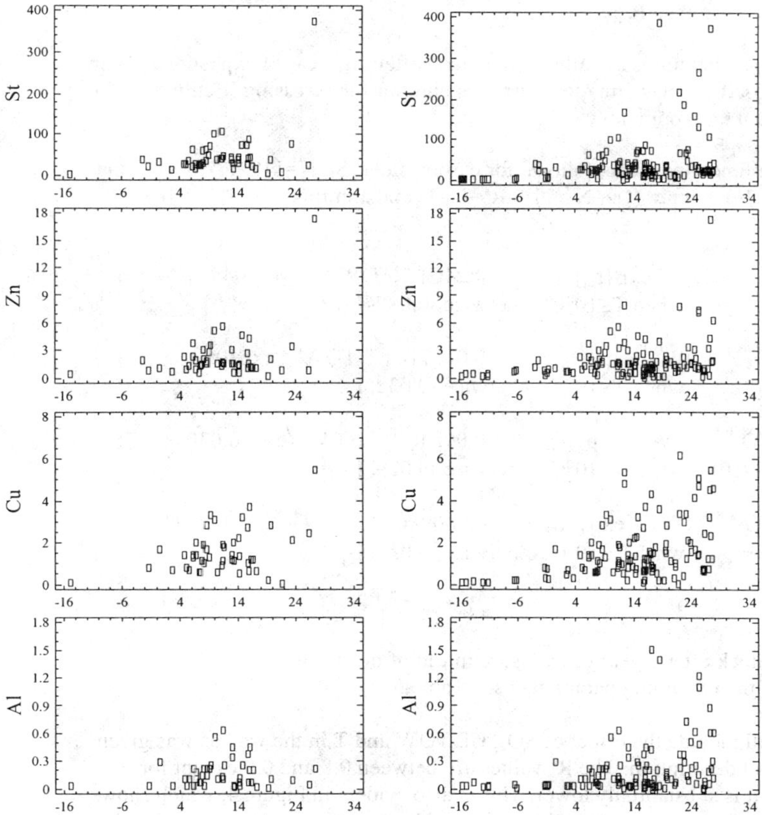

Figure 3 – *Yearly corrosion in μm/year of Carbon Steel, Zinc, Copper and Aluminium vs. average annual temperature (T) in °C: left – ISOCORRAG data, right – ISOCORRAG-MICAT and Russian data.*

Table 5 – *Parameters used in dose-response functions including symbol, description, interval measured in the programme and unit. All parameters are expressed as annual averages.*

Symbol	Description	Interval	Unit
T	Temperature	-17.1 – 28.7	°C
TOW	Time of wetness	206 – 8760	$h\ yr^{-1}$
SO_2	SO_2 deposition	0.7 – 150.4	$mg\ m^{-2}\ day^{-1}$
Cl	Cl^- deposition	0.4 – 699.6	$mg\ m^{-2}\ day^{-1}$

Derivation of dose-response functions

The dose-response functions are all based on data after one year of exposure and can therefore only be used for classification purposes and not for assessing lifetimes of the materials in different environments.

The dose-response functions are as follows for carbon steel (St, N=119, R^2=0.87), zinc (Zn, N=116, R^2=0.78), copper (Cu, N= 114, R^2=0.81) and aluminium (Al, N= 108, R^2=0.61):

$$C_{St} = 0.085SO_2^{0.56}TOW^{0.53}\exp\{f_{St}\} + 0.24Cl^{0.47}TOW^{0.25}\exp\{0.049T\} \quad (5)$$
$$f_{St}(T) = 0.098(T-10) \quad \text{when } T \leq 10°C\text{, otherwise } -0.087(T-10)$$

$$C_{Zn} = 0.0053SO_2^{0.43}TOW^{0.53}\exp\{f_{Zn}\} + 0.00071Cl^{0.68}TOW^{0.30}\exp\{0.11T\} \quad (6)$$
$$f_{Zn}(T) = 0 \quad \text{when } T \leq 10°C\text{, otherwise } -0.032(T-10)$$

$$C_{Cu} = 0.00013SO_2^{0.55}TOW^{0.84}\exp\{f_{Cu}\} + 0.0024Cl^{0.31}TOW^{0.57}\exp\{0.030T\} \quad (7)$$
$$f_{Cu}(T) = 0.047(T-10) \quad \text{when } T \leq 10°C\text{, otherwise } -0.029(T-10)$$

$$C_{Al} = 0.00068SO_2^{0.87}TOW^{0.38}\exp\{f_{Al}\} + 0.00098Cl^{0.49}TOW^{0.38}\exp\{0.057T\} \quad (8)$$
$$f_{Al}(T) = 0 \quad \text{when } T \leq 10°C\text{, otherwise } -0.031(T-10)$$

where

C_{Me} = Corrosion attack after 1 year of exposure in μm of metal Me
For explanation of the remaining parameters see Table 5.

The functions all include the effect of SO_2, Cl, TOW and T in the way as was given in the general model description. The R^2 values are between 0.8 and 0.9 except for aluminium, where it is substantially lower. This is also evident in Figure 4, which shows the observed and predicted values for all functions. Aluminium experiences localised corrosion but the corrosion attack is calculated as uniform corrosion, which might be one reason for the low value. As is specified in the present standard maximum pit depth is a better indicator of the potential damage but this characteristic can not be evaluated after the first year of exposure.

The temperature functions differ from metal to metal in that the functions for carbon steel and copper (f_{St} and f_{Cu}), in contrast to zinc and aluminium (f_{Zn} and f_{Al}), show stronger temperature dependence for the SO_2 term in the cold temperature region than what is implicitly included in TOW. This is in qualitative agreement with results from the UN ECE ICP Materials [*6*] where the coefficients in the temperature functions for zinc, 0.062(T-10), and aluminium, 0.031(T-10), were significantly lower than that in the function for copper, 0.083(T-10). It should be noted that the ICP Materials functions include RH instead of TOW, i.e., no implicit temperature dependence is included in the ICP Materials functions. For carbon steel no dose-response function including the effect of temperature has so far been presented from ICP Materials.

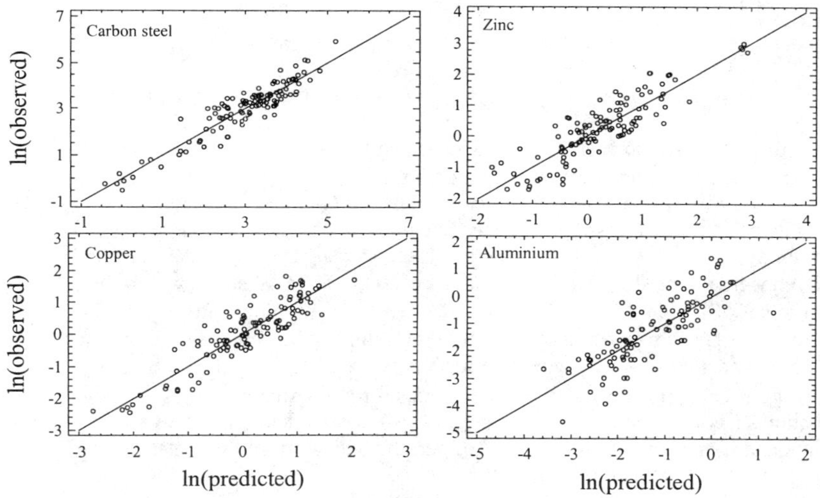

Figure 4 – *Observed vs. predicted values (logarithmic) for carbon steel (Eq. 5), zinc (Eq. 6), copper (Eq. 7) and aluminium (Eq. 8)*

The obtained dose-response functions can be used in several ways for the development of ISO 9223 but it is not in the scope of the present paper to discuss these options in detail and to recommend any procedure. Instead, different examples will be shown, with the main aim to illustrate the behaviour of the functions in graphical form since the functions in themselves are relatively complicated and non-descriptive.

Before illustrating the functions, the way of presenting the data, in particular the P and S categories should be described. Recalling Table 2, the SO_2 (P) and Cl (S) categories span a relatively large interval and with individual ranges for the different classes. If the data were to be presented using SO_2 and Cl as they are, for the intervals covering P_0-P_3 and S_0-S_3, respectively, the two upper categories (2-3) would completely dominate the picture since they represent more than 80% of the full intervals. Since the lower categories (0-1) are of equal interest this would be unsatisfactory. Instead, the transformation presented in Figure 5, were used in order to show the classes as continuos variables.

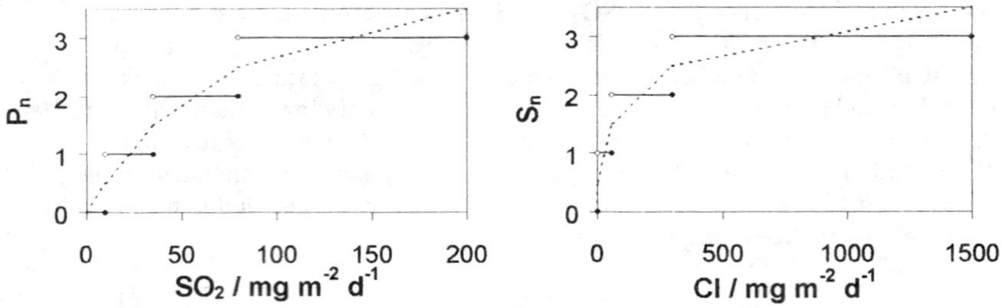

Figure 5 – *Definition of the 'continuos categories' used for illustration purposes in Figure 6 and 7 (-------) and the relation to the corrosivity categories defined in ISO 9223 (——).*

Figure 6 show calculated corrosivity categories for carbon steel based on Eq. 5. with TOW values restricted to τ_4 and T values restricted between 10 and 15 °C. As is seen from the figure the procedure of categorising environmental parameters introduces a large uncertainty in the estimation of corrosivity categories. Look for example in the box corresponding to P_1-S_1. There, the corrosivity category ranges from low (C_2) to very high (C_5) depending on different combinations of SO_2, Cl, TOW and T within the given restrictions P_1, S_1, τ_4 and $10°C<T<15°C$. Also worth mentioning is the area in the P_3-S_3 box above C_5. This corresponds to corrosion rates that are beyond the scope of the present standard. The area above C_5 would be further extended to lower P and S categories for the τ_5 category and to lower S categories for higher temperatures.

Figure 7 shows a comparison of the metals in the P-S domain taking the C_3-C_4 category boundary as an example. Except for copper, the behaviour of the metals are similar for this set of climatic conditions.

Last, the temperature effect is illustrated in Figure 8 taking zinc as an example. When Cl is low the temperature dependence shows a maximum. When Cl is high the increase of T leads to the increase of corrosion, confirming the high corrosion values in tropical marine environments [8]. In figure 8 the SO_2 and Cl levels are taken as constants while the TOW level varies according to the function given in Figure 9. The reason for using this curve as a basis for the illustration rather than a constant TOW level is that for low temperatures the temperature poses a restriction of the upper limit of TOW values [5]. The TOW function given in figure 9 represents average conditions of outdoor experimental data but the deviation from the curve can be substantial for an individual sites as can be seen in Figure 1 for relative humidity.

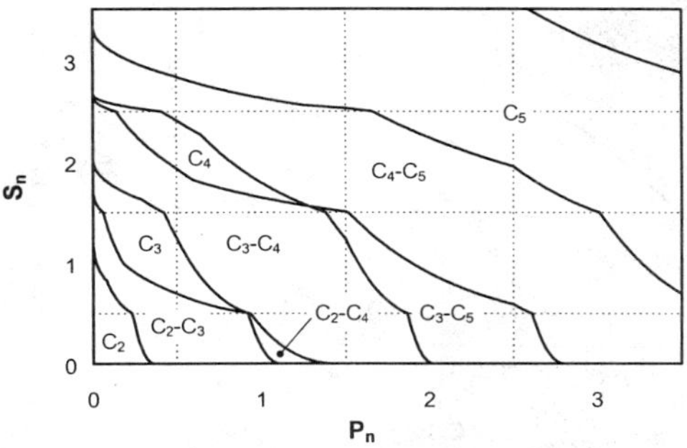

Figure 6 – *Calculated corrosivity categories for carbon steel (C_2-C_5) based on a dose-response function (Eq. 5) including SO_2 (P_0-P_3), Cl (S_0-S_3), TOW (τ_4) and T (10-15°C)*

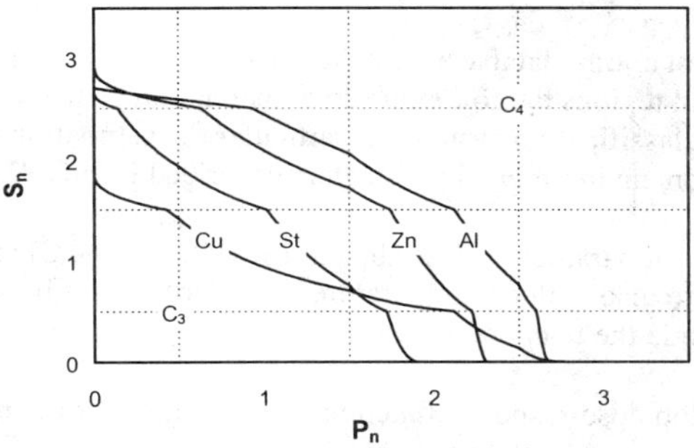

Figure 7 – *Calculated category boundaries C_3-C_4 based on dose-response functions (Eq. 5-8) and assuming constant values of TOW = 2500 h yr^{-1} and T = 10°C.*

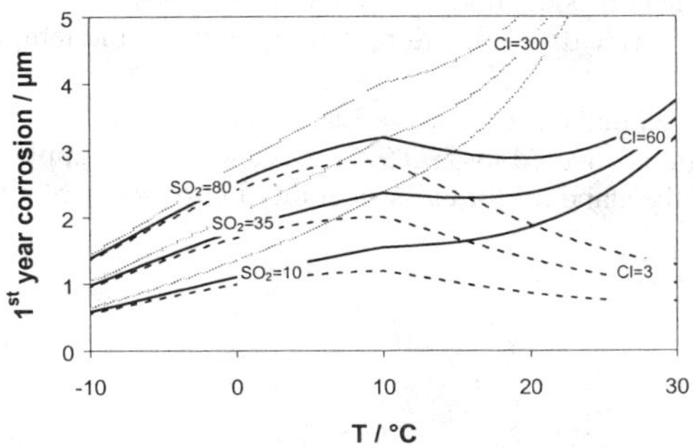

Figure 8 – *Calculated corrosion attack on zinc based on a dose-response function (Eq. 6) and different combinations of SO_2 and Cl as a function of temperature. TOW varies with T according Figure 9 for all curves.*

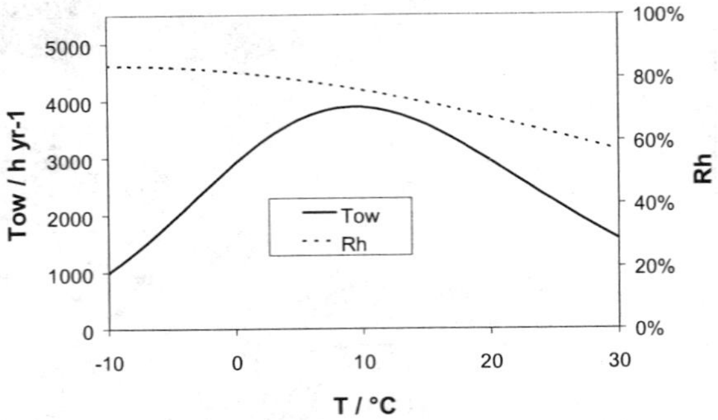

Figure 9 – *Average experimental data of RH and TOW as a function of T.*

Conclusions

Based on analysis of data from a joint database with data from ISOCORRAG, MICAT and some Russian sites in cold regions the following conclusions can be made regarding the improvement of the ISO classification system, in particular the estimation of corrosivity categories based on environmental parameters described in ISO 9223:

- The set of environmental parameters described in ISO 9223 (SO_2, Cl and TOW) needs to be extended to also include temperature, in order to describe the situation in areas outside the temperate region;

- It is essential to develop dose-response functions for calculation of corrosion attack in terms of the environmental parameters;

- Dose-response functions for carbon steel, zinc, copper and aluminium has been obtained describing the corrosion attack after 1 year of exposure as a function of SO_2 dry deposition, chloride dry deposition, time of wetness and temperature;

- An analysis of the developed dose-response functions in the range of environmental conditions covered by ISO 9223 has shown their applicability in the future revision of the standard for classification of corrosivity of atmospheric environments.

References

[1] Knotkova, D., "Atmospheric Corrosivity Classification. Results of the International Testing Program ISOCORRAG," *Corrosion Control for Low-Cost Reliability, 12th International Corrosion Congress, Vol. 2, Progress Industries Plant Operations,* NACE International, Houston, Texas, 1993, pp. 561-568.

[2] Dean, S.W., "ISO CORRAG Collaborative Atmospheric Exposure Program: A Preliminary Report," *Degradation of Metals in Atmosphere, ASTM STR 965,* S.W. Dean and T.S. Lee, Eds., American Society for Testing and Materials, West Conshohocken, PA, 1988, pp. 385-431.

[3] Knotkova, K., Kucera, V., Dean, S. W. and Boschek, P., "Classification of the Corrosivity of Atmosphere – Standardized Classification System and Approach for Adjustment", *Outdoor and Indoor Atmospheric Corrosion, ASTM STP 1421*, H. E. Townsend, Ed., American Society for Testing and Materials, West Conshohocken, PA, 2002.

[4] King, G.A., Duncan, J.R., "Some Apparent Limitations in using the ISO Atmospheric Corrosivity Categories," Corrosion & Materials Vol. 23, No. 1, 1998. Pp. 8-14 & 22-25.

[5] Tidblad, J. , A. A. Mikhailov and V. Kucera, "A model for calculation of time of wetness using relative humidity and temperature data.", Proceedings of the 14th International Corrosion Congress, Cape Town, South Africa, 1999.

[6] Tidblad, J., V. Kucera, A. A. Mikhailov, J. Henriksen, K. Kreislova, T. Yates, B. Stöckle and M. Schreiner, "Final dose-response functions and trend analysis from the UN ECE project on the effects of acid deposition.", Proceedings of the 14th International Corrosion Congress, Cape Town, South Africa, 1999.

[7] J. Tidblad, V. Kucera and A. A. Mikhailov "Mapping of Acid Deposition Effects and Calculation of Corrosion Costs on Zinc in China", *Water, Air and Soil Pollution*, accepted for publication.

[8] J. Tidblad, A. A. Mikhailov and V. Kucera, "Application of a Model for Prediction of Atmospheric Corrosion in Tropical Environments", in *Marine Corrosion in Tropical Environments, ASTM STP 1399*, S. W. Dean, G. H.-D. Delgadillo and J. B. Bushman, Eds., American Society for Testing and Materials, West Conshohocken, PA, 2000, pp. 18-32.

[9] "Functiones de Dano (Dosis/Respuesta)de la Corrosion Atmospherica en Iberoamerica," *Corrosion y Proteccion de Metales en las Atmosferas de Iberoamerica,* M. Morcillo, E. M. Almeida, B. M. Rosales *et al.* Eds., Programma CYTED, Madrid, Spain, 1998, pp. 629-660.

Carlos Arroyave,[1] Felix Echeverria,[1] Francisco Herrera,[1] Juan Delgado,[1] Daira Aragon,[1] and Manuel Morcillo[2]

NO_2 Measurements in Atmospheric Corrosion Studies

Reference: Arroyave, C., Echeverria, F., Herrera, F., Delgado, J., Aragon, D., and Morcillo, M., **"NO_2 Measurements in Atmospheric Corrosion Studies,"** *Outdoor Atmospheric Corrosion, ASTM STP 1421*, H. E. Townsend, Ed., American Society for testing and Materials International, West Conshohocken, PA, 2002.

Abstract: Recent trends of higher energy consumption, along with new combustion technologies and drastic reduction of SO_2 emissions, have led to an increased relevance of the effect of other atmospheric pollutants on material stability. As a significant constituent of this group of new contaminants, nitrogen oxides have been considered as the cause of several atmospheric corrosion problems. Consequently, measurement of the content of NO_2 in the atmosphere has become important. Looking for an easy analytical procedure to aid in conventional atmospheric corrosion studies, a technique based on the diffusion tube principles is introduced in this work. The results show good promise. Theoretical considerations, experimental results, their analysis and conclusions are included here, as a contribution to a better understanding of the atmospheric corrosion phenomena.

Keywords: Nitrogen dioxide, atmospheric pollution, diffusion tubes, pollution measurements

Introduction

The significant influence of some atmospheric pollutants on the stability of materials has led to corrosion studies aiming to obtain information regarding pollution levels of species such as chlorides and sulfur dioxide. In general, such pollution levels are measured as the amount of pollutant species deposited on a unit of surface in a unit of time, and the candle is the measuring method most employed for these species.

In the past few years, other pollutants have become important and consequently they have gained attention concerning their potential effects. Nitrogen oxides (especially nitrogen dioxide) are included in this group. They are emitted into the urban atmosphere as a by-product of combustion processes, primarily from vehicle exhausts [*1*]. Most of those emissions are nitrous oxide (NO), which then reacts with ozone (O_3) to form nitrogen dioxide (NO_2) [*2*]. Nevertheless, atmospheric corrosion studies have not yet included either the measurement or a systematic consideration of the influence of these

---

1Professor, lecturer, lecturer, lecturer and student respectively, Corrosion and Protection Group, University of Antioquia, P.O. Box 1226, Medellin, Colombia

2Director, CENIM, CSIC Av. Gregorio del Amo, 8. 28040, Madrid, Spain.

pollutant species. One of the main reasons is the use of a measuring procedure based on chemiluminiscence principles, which is, although well known, complicated and expensive.

Fortunately, a simple and low cost technique based on the Palmes' diffusion tubes has been developed. This method allows the measurement of the concentration of several individual species found in a gaseous mixture, such as the pollutant species in the atmosphere and particularly, nitrogen dioxide. Such measurements can be integrated in common studies of atmospheric corrosion, both in the laboratory and in the field. Consequently, the implementation of nitrogen dioxide measurements by the diffusion tube technique is proposed as a good practice in atmospheric corrosion studies, which could generate a better understanding of the mechanisms of material degradation resulting from exposure to the atmosphere. The effect of nitrogen oxides in atmospheric corrosion of metals and other materials has been studied by several authors [*3-7*]. Arroyave and Morcillo [*3*] reviewed the state of the art on structural materials, especially steel, copper and zinc. Nitrogen dioxide is believed to be the major precursor of nitric acid and nitrates species, which are well known to be involved in many chemical processes leading to material degradation.

Diffusion sampling has advantages in field studies, such as no need for calibration, sampling tubing, electricity, or technicians. In addition the samplers are re-usable, environmentally friendly and easy to carry and install. Studies employing diffusion tube samplers simultaneously with other measuring technique for NO_2 monitoring have shown good agreement between the two groups of data [*1,8-11*]. The present work presents the theoretical bases of the diffusion tube technique together with comparative results obtained in the Madrid atmosphere, and partial experimental results from an atmospheric corrosion study being carried out in the urban area of the city of Medellin. A statistical treatment of the data helps a better comprehend the results and to achieve more accurate conclusions.

Fundamentals

Deposited Amount

The method is based on the establishment of a concentration gradient along a cylinder, between a representative point in the atmosphere, and a sink for the gas under consideration, on the opposite side of the tube[*12*].

Assuming that the situation along the cylinder is exclusively related to one-dimensional transfer of NO_2 in the air (there are not changes in concentration with time for specific point), there will be a gas flux between the two ends of the tube, according to Fick's first law

$$F = -DdC/dz \tag{1}$$

where:

F = NO_2 flux (mol/cm^2•s),
D = NO_2 diffusion coefficient in air (cm^2/s),

C = NO_2 concentration (mol/cm³), and
z = Distance in the diffusion direction (cm).

The amount Q (mol) of NO_2 passed between the ends of the tube of radius r (cm), in a time t (s), is

$$Q = F(\pi r^2)t \tag{2}$$

Replacing

$$Q = (-Ddc/dz)(\pi r^2)t \tag{3}$$

And, consequently

$$dc/dz = -(C-C_0)/z \tag{4}$$

where

C = NO_2 concentration to be measured in the atmosphere and
C_0 = NO_2 concentration on the sink surface.

According to that

$$Q = D(C-C_0)(\pi r^2)t/z \tag{5}$$

Furthermore, if the sink surface has a 100% NO_2 absorbent substance, $C_0 = 0$, and

$$Q = DC(\pi r^2)t/z \tag{6}$$

Diffusion Coefficient

An approximate value of the molecular diffusivity for nitrogen dioxide in the air can be obtained from the Equation [*13*]

$$D = 0.0018583T^{3/2}[(M_1 + M_2)/M_1M_2]^{1/2}/P(\sigma_{12})^2\Omega_{12} \tag{7}$$

where

T= Temperature (K),
P= Total pressure (atm),
$M_{1,2}$= NO_2 and air molecular weights,
σ_{12}= Lennard-Jones forces constant, measuring of the molecular distance between the two colliding species
Ω_{12}= Colliding integral, function of K_BT/ε_{12}, where K_B is the Boltzmann's constant and ε_{12} is the Lennard-Jones forces constant, measuring the maximum approaching energy between the two molecular species (non-dimensional).

The approximate molecular weight for dry air can be calculated from its fractional composition, in volume, and for practical effects, can be assumed as 29 g/mol. The force constants can be calculated from

$$\sigma_{12} = (\sigma_1 + \sigma_2)/2 \quad (8)$$

$$\varepsilon_{12} = (\varepsilon_1 \varepsilon_2)^{1/2} \quad (9)$$

And the best path for the estimation of the individual force constants is from viscosity data for several temperatures, simultaneously resolving two equations like[*14*]

$$\eta \, x \, 10^7 = 266.93(MT)^{1/2}\Omega/\sigma \quad (10)$$

where

η = Viscosity (g/cm.s) and
Ω = Intrinsic colliding integral, function of $K_B T/\varepsilon$.

Thus, the following values for NO_2 in air, at 21.1°C, have been found [*15*]: σ_{NO2} = 3.765 Å; ε_{NO2}/K_B = 210 K, σ_{AIR} = 3.682 Å and ε_{AIR}/K_B = 84.6 K. As a result, the best estimation for the NO_2 diffusion coefficient in air, at 21.1°C and 1 atm. of pressure, is 0.154 cm^2/s. From this value, the diffusivity for the same system at another temperature T_2, can be obtained from the following relationship [*16*]:

$$D_{T_2} = 0154(T_2/294.3)^{3/2}(\Omega_{12})_{294.3}/(\Omega_{12})_{T_2} \quad (11)$$

Table 1 shows some approximate values for the colliding integral, at several temperatures.

Table 1 - *Approximate values of the colliding integral for the NO_2-air system, at several temperatures[16].*

T (K)	Ω
279.0	1.128
286.9	1.116
294.9	1.105
302.9	1.094
310.8	1.084

Chemistry of the Absorption Process and the Analysis Procedure

Levaggi et. al. [*17*] found that triethanolamine (TEA) fixed all the contacted NO_2 and, taking advantage of that situation, proposed the use of TEA for the fixation of the atmospheric nitrogen dioxide. Then, employing laboratory analysis, they obtained the pollutant concentrations. During absorption, the TEA passes to nitrosodiethanolamine [*18*] and later during analysis the nitrosamine hydrolizes to turn into nitrite [*19*]. Hereby, nitrite can be easily and accurately measured by ion chromatography [*20*] or conventional spectrophotometric analysis, by estimation of the diazotization effect of sulfanilamine on nitrite. In this process, the diazonium salt produced reacts with N-1-naphthylethylene-diamine dihydrocloride (NEDA) changing into a purple azo dye. Consequently, the concentration of this latest species can be measured by a spectrophotometric procedure.

Ion chromatography allows a determination limit about two orders of magnitude below the limit obtained in a spectrophotometer. Using IC allows the determination of less than 10^{-10} moles of nitrites in 5 mL sample using electrochemical suppression and a concentrator column [*10*]. Krochmal and Kalina [*10*] determined that the value and variation of analyte concentration in the blanks has an important influence in the precision of the technique. Thus, a ratio of the mass of absorbed analyte to the blank value should be as high as possible in order to obtain better precision in the measurement. The implication of this is that the rate of sampling of the diffusion tube should be in close relationship with the actual concentration of pollutant species being monitored. The sampling rate can be selected by variation of the sampler relative geometry as explained below.

Geometric Considerations

From Equation 6, it can be inferred that the use of a simple relationship between the area and the cylinder length will make calculations easier, and therefore the significance of the phenomena under study is more clearly understood. Usually, a relation 1 : 10 is considered, and Equation 6 for NO_2 at 1 atm. and 21.2°C for t hours, becomes as follows

$$Q = 55.44Ct \tag{12}$$

According to this expression, during one hour the tube absorbs 55.44C moles of NO_2. In other words, during one hour the tube absorbs the amount of NO_2 contained in 55.44 cm^3 of air. Rearranging Equation 12, the concentration of NO_2 can be calculated from the following formula, in which C is in $\mu g/m^3$, Q in micrograms and t in hours

$$C = 18038Q/t \tag{13}$$

Obviously, changing either the length of the cylinder or the sampling area, the coefficient in Equation 12 is altered, making the sampler more or less sensitive to pollutant concentration; a longer and narrower sampling tube is suitable for measurements in highly polluted atmospheres or during long term exposures and vice versa.

Shooter [*19*] reports variations of about 15% comparing diffusion tube measurements with conventional procedures, and a lower detection limit corresponding to a measure during one week in an atmosphere with 0.2 ppb NO_2. Similarly a 15% variation was obtained by Ferm and Svanberg [*9*] in a study where the passive or diffusive method was compared with a volumetric or active technique in which the air is pumped through the samplers. On the other hand, Shooter [*19*] reports other incident factors that must be taken into account for more accurate assessments and deeper studies: They are temperature, relative humidity, specific characteristics of the sampler, and wind speed. A more recent work by Gair and Penkett [*21*] reported a very significant effect of turbulence and wind speed, showing important reductions in the NO_2 values when these climatic factors increase. Ferm and Svanberg [*9*] also studied the wind effect on the concentration values obtained with diffusive samplers, concluding that when protecting the open end of the sampler with a fine stainless steel screen allows the NO_2 concentration data obtained with diffusion tubes placed outdoors to closely agree with the values from indoors sampling. The wind turbulence results in a decrease of the effective diffusion length of the tube [*21*].

Experimental Procedure

Samplers

Cylindrical diffusion samplers of a commercial poly-methyl-methacrylate tube piece, 10.0 mm inner diameter and 79.0 mm length, were cut. Consequently, a cross area/length ratio equal to 1.0 mm is obtained. A fixed PVC cap was used on the sink end and a removable one at the other end of the tube. Between the fixed cap and the acrylic tube were sandwiched two stainless steel grids (BS410 N°44), 11.0 mm diameter.

All components were separately washed with ultrasonic agitation in a detergent solution. Thereafter, the elements were rinsed and kept in high purity water until the final assembling of the sampler. It should be noted that all parts were dried in a hot air stream before assembling.

The steel meshes were impregnated directly with 40 μL of the absorbent solution (20 percent (volumetric percentage) aqueous solution of TEA, that contains 80 μL per 10 mL of a 10 percent polyoxyethylene lauryl ether wetting solution). Then, the second caps were placed, and the samplers put into a hermetically sealed polyethylene bag, and refrigerated until the time of exposure.

Exposure

The samplers were held in a vertical position, with the open end facing down, avoiding direct water and particulate material deposition. The tubes were fastened to a rigid structure using wire clips and allowing the free entrance of air into the tube. No provision was taken to protect the samplers from direct sunlight, rain or wind effects. An accurate measurement of the time of exposure was done by registering the exact starting and finishing time, with an accuracy of a few minutes. In the present work two groups of data are reported. In Group 1, the usual period of exposure of one week was employed and two places were selected for monitoring purposes, both on the top of two buildings at the campus of the Complutense University of Madrid, 500 m from each other, one of them corresponding to a conventional pollution measuring station, including a NO_2 chemiluminicence instrument (Building 2).

The second group (Group 2) of data has been obtained by exposure of diffusion tube samplers in ten different locations in the urban area of the city of Medellin, as shown in the map of Figure 1. At each one of the ten stations three samplers were installed around lamp posts at about 3 to 4 meter height, with an angular separation between each other of about 90 degrees and a radial distance from the post of approximately 40 cm. The selection of the ten different locations was based on two main aspects: characteristics of the place itself and statistical distribution of the station according to the sort of activity of the area, that is whether the area is mainly rural, residential, commercial or industrial. For the first decision parameter the comments of Croxford and Penn [*22*] about the siting of urban pollution monitors were followed. Statistical considerations were the second decision parameter, mainly considering the weight of a given economic area on the emission of pollutant species into the environment and its fraction out of the whole city area. In this sense, the calculation of the number of stations for each one of the zones, was based on information of SO_2 pollutant levels in Medellin and applying a randomic, stratified sampling design. Thereby, four stations were located at industrial areas, three stations sited in residential zones, two in commercial areas and one at a rural site. Each batch of diffusive samplers were exposed for a period of about a month. According to the results of Heal et al. [*1*], this exposure period appears to make a more accurate monitoring of NO_2 in the atmosphere than weekly exposures. Data obtained during the first seven periods of exposure are presented and analyzed here.

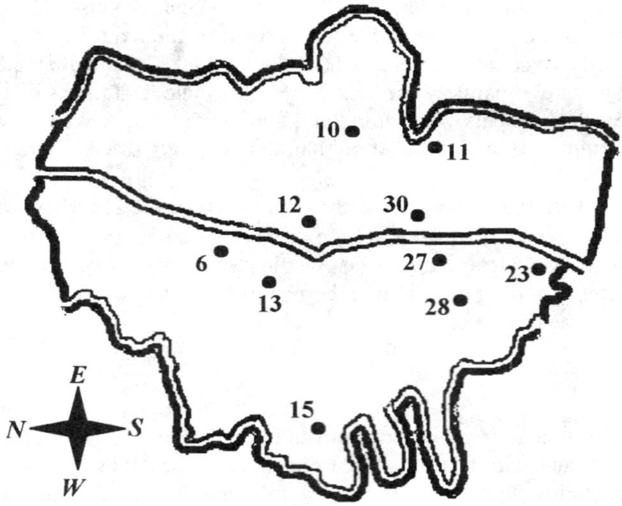

Figure 1- *Diagram of Medellin city area showing the spatial distribution of the ten NO_2 monitoring stations. The numbers refer to conventions employed in the actual research project being carried out across the city.*

Spectrophotometric Analysis

The samplers were held vertically with the fixed cap at the bottom end and 1.0 mL of high purity water was added into the tube, followed by 1.0 mL of solution No. 1 (250 mL of solution with 5 g of sulfanilamide and 15 mL of orthophosforic acid) and 0.1 mL of solution No. 2 (250 mL of solution containing 0.35 g of NEDA). Replacing the cap at the top end, the sampler was shaken vigorously and then left to stand for about 15 minutes. Later on, the whole amount of solution from each one of the tubes was taken to the spectrophotometer cell and the absorbance measured at a wave length of 540 nm.

Results

Experiments, Group 1

Table 2 shows the first group of data obtained at the campus of the Complutense University of Madrid, including both the concentration values measured by using diffusion tubes, and the accumulated data from the daily measurements by conventional chemiluminiscence, for the same sampling period (i.e. one week).

Table 2 - *NO_2 concentrations monitored at the campus of the Complutense University of Madrid.*

Concentration of NO_2 ($\mu g/m^3$)		
Diffusion tubes		**Conventional**
Building 1	**Building 2**	**Chemiluminiscence**
28		35
53		24
44		3
	32	22
	40	34

The data in Table 2 show two features. Firstly there is little agreement between the diffusion tubes measurements in Building 1, and second, overestimation of the NO_2 concentrations given by samplers sited in Building 2, all of this compared with the chemiluminiscence readings.

Experiments, Group 2

Given the amount of data for this group of experiments, the results are more clearly presented as XY charts, representing the relationships of the different variables with the NO_2 concentrations measured. The variables considered are exposure period, site of exposure, and sort of area in which the station is located. The exposure period was about a month, with the first batch of samplers installed in the middle of May 2000. In the figures shown below, the exposure periods have been designated with the numbers 1 to 7, corresponding to first, second, third, etc., month of exposure. Regarding the exposure site, the different stations have been identified by a number as indicated in the map of Figure 1. Stations 6, 23, 27, and 30 are situated at industrial areas. Stations 10, 15, and 28 are located in areas primary residential. Stations 12 and 13 are located in zones of strong commercial activity and finally, Station 11 was installed in a rural borderland area of the city.

Variation of Pollutant Concentration for each Station along the Exposure Period - Figure 2 shows the variation of the NO_2 concentration along the total exposure period. The concentrations of pollutant measured vary continuously from month to month in every station with some stations presenting stronger variations than others. For instance, the concentration values obtained for Station 6 change a lot more than the corresponding values for station 15. Comparison of stations located in areas with similar characteristics show similar trends along the seven exposure periods. This statement is valid for any one of the four different zones. More drastic variations are expected in stations located at industrial sites as pollutant concentrations in these areas respond to changes in the level of activity of the factories surrounding the station. There are a few things to be highlighted, such as the peak presented in the concentration value of Station 6 that corresponded to the monitoring of the fourth month of the experiment. Another one is the small variation shown by the samplers of Stations 11, 12, and 15. They all belong to a different sort of human activity; rural, commercial, and residential, respectively. The levels of pollutant concentration given by Stations 11 and 15 are relatively low and quite similar between one another, whereas Station 12 presents in most periods the highest levels of NO_2 pollution. Of particular interest are the high NO_2 concentrations measured in Station 10 for the four final periods of monitoring, given that this sampling station is located in a residential zone. Station 28 similarly shows the two last periods of increasing pollutant concentration. An overview of the curves for all the stations appears to have a common feature; the NO_2 concentration values have a trend to increase starting in the second month of exposure and reaching a maximum between the fourth and fifth month. This is not exactly correct for all stations but several of them present such a behavior.

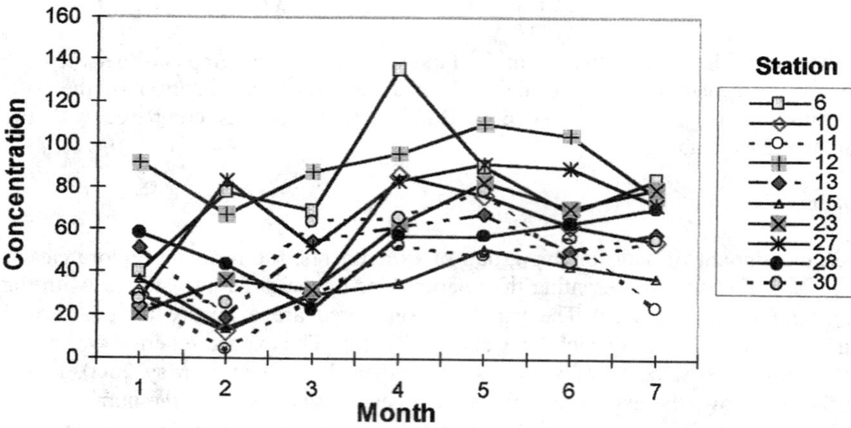

Figure 2 - *Variation of the NO_2 concentration for each station as a function of the period of exposure. Concentration scale is given in μg NO_2/m^3. Industrial: Stations 6, 23, 27, and 30; commercial: Stations 12 and 13; residential: Stations 10, 15 and 28; rural: Station 11.*

Average Variation of Pollutant Concentration for each Monitoring Station - Figure 3 presents the average values for each station in the whole exposure period. Stations 6 and 12 have the higher mean values of pollutant concentration. The next highest value is found for Station 27. Stations 6 and 27 are located at industrial zones while Station 12 is sited at the city center of Medellin, an area of strong commercial activity. The lowest values were obtained for Stations 11 and 15, which are rural and residential areas, correspondingly. The remaining stations show average values quite similar to each other.

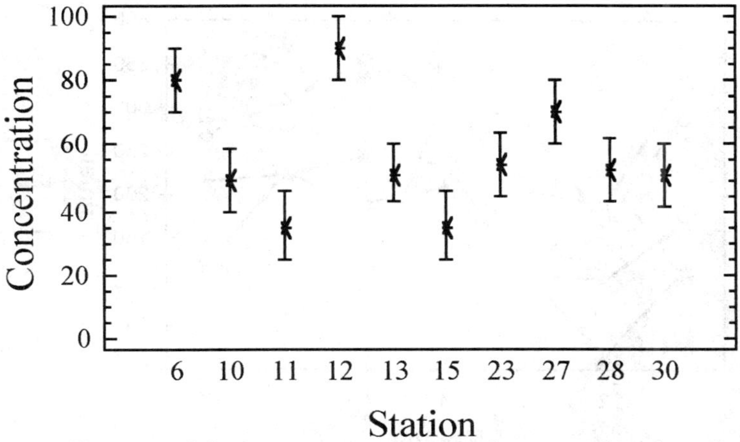

Figure 3 - *Variation of the average concentration of nitrogen dioxide in the total experimental period (about seven months). Concentration scale is given in $\mu g\ NO_2/m^3$.*

Variation of Pollutant Concentration for each Zone along the Exposure Period - Figure 4 represents the changes in NO_2 concentration during the whole period of experimentation for each individual zone. The curves in this figure clearly reveal a general increase in pollutant concentration starting from the second or third month of exposure and reaching top values at about the fourth or fifth month of monitoring. In addition, it appears that the concentration of pollutant is in two levels, the higher corresponding to stations located in industrial and commercial zones and a lower one for stations located at residential and rural areas although the values for the rural station are in most cases the lowest. The atypical average measurement for the industrial zone in Month 1 closely reflects the values obtained for each one of the stations sited at industrial areas, as can be observed in Figure 2.

Relationship between Pollutant Concentration and Climatic Changes - Figure 4 additionally shows the variations of rainfall levels in the urban area of the city of Medellin during the same periods of exposure of the NO_2 samplers. This curve reveals some agreement with the variation trend of the pollutant concentration curves, namely a high level of rainfall corresponds to high NO_2 average concentrations for commercial, rural, and residential zones. In the following exposure period, both rainfall levels and nitrogen dioxide concentrations for the same areas decrease and later increase again during the third and fourth exposure periods. In the three final periods of experimentation a net slight decrease for all four zones is observed corresponding to an important drop in rainfall. The rainfall level for each period was calculated adding the daily readings obtained by the local environmental authority. It must be highlighted that the rainfall readings are average data for the whole urban area of Medellin and given geographic characteristics of the urban area the rainfall levels could vary across different parts of the city.

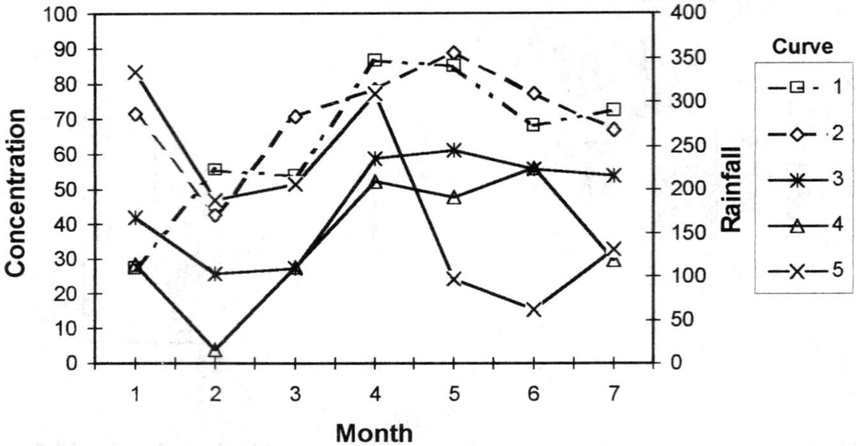

Figure 4 - *Average concentration changes of nitrogen dioxide for each zone as a function of the exposure period (Left hand side scale, concentration in μg NO_2/m^3). Additionally, this figure shows the changes in average rainfall levels for the same periods (Right hand side scale in mm). 1-Industrial, 2-Commercial, 3-Residential, 4-Rural and 5-Rainfall.*

Average Variation of Pollutant Concentration for each Month of Exposure - Figure 5 presents the variation of the average concentration for each monitored month. This representation of the data confirms that, in general, the levels of NO_2 concentration increased after the second month of exposure taking the higher values at the fourth and fifth months, to finally decrease slightly during the sixth and seventh months. The lowest average value according to the data obtained took place during the second month of the experiment.

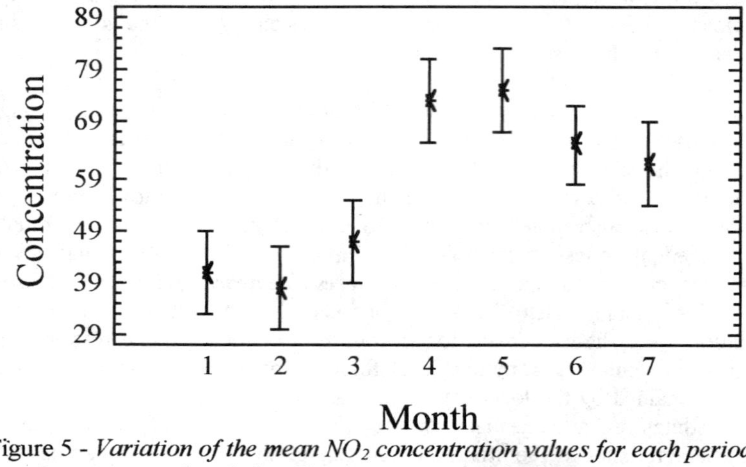

Figure 5 - *Variation of the mean NO_2 concentration values for each period of experimentation and including the measurements of all the monitoring stations. Concentration scale is given in μg NO_2/m^3.*

Discussion

Regarding the data of the Group 1 experiments, it can be noted that the chemiluminicence equipment is located at Building 2 and therefore the measurements obtained for that technique are more representative of the pollution condition around Building 2. Thus the data obtained employing the two measuring techniques for the concentration of NO_2 in Building 2 indicates that passive diffusion tubes are a useful, low cost and simple tool to measure nitrogen dioxide content in the atmosphere, either for air quality control or during studies of atmospheric corrosion.

Overestimation of nitrogen dioxide contents is believed to be caused by two main factors; first the effect of wind turbulence in the tube, effectively reduces the diffusion path, and second the relationship between NO_2 photolysis and the generation of NO_2 due to chemical reaction between NO and O_3. The effect of wind turbulence is known to be dependent on tube geometry; shorter tubes are more affected than longer ones [*21*]. In addition, wind effects can be diminished by placing a grid at the open end of the sampler [*9*]. Regarding the occurrence of chemical reaction on the inside of the tube, it has been reported that the net effect is to increase the actual concentration of nitrogen dioxide. However, measurements obtained with longer monitoring periods (one month instead of one week) are more accurate, possibly due to the existence of a chemical process at the absorbent surface [*1*], which has a net balancing effect in the nitrogen dioxide measurements.

Considering that combustion processes, primary vehicle emissions, are the main source of NO_2 pollution, it could be expected that both commercial and industrial sites present the higher values of pollutant concentration. The data obtained for Stations 12 and 6 are similar in average, although the characteristics of these two locations are different. In the first case, the samplers are located at the center of the city, an area with strong commercial activity with constant presence of slow moving traffic. On the other hand, Station 6 is sited at an area of industrial activity with a main cross roads nearby. Therefore, the environment around Station 6 is affected by the emissions of a big volume of fast moving traffic. Although a less critical case, Station 27 is also located near a cross roads with heavy traffic, especially at rush hours.

As expected according to the characteristics of the monitored areas, Stations 11 (rural) and 15 (residential) appear to be the least polluted; neither place presents evident sources of pollution with apparently low traffic flux. A statistical test of multiple comparisons indicates that it is possible to rank the stations in five groups, according to the average pollutant concentrations, as follows: Low pollutant concentration: stations 11 (34.86 $\mu g/m^3$) and 15 (35.11 $\mu g/m^3$). Medium pollutant concentration: Stations 10 (49.73 $\mu g/m^3$), 13 (51.58 $\mu g/m^3$), 23 (54.40 $\mu g/m^3$), 28 (52.95 $\mu g/m^3$) and 30 (51.25 $\mu g/m^3$). High medium pollutant concentration: Station 27 (70.82 $\mu g/m^3$). Lower high pollutant concentration: Station 6 (80.31 $\mu g/m^3$). High pollutant concentration: Station 12 (90.17 $\mu g/m^3$).

The pollutant concentrations reported in the present work are within the ranges for NO_2 contents in urban atmospheres, which is 1.6 to 121.0 $\mu g/m^3$[3] whereas the estimated minimun concentration inducing metal corrosion is believed to be about 30 $\mu g/m^3$[3]. The NO_2 levels related to health problems are well above either one of these figures, the lower toxic limit is about 9400 $\mu g/m^3$[3]. Kassomenos et al.[23] have reported data for the annual mean NO_2 concentration in several sites in the Athens metropolitan area. Comparison of this data with the results of the present work, indicates close agreement for sites of similar characteristics, namely stations located close to heavy traffic roads measure pollutant concentrations of about 80 to100 $\mu g/m^3$ similar to what is indicated by the diffusion samplers here for Stations 12 and 6. Sites with medium traffic in the work of Kassomenos et al.[23] present NO_2 concentrations ranging from 40 to 60 $\mu g/m^3$ in agreement with the results of the present work for stations classified as having medium pollutant concentrations, and finally the reported pollution levels for NO_2 in rural and residential areas of Athens are about 20 to 40 $\mu g/m^3$, which is again in agreement with the results obtained here for Stations 11 and 15 with low traffic flux. It is important to note that the comparison is based mainly on the characteristics of traffic in the monitored zones without considering other factors that could have important effects, such as the climatic factors.

A statistical test of multiple comparisons for the different periods (months) of exposure indicates that the data can be divided in three classes, based on the average values for all stations in the same month. Months 1, 2 and 3 show average low pollution levels (ranging from about 37 to 47 $\mu g/m^3$), the following two exposure periods (4 and 5) presented the higher concentrations of nitrogen dioxide in the atmosphere (about 73 to 75 $\mu g/m^3$) and finally measurements during Months 6 and 7 (about 60 to 65 $\mu g/m^3$) indicate an average slight reduction of NO_2 content for the urban area of Medellin. Some explanation to the variation of nitrogen dioxide content, across the whole period of experimentation, can be found looking at rainfall data during the same time (see Figure 4), however, the behavior of other climatic factors such as wind, temperature, and sun radiation, which are believed to influence air quality[*19*], must be investigated.

The analysis of the average NO_2 concentration data for each one of the zones with a statistical test of multiple comparisons reveals, as expected, that the rural zone contains the lower amounts of pollutant with a mean value of about 35.11 $\mu g/m^3$; next higher is the residential zone with a mean concentration value of about 46.34 $\mu g/m^3$; and the most polluted areas are the industrial and the commercial with average values of 64.03 $\mu g/m^3$ and 70.87 $\mu g/m^3$, respectively. The curves in Figure 4 additionally indicate that the above ranking of the different zones is valid for almost every month of experimentation, with only one exception, which is the atypical data, both for the average value and the individual measurements, for the industrial sites during the first month of exposure.

Considering the important role of NO_2 on atmospheric corrosion processes for various materials, as reviewed by Arroyave and Morcillo [*3*], it appears very appealing to employ passive diffusion tubes during studies of atmospheric corrosion. The characteristics of the diffusive samplers make them quite suitable to be installed alongside corrosivity specimens. Since the passive diffusive tubes are accumulative and are a continuous measurement technique, similar to the effect of the atmosphere on a metallic surface, the data generated are expected to be more closely related to the corrosion process than discrete measurements. Furthermore, the discussion of results presented in the paragraphs above shows how diffusive samplers monitor pollutant concentrations with both temporal and spatial resolution. On the other hand, given the low cost and small size of the samplers it is feasible to install as many tubes as wanted in every monitoring station without interfering at all with the atmospheric corrosion specimens installed at the same site.

Conclusions

It was possible to arrange a systematic and coherent procedure for the monitoring and measurement of NO_2, as a very useful complement to the conventional studies on atmospheric corrosion.

The described technique has attractive advantages, including very simple samplers, short periods of time for adequate answers, and easy analytical procedures, with a resulting extremely low cost. In addition, the capabilities of the passive diffusion tubes for both spatial and temporal resolution are the main advantages of this technique as a complement to other atmospheric corrosion samplers.

The changing effects of human activities on the atmosphere has brought a growing impact of NO_2 on corrosion problems, and experimental results indicate that passive diffusion tube practice is an important alternative for gas monitoring. In this work, a good correlation has been shown between the presence of the major source of pollution and the readings extracted from the diffusive samplers.

It is necessary to underline that better understanding of the climatic conditions of the area under study will result in more accurate concentration calculations and consequently better understanding of the NO_2 corrosive activity over metals and other materials.

There are some climatic factors influencing the ideal aerodynamic and chemical conditions in the passive diffusion sampler, such as wind and within-tube chemical reactions, that require deeper understanding in order to improve the accuracy of the technique.

References

[1] Heal, M. R., O'Donoghue, M. A. and Cape J. N., "Overestimation of Urban Nitrogen Dioxide by Passive Diffusion Tubes: a Comparative Exposure and Model Study," *Atmospheric Environment*, Vol. 33, 1999, pp. 513-524.

[2] Gardner, M. W. and Dorling, S. R., "Neural Network Modelling and Prediction of Hourly NO_x and NO2 Concentrations in Urban Air in London," *Atmospheric Environment*, Vol. 33, 1999, pp. 709-719.

[3] Arroyave, C. and Morcillo, M., "The Effect of Nitrogen Oxides in Atmospheric Corrosion of Metals," *Corrosion Science*, Vol. 37, No. 2, 1995, pp. 293-305.

[4] Hannef, S. J., Johnson, J. B., Dickinson, C., Thompson, G. E. and Wood. G. C., "Effect of Dry Deposition of NO_x and SO_2 Gaseous Pollutants on the Degradation of Calcareous Building Stones," *Atmospheric Environment*, Vol. 26A, No. 16, 1992, pp. 2963-2974.

[5] Graedel, T. E., "Concentrations and Metal Interactions of Atmospheric Trace Gases Involved in Corrosion," *Proceedings of the 9th International Corrosion Congress*, 1984, pp. 396-401.

[6] Ahmed, Z. and Al-Sulaiman, F. A., "Corrosion of Cars in Eastern Coastal Environment of Saudi Arabia," *British Corrosion Journal*," Vol. 28, No. 2, 1993, pp. 112-116.

[7] Takazawa, H., "Effect of NO_2 on the Atmospheric Corrosion of Metals," *Boshoku Gijutsu*, Vol. 34, 1985, pp. 612-617.

[8] Heal, M. R. and Cape, J. N., "A Numerical Evaluation of Chemical Interferences

in the Measurement of Ambient Nitrogen Dioxide by Passive Diffusion Samplers," *Atmospheric Environment*, Vol. 31, 1997, pp. 1911-1923.

[9] Ferm, M. and Svanberg P., "Cost-Efficient Techniques for Urban and Background Measurements of SO_2 and NO_2," *Atmospheric Environment*, Vol. 32, No. 8, 1998, pp. 1377-1381.

[10] Krochman, D. and Kalina, A., "A Method of Nitrogen Dioxide and Sulphur Dioxide Determination in Ambient Air by Use of Passive Samplers and Ion Chromatography," *Atmospheric Environment*, Vol. 31, No. 20, 1997, pp. 3473-3479.

[11] Kasper-Giebl, A. and Puxbaum, H., "Deposition of Particulate Matter in Diffusion Tube Samplers for the Determination of NO_2 and SO_2," *Atmospheric Environment*, Vol. 33, 1999, pp. 1323-1326.

[12] Palmes, E.D. and Gunnison, A. F., "Personal Monitoring Device for Gaseous Contaminants," *American Industrial Hygiene Association Journal*, Vol. 34, 1973, pp. 78-81.

[13] Reid, R. C. and Sherwood, T. K., *The Properties of Gases and Liquids*, 2nd Edition, Mc Graw-Hill, New York, 1966.

[14] Wilke, C. R. and Lee, C. Y., "Estimation of Diffusion Coefficients for Gases and Vapors," *Industrial and Engineering Chemistry*, Vol. 47, No. 6, 1955, pp. 1253-1257.

[15] Palmes, E. D., Gunnison, A. F., Dimattio, J. and Tomczyk, C., "Personal Sampler for Nitrogen Dioxide," *American Industrial Hygiene Association Journal*, Vol. 37, 1976, pp.570-577.

[16] Skelland, A. H. P., *Diffusional Mass Transfer*, Jonh Wiley & Sons, New York, 1974.

[17] Levaggi D. A., Siu, W. and Feldstein, M., "A New Method for Measuring Average 24-hour Nitrogen Dioxide Concentrations in the Atmosphere," *Journal of the Air Pollution Control Association*, Vol. 23, No. 1, 1973 pp.30-33.

[18] Aoyama, T. and Yashiro, T., " Analytical Study of Low-Concentration Gases: IV. Investigation of the Reaction by Trapping Nitrogen Dioxide in Air Using the Triethanolamine Method," *Journal of Chromatography*, Vol. 265, 1983, pp. 69-78.

[19] Shooter, D., "Nitrogen Dioxide and its Determination in the Atmosphere: A Simple Method for Surveying Ambient Pollution Concentrations," *Journal of Chemical Education*, Vol. 70, No. 5, 1993, pp.A133-A140.

[20] Gair, A. J., Penkett, S. A. and Oyola, P., " Development of a Simple Passive Technique for the Determination of Nitrogen Dioxide in Remote Continental Locations," *Atmospheric Environment*, Vol. 25A, No. 9, 1991, pp.1927-1939.

[21] Gair, A. J. and Penkett, S. A., "The Effects of Wind Speed and Turbulence on the Performance of Diffusion Tube Samplers," *Atmospheric Environment*, Vol. 29, No. 18, 1995, pp. 2529-2533.

[22] Crosford, B. and Penn, A., "Siting Considerations for Urban Pollution Monitors," *Atmospheric Environment*, Vol. 32, No. 6, 1998, pp. 1049-1057.

[23] Kassomenos, P., Skouloudis, A. N., Lykoudis, S. And Flocas, H. A., "Air-Quality Indicators for Uniform Indexing of Atmospheric Pollution Over Large Metropolitan Areas," *Atmospheric Environment*, Vol. 33, 1999, pp. 1861-1879.

Le Thi Hong Lien[1] and Pham Thy San[1]

The Effect of Environmental Factors on Carbon Steel Atmospheric Corrosion; The Prediction of Corrosion

---

Reference: Le Thi Hong Lien and Pham Thy San, **"The Effect of Environmental Factors on Carbon Steel Atmospheric Corrosion; The Prediction of Corrosion,"** *Outdoor Atmospheric Corrosion, ASTM STP 1421,* H. E. Townsend, Ed., American Society for Testing and Materials International, West Conshohocken, PA, 2002.

Abstract: Carbon steel was exposed at atmospheric testing sites all over Vietnam in the period 1995-2000. The corrosion rate was determined by the weight loss method. The environmental data were collected at the same time of the exposures. The effects of the environmental factors on the corrosion of carbon steel were investigated. It showed that the corrosion rate increases with increasing relative humidity, TOW, chloride concentration in the air and decreasing air temperature. Based on the data of observed corrosion and the environmental factors the linear regression equation of one-year corrosion is given for carbon steel in coastal and continental area.

Keywords: atmospheric corrosion, air temperature, relative humidity, time of wetness (TOW), chloride deposition rate, linear regression equation.

Introduction

Vietnam is situated in the tropical zone. The relative humidity are high all year, the time of wetness (TOW) of the atmosphere in Vietnam is often about 4500 – 6500 h/year. These factors accelerate the atmospheric corrosion of metals, although the air salinity is not very high and the SO_2 concentration is low.

In this paper the effects of the environmental factors on the corrosion of carbon steel are presented. Linear equation is given to calculate the corrosion rate of carbon steel from environmental parameters.

Experimental Method

Carbon steel specimens, 150x100x2mm, were exposed at test sites from the north to the south of Vietnam. The positions of these sites are shown in the map (Figure 1) and Table 1.

The specimens were placed on racks at 45°, facing to the sea at the marine test sites and to the south east at the others. The experiments were repeated three times from July 1995 to July 1998. The specimens were taken off after each one year of exposure.

---

[1]Research scientists. Corrosion Research Center. Institute for Material Science. NCST of Vietnam. Hoang Quoc Viet Rd.. Cau Giay Dist. Hanoi. Email: honglien@ims.ncst.ac.vn or PtSan @Hotmail.com.

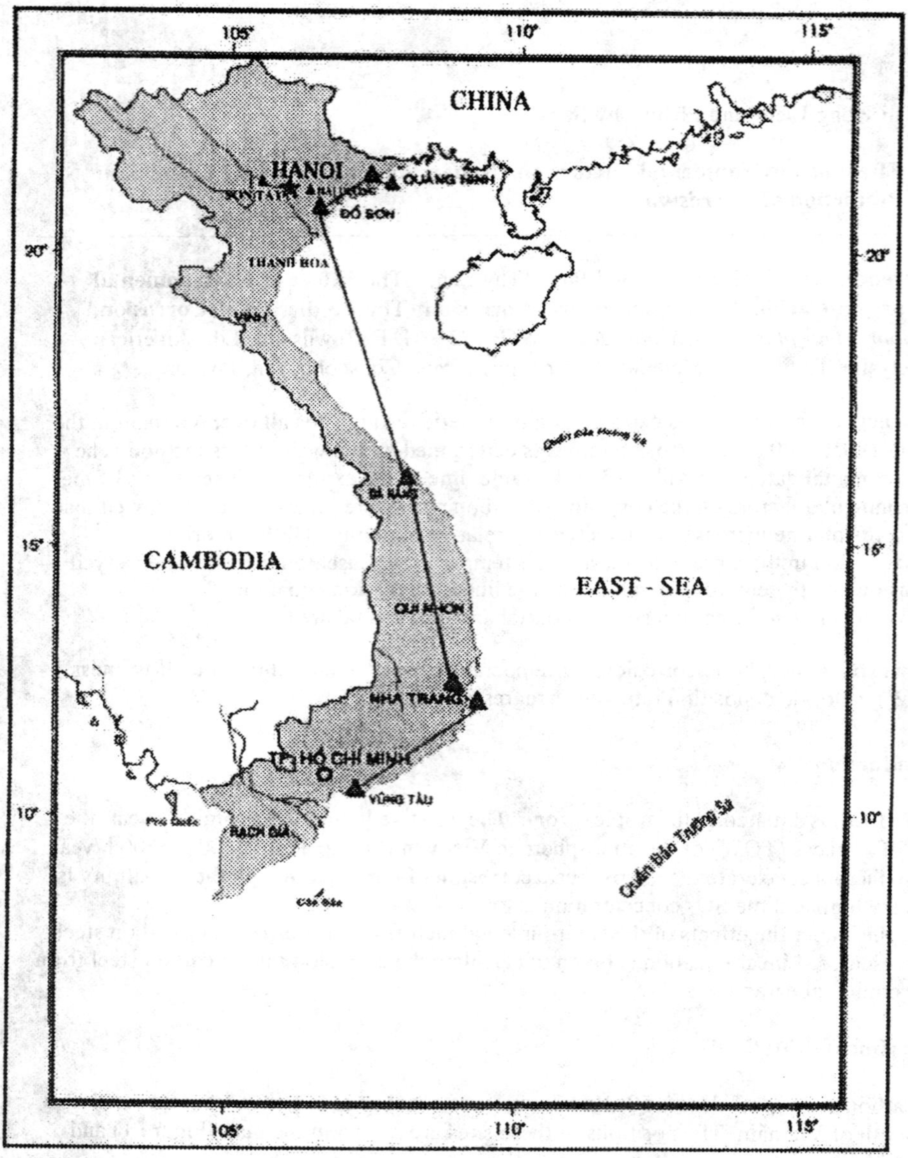

Figure 1 - *The position of test sites.*

The corrosion products were removed by chemical cleaning in accordance with ISO 8407: Metals and Alloys - Procedure for removal of corrosion products from corrosion test specimens, using the following solution: 500mL hydrochloric acid (HCl) (ρ=1.19g/cm^3); 3.5g hexamethylenetetramine, distilled water to make 1000mL at 25°C.

The corrosion rates were evaluated from the weight losses of specimens according to ISO 9226/1992: Corrosion of Metals and Alloys - Corrosivity of atmospheres - Method of determination of corrosivity using the corrosion loss standard specimens.

Environmental data were collected at the same time as the exposures: Air temperature and relative humidity obtained from Meteostations located at the test sites were used to calculate TOW. The chloride deposition rates were assessed by the "dry gauze" method in accordance with GOST 9039-1974 ECZKC.

The statistical treatment of collected data was performed with the aid of STATITCF software.

Table 1- *Characteristics of test sites.*

Test sites	Height above sea-level, m	Distance from the sea-shore, km	Type of atmosphere
Hanoi	5	90	Urban in the North
Doson	5	0.03	Marine in the North
Danang	6	0.5	Urban in the Central
Nhatrang	16	0.05	Marine in the Central
HoChiMinh City	6	90	Urban in the South
Vungtau	4	1.0	Marine in he South
Sontay	6	120	Rural in the North

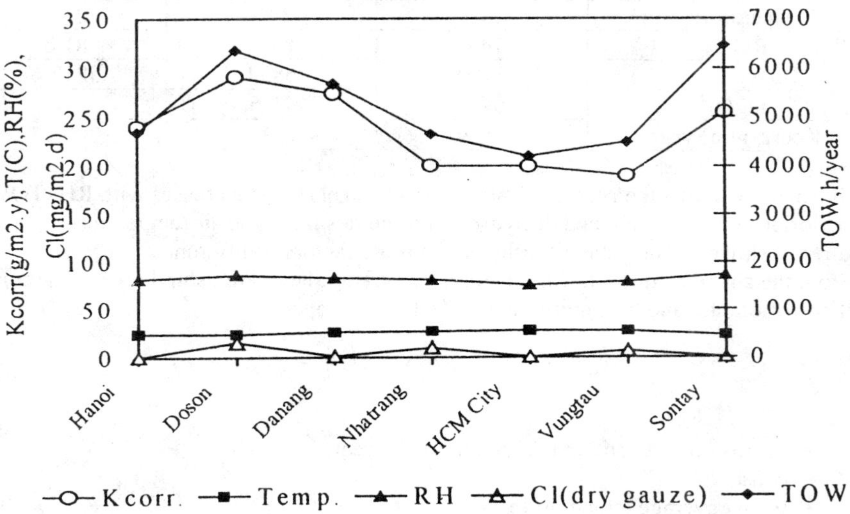

Figure 2 – *Corrosion rate depends on the environmental parameters.*

Results and Discussion

Table 2 gives the data of climate and salinity of test sites, collected during the exposure times.

Table 2 - *Environmental data during the exposure time. (Average annual values from 7 1995 to 6 1998).*

Testing sites	Average temp., °C	RH,%	TOW h/year	Cl^- mg/m^2day (1)	Cl^- mg/m^2day (2)	Cl^- ISO category
Hanoi	24.1	81.3	4697	0.32	0.64	S_0
Doson	23.6	86.1	6359	15.25	30.5	S_1
Danang	25.9	83.1	5672	1.54	3.08	S_1
Nhatrang	26.7	80.8	4651	9.65	19.3	S_1
HCM City	27.7	75.6	4205	0.61	1.22	S_0
Vungtau	27.5	79.3	4495	6.79	13.58	S_1
Sontay	23.4	86.3	6451	0.28	0.56	S_0

(1) - Data collected by " dry gauze" method (C_{dry}),
(2) - Data calculated by formula: $C_{wet} = 1.9\ C_{dry}$ or $C_{wet} = 2.1\ C_{dry}$ [*1,2*].

Table 3 - *The environmental and corrosion data used in the regression treatment of test sites.*

	Min	Max	Mean
Temperature, °C	23.1	28.2	25.6
RH,%	74.0	87.5	81.8
TOW, h/year	3970	6834	5218.6
Cl^-, mg/m^2day	0.27	25.6	4.92
Kcorr, $g/m^2.year$	175.80	319.1	239.37

It can be seen that the corrosion rate of carbon steel (K_{corr}) increases with RH, TOW and chloride concentration, and decreases with the raising of the air temperature (Figure 2). However, the level of influence of these factors is different.

From the above results a linear equation expressing the relationship between carbon steel corrosion rate and the environmental factors is proposed

$$K_{corr} = a_1T + a_2RH + a_3TOW + a_4[Cl] + a_5 \quad (1)$$

where:

K_{corr} = corrosion after one year exposure, $g/m^2.year$,
a_i = const,
T = annual average temperature, ^0C,
RH = annual average relative humidity, %,
TOW = time of wetness, h/year, and
$[Cl]$ = chloride concentration, $mg/m^2.day$.

The regression treatment was done using 19 measured data of carbon steel corrosion rate and the respective environmental parameters at the same time of exposure (Table 3). The regression equation for corrosion of carbon steel after one year was found as follows

$$K_{corr} = -8.78T + 5.25RH + 0.0081\ TOW + 0.77[Cl] - 10.228 \qquad R = 0.94 \qquad (2)$$

It is noted that the air temperature, relative humidity and TOW more strongly influence the corrosion rate of carbon steel than the chloride ion. This may be because the chloride concentrations are not very high (equivalent to S_0 and S_1 of ISO category).

Table 4 - *The comparison between corrosion rates by experiment and calculation.*

Test sites	Obs. corr. rate	Cal. corr. Rate[1]	Errors	Cal. corr. Rate[2]	Errors	Cal. corr. Rate[3]	Errors
Hanoi	240.36	243.291	1%	243.044	1%	230.668	-4%
Doson	290.7	297.839	2%	286.096	-2%	281.994	-3%
Danang	264.64	245.774	-8%	244.588	-8%	231.484	-14%
Nhatrang	254.23	224.649	-13%	217.219	-15%	213.607	-19%
HCM city	191.92	177.996	-8%	177.526	-8%	170.696	-12%
Vungtau	191.23	206.284	7%	201.056	5%	192.049	0%
Sontay	289.74	255.15	13.5%				

[1] Using the average environmental data 7/1995 – 7/1998 (during the exposure time),
[2] As above but without chloride concentration,
[3] Using the annual average environmental data 1/1996 – 12/1999.

Table 4 gives the corrosion rates of carbon steel calculated by Equation (2) using the average environmental data during the exposure time 7/1995 –7/1998 (Table 2). The errors are ≤ 13% compared with the observed average corrosion rates. When calculating the corrosion rate based on the annual average environmental data obtained by Meteostations (1/1996 –12/1999); the errors are not high and can be accepted. If calculating corrosion rate without using the chloride concentration, the errors change negligibly (Table 4), therefore, the corrosion rate of carbon steel can be approximately calculated from only the climatic data for continental areas

$$K_{corr} = -8.78T + 5.25RH + 0.0081\ TOW - 10.228 \qquad (3)$$

Conclusions

1. In Vietnam, the corrosion rate of carbon steel increases with RH, TOW and Cl, but the effect of chloride is not clear because the chloride concentrations are not very high at the test sites.

2. The corrosion rate of carbon steel can be approximately calculated according to the linear regression Equation (2), depending on the environmental data with accepted errors. In the continental areas (far from the sea) the corrosion rate can be calculated without using the salinity data, following Equation (3).

3. However, the given equations should be continued improving by the addition of the SO_2 concentrations in the urban atmospheres and should be verified at other sites.

References

[1] Strekalov, P. V., Panchenko, Y. M., Zachita metallov, 1990, Vol. 26, No 6, p. 883 (Russian)

[2] Strekalov, P. V., Panchenko, Y. M., Zachita metallov, 1992, Vol. 28, No 2, p. 269 (Russian).

Dagmar Knotkova,[1] Vladimir Kucera,[2] Sheldon W. Dean,[3] and Petr Boschek[4]

Classification of the Corrosivity of the Atmosphere – Standardized Classification System and Approach for Adjustment

Reference: Knotkova, D., Kucera, V., Dean, S. W. and Boschek, P., **"Classification of the Corrosivity of Atmosphere – Standardized Classification System and Approach for Adjustment,"** *Outdoor Atmospheric Corrosion, ASTM STP 1421*, H. E. Townsend, Ed., American Society for Testing and Materials International, West Conshohocken, PA, 2002.

Abstract: Atmospheric corrosivity information is essential for selection of materials of construction, protective coatings and other measures used to prevent corrosion. Standards ISO 9223 – 9226 elaborated in ISO/TC 156/WG 4 are well accepted, used by corrosion engineers and introduced in other normative documents. Work on adjustment of the atmospheric corrosivity classification system is going on in the WG 4 with the aim to cover some gaps in this system. Statistical treatment of the ISOCORRAG program data and other data makes it possible to obtain a better definition of environmental influencies and specification of standard specimens. More sensitive procedures for corrosivity derivation for low aggressive indoor atmospheres have been recommended.

Keywords: atmospheric corrosivity, standards, standard specimen, environmental influence, dose-response function, indoor corrosivity.

Introduction

Within the work of ISO/TC 156/WG 4 ATMOSPHERIC CORROSION TESTING AND CLASSIFICATION OF CORROSIVITY OF ATMOSPHERE standards for corrosivity classification were elaborated. The purpose was to elaborate standards convenient for needs of corrosion engineers and users of technical products. Standards are well accepted and introduced in other technical committees (ISO/TC 107, ISO/TC 35/SC 14, CEN/TC 262, CEN/TC 240). The engineering application of the classification system will be supported by guidance formulated in ISO/DIS 11303, Corrosion of metals and alloys; Guidelines for selection of protection methods against atmospheric corrosion.

[1] Dr., convenor of ISO/TC 156/WG 4, senior scientist, SVUOM Ltd., U Mestanskeho pivovaru 4, 170 00 Praha 7, Czech Republic.
[2] Dr., senior scientist, SCI, Kräftriket 23, SE-104 05, Stockholm, Sweden.
[3] Dr., senior scientist, Air Products and Chemicals, Inc., 7201 Hamilton Boulevard, Allentown, PA, 18195- 1501, USA.
[4] Dr., senior lecturer, Charles University, Celetna 20,110 00 Praha 1, Czech Republic.

supported by guidance formulated in ISO/DIS 11303, Corrosion of metals and alloys; Guidelines for selection of protection methods against atmospheric corrosion.

Despite wide acceptance, the standards have also been subject to criticism mainly concerning the difference in corrosivity categories obtained by exposure of standard specimens and from available environmental data. Concern has also been raised on the effect of composition and thickness of the standard plate specimens of carbon steel and zinc on the corrosion rate.

Work on adjustment of the atmospheric corrosivity classification systems is going on in the WG 4 during recent years mainly in cooperation by the Czech, Swedish, American, Russian, Spanish and Australian members.

The fundamental contribution to the improvement of the corrosivity classification system was a systematic evaluation and adjustment, including statistical processing of the ISOCORRAG program [*1, 2*]. This program was carried out at more than 50 sites in 13 countries and was designed to follow the ISO classification system both by its methodology and structure. Also, the MICAT program [*3*] was organized in a similar manner later on.

This paper indicates the drawbacks of the current system in a simplified manner and defines new approaches in revision and creation of new standards for classifying corrosivity of atmospheres.

Adjustment of the Classification System Presented in ISO 9223

Characterization of the Standardized System

The standardized system is presented schematically in Figure 1. It covers standards:

- ISO 9223:1992 Corrosion of metals and alloys. Corrosivity of atmospheres. Classification,
- ISO 9224:1992 Corrosion of metals and alloys. Corrosivity of atmospheres. Guiding values for the corrosivity categories.,
- ISO 9225:1992 Corrosion of metals and alloys. Corrosivity of atmospheres. Measurement of pollution and
- ISO 9226:1992 Corrosion of metals and alloys. Corrosivity of atmospheres. Determination of corrosion rate of standard specimen for the evaluation of corrosivity.

Description of the ISO 9223 (parameters, levels, ranges for corrosivity categories) and the general approach for the atmospheric corrosivity classification are object of an other publication [*4*].

A normalized classification system is a simple system with a consistent structure, which basically meets the requirements for systems designed for wide engineering use. The basic assumptions required for wide applicability of the system also represent weak points in some cases:

- possibility of a comparable derivation of corrosivity from annual averages of three decisive factors of the environment and from results of an annual corrosion test,

- possibility to associate long-term and steady state corrosion rates with a corrosivity category derived from annual environmental or corrosion values,
- comparability of annual corrosion loss determined using standard flat specimens or helix specimens,
- low cost and easy-to-use methods of measurements of pollution deposition are standardized, that, however, do not fully exploit possibilities in the field,
- the standardized approach for determination of the one-year corrosion loss is not sensitive enough for determination of corrosion attack at C1 corrosivity category, and
- the fact that the system does not show corrosivity information for values over C5 corrosivity category does not mean that higher corrosion losses cannot be reached, but this limitation must be explained within the standard so that the applicability of the standard is better defined.

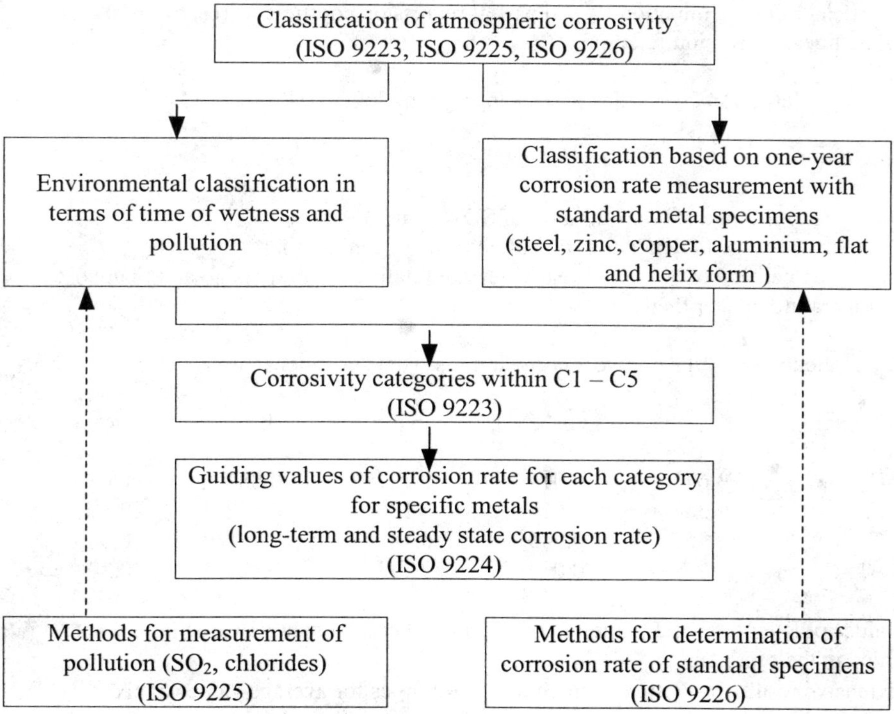

Figure 1 - *Scheme for classification of atmospheric corrosivity approach in ISO 9223-9226.*

Application of the ISOCORRAG Program Results on the Procedures Defined in ISO 9223 and ISO 9224 - Verification of the System, Definition of Gaps.

Papers focusing on analyzing partial relations are subject to individual publications [*5*]. The main objectives of the carried-out evaluations are:

- comparison of corrosivity categories derived on different ways for four metals,
- assessment of relation between corrosivity categories derived from annual corrosion losses and corrosion rate during eight-year exposure in relation to standard corrosion values according to ISO 9224.

The final version of the ISOCORRAG database was used to derive regression functions. These functions derived from the ISOCORRAG data were applied in different ways of corrosivity and guiding corrosion values estimation according to procedures shown in ISO 9223 and ISO 9224. The aim of these analyses was to contribute to the validation of the classification system.

Coefficient of determination R^2 and partial regression coefficients (b_0, b_1, b_2, b_3) from the linear regression function

$$\ln(\text{corr}) = b_0 + b_1.[SO_2] + b_2.\ln[Cl] + b_3.\ln[TOW] \qquad (1)$$

where

$[SO_2]$ = yearly average of concentration of SO_2 ($\mu g.m^{-3}$),
$[Cl]$ = yearly average of deposition rate of chloride ($\mu g.m^{-2}.day$),
$[TOW]$ = per centage of hours per year when the relative humidity is greater than 80% at a temperature greater than 0°C (%),

used for calculation of the one-year corrosion loss (flat specimens) are as follows:

Metal	b_0	b_1	b_2	b_3	R^2
Fe	3.647	0.011	0.137	0.883	0.63
Zn	0.388	0.010	0.126	0.552	0.49
Cu	0.354	0.005	0.148	0.702	0.58
Al	-1.972	0.014	0.233	0.225	0.39

Values of R^2 correspond to the fact that many other atmospheric variables are not included in the function.

Extensive results [*6-7*] are documented by examples for average values of six repeated one-year exposures. Calculations and derivations are summarized in Table 1.

The set of derived corrosivity categories for steel and zinc is relatively consistent. The set of derived corrosivity categories for copper is changeable with higher differences for higher mass loss values (measured and calculated too). It is a principal disproportion for aluminum between corrosivity categories derived from classified environmental characteristics and from measured mass losses.

Table 1 – *Comparison of corrosivity categories derived on different way (from environmental categories, estimated and calculated corrosion loss).*

test site	classified environmental categories			corrosivity category		
	τ	P	S	C 1)	C 2)	C 3)
steel						
ARG5 - Jubay	τ4	P0	S1	C3	C3	C2
CS2 – Praha	τ3	P2	S1	C4	C3	C3
CS3 – Kopisty	τ3	P2	S0	C3	C4	C3
SF2 – Otaniemi	τ4	P1	S0	C3	C3	C2
S3 – Kvarvik	τ4	P0	S3	C5	C4	C4
US3 – Panama	τ5	P2	S3	C5	C5	C5
zinc						
ARG5 - Jubay	τ4	P0	S1	C3	C3	C3
CS2 – Praha	τ3	P2	S1	C3	C4	C3
CS3 – Kopisty	τ3	P2	S0	C3	C4	C3
SF2 – Otaniemi	τ4	P1	S0	C3	C3	C3
S3 – Kvarvik	τ4	P0	S3	C5	C3	C4
US3 – Panama	τ5	P2	S3	C5	C5	C5
copper						
ARG5 - Jubay	τ4	P0	S1	C3	C4	C3
CS2 – Praha	τ3	P2	S1	C3	C4	C3
CS3 – Kopisty	τ3	P2	S0	C3	C5	C3
SF2 – Otaniemi	τ4	P1	S0	C3	C3	C3
S3 – Kvarvik	τ4	P0	S3	C5	C4	C4
US3 – Panama	τ5	P2	S3	C5	C5	C5
aluminum						
ARG5 - Jubay	τ4	P0	S1	C3	C3	C2
CS2 – Praha	τ3	P2	S1	C3	C3	C2
CS3 – Kopisty	τ3	P2	S0	C3	C3	C2
SF2 – Otaniemi	τ4	P1	S0	C3	C2	C2
S3 – Kvarvik	τ4	P0	S3	C5	C2	C2
US3 – Panama	τ5	P2	S3	C5	C2	C3

where

- environmental characterization is expressed in classification categories for TOW, SO_2 concentration and salinity deposition rate (Table 1, 2 and 3 from ISO 9223);
- corrosivity category $C^{1)}$ is derived from classified environmental information (Table 6 from ISO 9223);

- corrosivity category C[2)] is derived from one-year mass loss values (Table 5 from ISO 9223); and
- corrosivity category C[3)] is derived from calculated one-year mass loss values. Dose-response functions (1) were used for this calculation.

On a general level, the dose-response functions derived for annual corrosion results were applied to the whole extent of combinations of pollution levels and time of wetness in order to determine the extent of an interval of overlapping corrosivity categories. The calculation was performed for the border limits of the classified intervals (example see Table 2).

Table 2 - *Prediction of standardized corrosivity categories based on equation (1) – examples of calculated results for limited number of combination of environmental characteristics.*

classified environmental caterogy			calculated corrosion loss, μm/a		corrosivity categories	
P	S	τ	min	max	covered	mean
calculation for: steel						
P0	S0	τ12	0.3	9.3	[1 2]	C2
P0	S0	τ3	4.6	20.4	[2]	C2
P0	S0	τ4	10.1	26.6	[2 3]	C2
P0	S0	τ5	13.2	31.1	[2 3]	C2
P0	S3	τ12	1.3	27.6	[2 3]	C2
P0	S3	τ3	18.6	60.1	[2 3 4]	C3
P0	S3	τ4	40.6	78.4	[3 4]	C4
P0	S3	τ5	53.0	91.3	[4 5]	C4
P1	S3	τ12	1.5	35.8	[2 3]	C2
P1	S3	τ3	20.8	78.0	[2 3 4]	C3
P1	S3	τ4	45.4	101.9	[3 4 5]	C4
P1	S3	τ5	59.3	119.3	[4 5]	C5
calculation for: zinc						
P0	S1	τ12	0.10	0.35	[1 2]	C2
P0	S1	τ3	0.21	1.36	[2 3]	C3
P0	S1	τ4	0.84	2.18	[3 4]	C3
P0	S1	τ5	1.34	2.87	[3 4]	C4
P1	S1	τ12	0.10	0.45	[1 2]	C2
P1	S1	τ3	0.24	1.78	[2 3]	C3
P1	S1	τ4	0.94	2.84	[3 4]	C3
P1	S1	τ5	1.50	3.75	[3 4]	C3

P2	S3	τ12	0.01	1.09	[1 2 3]	C2
P2	S3	τ3	0.55	4.28	[2 3 4 5]	C4
P2	S3	τ4	2.17	6.84	[4 5]	C5
P2	S3	τ5	3.48	9.02	[4 5]	C5

where

- P, S and τ categories of environmental parameters (Tables 1,2, and 3 from ISO 9223)
- minimum and maximum calculated corrosion loss derived for a given combination of the environmental categories
- corrosivity categories according ISO 9223 (Table 5 of this Standard) covered by this minimum and maximum corrosion loss and
- mean corrosivity category derived from the mean of the minimum - maximum corrosion loss interval.

Conclusions to the full treatment are presented in WG 4 document [*6*]. As a result of the high variability in the input data, general characteristics of the estimated regression model and the span in the classified intervals the estimated ranges of corrosion losses (corrosivity categories) are rather large. The mean corrosivity category is in most cases not in principal contradiction to the classified system (Table 6 in ISO 9223). A similar calculation based on smaller (more narrow) intervals of individual environmental categories was performed too.

Estimated corrosivity ranges do not cover so many corrosivity categories for the individual environmental category combinations. This logical approach is not the best for the technical application of the corrosivity system. Corrosion engineers and designers need relatively simple general information, which can be used for the selection of materials of construction or other methods of corrosion protection.

The corrosion of aluminum is a complicated issue. The ISOCORRAG results showed the greatest variation of rates with this metal. Furthermore, the damage suffered by aluminum tends to be pitting and other forms of localized attack. Using mass loss to calculate thickness loss is obviously incorrect and tends to minimize the severity of the problem. One-year mass loss does not give accurate information to determine guiding values from ISO 9224. However, the use of corrosion classification is valuable as a qualitative indicator of the problem aluminum components may have in the atmosphere. Therefore, it is recommended that aluminum be singled out with a warning statement that the corrosion rate values should not be used as a quantitative measure of the damage that may occur but rather as a guide or lower limit.

Application of Flat and Helix Specimens for the Determination of Atmospheric Corrosivity

ISO standard 9226 provides for two types of specimens for the purpose of determining atmospheric corrosivity by mass loss measurement after a one-year exposure. These specimens are either flat panels, 100x150x203 mm, or 1 m long wire,

2-3 mm in diameter wound into an open helix. The latter specimen was included because some countries have used it and it was claimed to be insensitive to orientation issues.

When the ISOCORRAG program was formulated, both types of specimens were included in the program so that we could evaluate their relative performance. As the program continued, it became apparent that the wire helices tended to corrode more rapidly than the corresponding flat panels. Review of the literature for zinc specimens [*8*] suggested that zinc wires tend to show higher atmospheric corrosion rates than flat panels with the rate increasing, as the diameter is smaller. However, no explanation was provided for this observation.

A comprehensive review of the ISOCORRAG data confirmed that the helices showed significantly higher corrosion rates than the flat panels for all four metals in this program, and the smaller diameter wires showed proportionately higher rates than the larger diameter wires [*9*]. This study also showed that sulfur dioxide accelerated the corrosion rate of zinc wires more than flat panels, and chloride deposition accelerated the rate of copper corrosion more than flat panels, while both of these environmental factors accelerated the rate of aluminum wires as compared to flat panels. These effects were attenuated in the multi-year exposures and, in every case, the wire rates began to approach the flat panel rates after eight years of exposure.

The explanation of these effects appears to be related to the fact that wires are more effective at collecting both gaseous and particulate corrosive compounds than flat panels. These increases in collection efficiency are dependent upon wire diameter and can vary from a factor of two to many times depending upon the particulate diameter and wind velocity. In the case of a 3 mm wire, the adsorption of sulfur dioxide will be approximately three times higher on the flat panel.

Because of these unexpected factors that cause wire specimens to have systematically higher corrosion rates in atmospheric corrosion, it was decided to eliminate this type of specimen from the standard. However, those findings serve as a cautionary note that flat panel results may underestimate the corrosion rates that wires experience in service.

Adjustment of the Procedure for Determination of One-Year Corrosion Rates on Standard Specimens

The adjustment consists of a more precise definition of the composition of metals for standard specimens and in specification of the thickness for specimens of steel and zinc. The proposals are based on the work performed by G. King [*10*].

The composition limits placed on both steel and zinc in ISO 9226 to serve as standard materials for measuring of atmospheric corrosivity are inadequate to ensure reproducible batch-to-batch performance. The ASTM standardized practice partly overcomes this problem for steel by recommending that copper-bearing structural carbon steel (such as A36 with 0.2% copper minimum) be used for characterizing corrosivity with respect to steel. However, this material may be difficult to obtain considering the steelmaking practices used today.

The thickness of standard corrosion specimens needs to be specified better for both steel and zinc. An influence of thickness on the corrosion rate of zinc has been demonstrated in limited studies [*10*]. Marked differences in the corrosion rates of zinc have been measured for different seasons and initiation times of exposure [*10*], and this issue is one that should also be noted for the users of ISO Standards 9223 and 9226.

Within the ISO/TC 156 WG 4 a special comparative exposure program is under preparation for judgement of relevant quality and thickness of standard specimens of zinc and steel. The final recommendation will be supported by the results of this program.

Another problem is connected with corrosion of aluminum. The one-year corrosion loss evaluation as uniform attack is not correct and can give misleading information about the corrosion behavior of aluminum in a given atmosphere. More warning statements in ISO 9226 are necessary.

Approach for Atmospheric Corrosivity Derivation Based on Application of Dose-Response Functions

This chapter presents general possibilities but also some of the problems resulting from this approach. The actual selection of optimum functions and work with it are presented in detail in a separate paper [*11*]. This paper includes analysis of ways for characterization of the temperature – humidity complex. The approach of application of TOW value in the revised version of the classification standard will not be changed. The limitations related to TOW definition will be mentioned.

There are two ways of application of dose-response function in revision of ISO 9223:

- improved association of corrosivity category with combinations of intervals of classified environmental characteristics. This approach cannot be recommended according to relatively wide intervals of parameters (see Table 2).
- direct application of function for measured environmental characteristics for the corrosivity category derivation.

If these functions are to be used, the inaccuracies of the calculated value must be statistically defined in a professional manner. Confidence limits should be defined. One-sided confidential regions for an individual value of corrosion loss should be sufficient for practical use because users are interested in the upper reliability of the predicted value.

Evaluation of Long-term Corrosion Rates from Site Corrosivity Determination

The ISO 9224 standard provides for an estimation of corrosion rates for both intermediate and long-term exposures based on site corrosivity classification (guiding values). These estimations are based on the known behavior of engineering metals in atmospheric exposures where the buildup of corrosion product layers gradually reduces the rate of corrosion. However, the results in the ISOCORRAG program have given us a more definitive data set from which the corrosion damage can be estimated with greater accuracy. An analysis of these results is in another publication [*12*].

The approach used was to employ the kinetic model that has been widely used for atmospheric corrosion:

$$ML = aT^b \quad (2)$$

or

$$\log (ML) = a^1 + b \log T \quad (3)$$

where

ML = mass loss per unit area,
T = exposure duration (years),
a = mass loss per unit area in one year ($a^1 = \log a$), and
b = time exponent, usually less than one.

The values for *a* or a^1 have been measured and, by means of a variety of dose-response functions, it is possible to estimate these values with reasonable accuracy for the steel, zinc, and copper. The results for aluminum show much larger variations, probably because the mechanism of aluminum corrosion is a pitting mechanism rather than general corrosion.

Regression analyses of the *b* values versus climatic and corrosive component levels showed much less variation. In the case of steel, only time of wetness and, to a much smaller extent, chloride affected the *b* value. Both of these parameters increased the *b* value indicating that they made the rust layer less protective. Sulfur dioxide and chloride affected the *b* value for copper but chloride tended to reduce the *b* value while sulfur dioxide increased it. This surprising result is in contrast to the effect chloride deposition has on the initial rate of attack, the *a* value, where it increased the initial rate. As a result, the initially higher rate decreases faster and, ultimately the mass loss falls below where it would be without any chloride deposition. This can require 3-10 years to occur with higher chloride deposition rates requiring longer times. In the case of both zinc and aluminum, the *b* values are not dependent upon the atmospheric variables.

If one uses equation (2) to estimate intermediate or long-term corrosion losses, much greater accuracy can be achieved than with a less specific corrosion category estimation. The best results are obtained in cases where one-year corrosion losses are measured. In that case, it is only necessary to estimate the *b* value and obtain the corrosion loss at any subsequent time. By adding two standard deviations to the estimated *b* value, an upper bound that incorporates 97.5% of the uncertainty can be determined. In the cases where it is necessary to estimate both the *a* and *b* values, the errors become significantly larger. However, in most cases, it is possible to predict to within a factor of two the corrosion losses for long-term exposures. These approaches should make the new revised standard more accurate in predicting the performance of metals in outdoor service.

Figure 2 shows a comparison between the eight-year predictions for four metals compared to the measured corrosion losses in the ISOCORRAG program. The values for the ISO 9224 predictions were based on the maximum rates for the predicted corrosion category. The categories were estimated either from the environmental information or from the one-year corrosion losses as described in ISO 9223. The projections from equation (3) include a two standard error increment to make the results conservative. The results shown in Figure 2 clearly demonstrate the greater fidelity and smaller tendency to underestimate the corrosion damage of the equation (3) approach. The large scatter in the predicted results may be caused by the fact that many

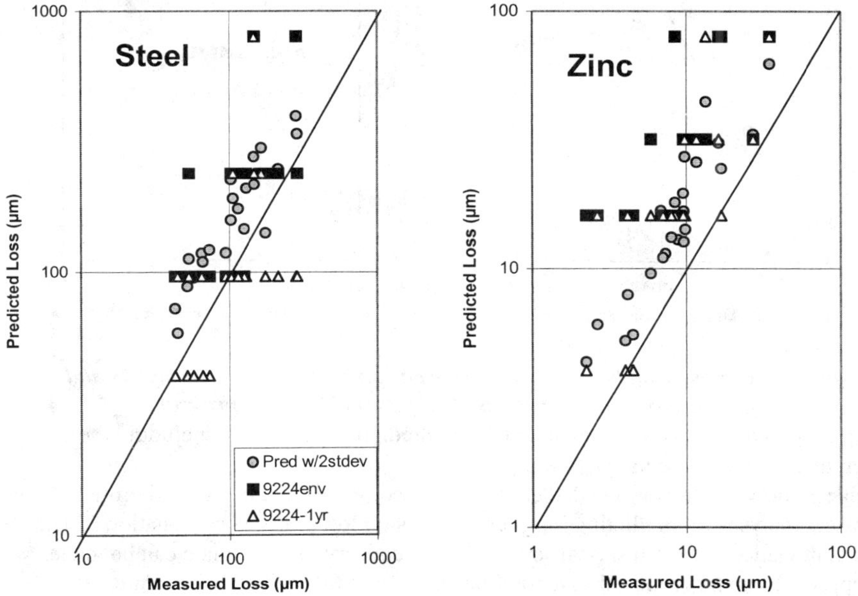

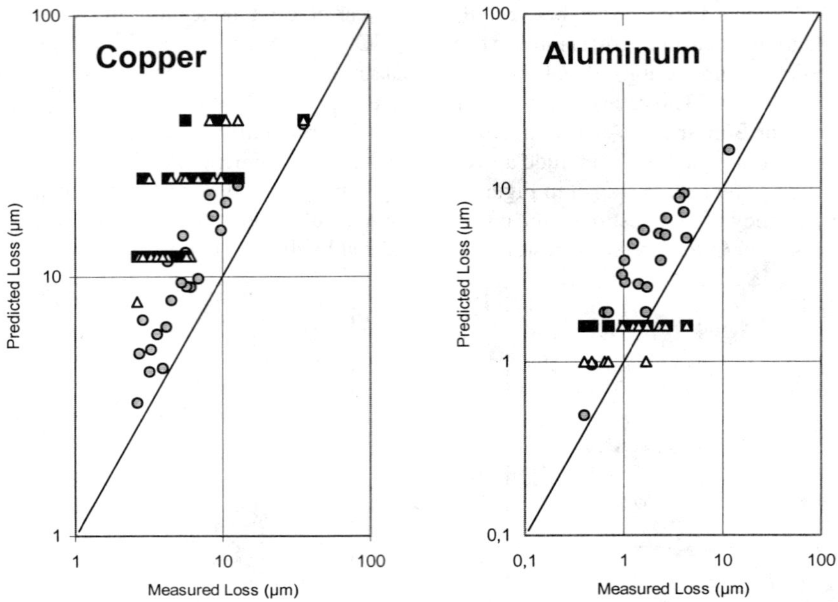

Figure 2 – *Comparison between the eight-year predictions for four metals and the measured corrosion losses in the ISOCORRAG program.*

atmospheric variables were not included in the prediction equations, including the temperature of the environment.

Table 3 shows the recommended equations based on minimum standard error regression analyses for predicting long-term corrosion losses based on equation 3. The standard deviations are also shown so that a more conservative estimate can be made. This approach can be proposed as a possible revision to the ISO 9224 standard.

Table 3 - *Recommended regression results to predict time exponents for long-term corrosion losses*[1]

Metal	Intercept	TOW Coef[2]	Cl Coef[3]	SD
Fe	0.348	5.20	--	0.110
Zn	0.786	--	3.10	0.137
Cu	0.537	4.78	- 4.08	0.133
Al	0.728	--	--	0.181

[1] b = intercept + (TOW coef) TOW + (Cl Coef) Cl average value
For 97.5% confidence limit value add 2 SDs
[2] TOW = time of wetness, RH>80%, t >0°C hrs per year x 10^{-5}
[3] Cl = chloride deposition in mg/m^{2}day x 10^{-4}

Corrosivity Classification for Indoor Atmospheres with Low Corrosivity

Procedures prescribed in ISO 9223 provide reasonably good results in derivation of corrosivity categories for outdoor atmospheric environments. This classification system is too course for indoor environments with low corrosivity. More sensitive procedures have to be chosen for derivation of corrosivity categories in less aggressive indoor environments, such as places where electronic devices, sophisticated technical products or works of art and historical objects are stored. A new system for classification of corrosivity of indoor atmospheres has been developed and the following 3 documents are prepared for the first voting:

- ISO/CD 11844 Part 1: Classification of corrosivity of indoor atmospheres. Determination and estimation of indoor corrosivity.
- ISO/CD 11844 Part 2: Classification of indoor atmospheres. Determination of corrosion attack in indoor atmospheres.
- ISO /CD 11844 Part 3: Classification of indoor atmospheres. Measurement of environmental parameters affecting indoor corrosivity.

The aim of the new elaborated proposals of standards is to provide a consistent method of indoor corrosivity classification with improved procedures for determination, based on evaluation of corrosion attack, and estimation, based on environmental characterization, of indoor corrosivity categories. A scheme representing the approach for the indoor corrosivity classification is presented in Figure 3.

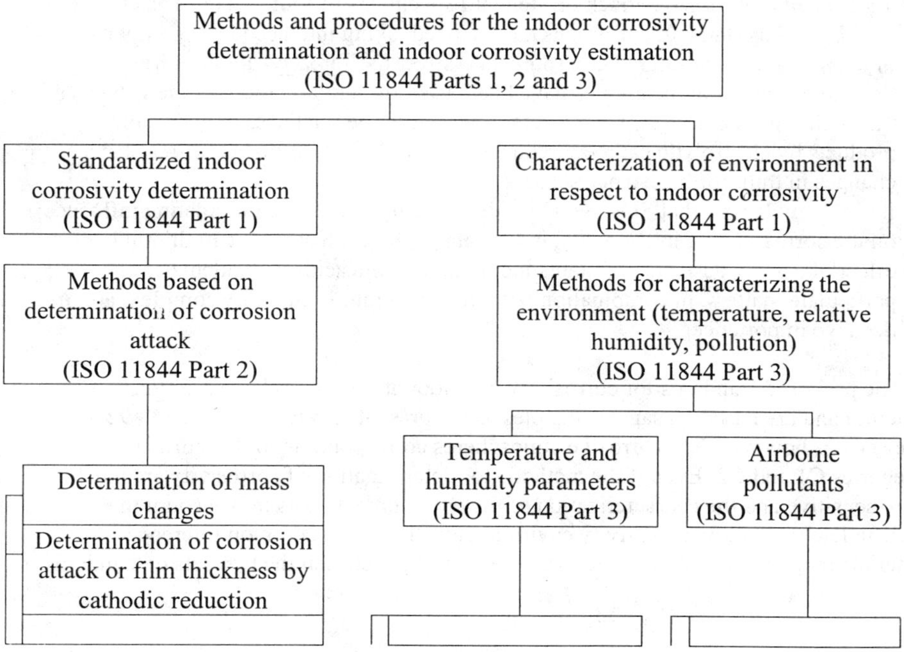

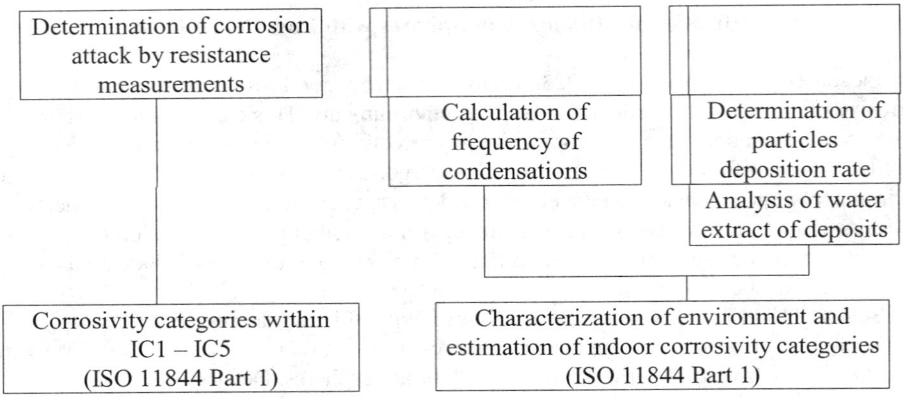

Figure 3 - *Scheme for classification of corrosivity in indoor atmospheres.*

The new proposals are based on results from systematic investigations of parameters affecting corrosion performed both in climate chambers and in extensive field exposures in widely differing types of indoor environments. In the investigations different sensitive techniques were used for the determination of corrosion attack and different methods for monitoring and measurement of environmental parameters [*13*]. The main aspects of the new approach are as follows:

- The determination of corrosivity categories of indoor atmospheres is based on the assessment of corrosion attack on standard specimens of four metals. The choice of metals used as standard specimens has changed taking into account the importance of occurrence in considered environments (silver has replaced aluminum).
- Sensitive evaluation techniques have been used for determination of mass changes using ultramicrobalances, for the determination of the thickness of corrosion products by electrolytic cathodic reduction or for the determination of resistance changes in thin film resistance sensors.
- It is acknowledged that SO_2 and chloride levels do not have a dominating influence on the corrosion rate in the actual indoor atmospheres. Instead the multipollutant effect of several gaseous pollutants including organic acids and aldehydes and particulate matters, in combination with the temperature-humidity complex, are of decisive importance.

The proposed standards for corrosivity in indoor atmospheres have a similar structure and are based on same principles as the present standards ISO 9223 –9226. They cover, however, less corrosive atmospheres corresponding to the corrosivity categories C1 and C2. Even if the methods for determination of corrosivity are much more sensitive and the characterization of the environment takes into account the specific features of low corrosivity environments, the use of common reference materials (copper, zinc, carbon steel) creates a bridge between the two sets of standards.

Conclusion

The classification system, as defined in ISO 9223 - 9226, is widely used in practice and was evaluated using results of ISOCORRAG and other programs. These experiences and the overall development of the knowledge in the field are raising the need to revise the system in a way that will consider generally valid principles, but will restrict sources of inaccuracies and will introduce new procedures for derivation of corrosivity from environmental data. This should lead to a much higher comparability of corrosivity categories derived from corrosion losses and environmental characteristics.

The original classification system does not cover conditions of low-corrosivity indoor atmospheres in a sufficient manner even after introducing the proposed changes, therefore proposals of special standards with a higher sensitivity for assessing these conditions in relation to corrosivity were developed.

References:

[1] Knotkova, D., Boschek, P., and Kreislova, K., "Results of ISO CORRAG Program: Processing of One-Year Data in Respect to Corrosivity Classification," *Atmospheric Corrosion, ASTM STP 1239,* W.W.Kirk, and H.H.Lawson, Eds., America Society for Testing and Materials, West Conshohocken, PA, 1995.

[2] Boschek, P. and Knotkova, D., "Program ISOCORRAG – Repeated One-Year Exposures. Optimal Regression Function for Adjustment of Corrosivity Categories," Proceedings of the UN/ECE Workshop on Quantification of Effects of Air Pollutants on Materials, Berlin, May 24-27, 1998.

[3] Morcillo, M., "Atmospheric Corrosion in Ibero-America: The MICAT Project," *Atmospheric Corrosion, ASTM STP 1239*, W.W. Kirk and Herbert H. Lawson, Eds., American Society for Testing and Materials, Philadelphia, 1995

[4] Vrobel L. and Knotkova D., "Using the Classification of Corrosivity of Atmospheres to Extend the Service Life of MAterials, Structures, and Products," *The Degradation of Metals in the Atmosphere, ASTM STP 965,* S.W.dean and T.S.Lee, Eds., American Society of Testing and MAterials, Philadelphia, 1988

[5] Dean, S.W. and Reiser, D.B., "Analysis of Data from ISOCORRAG Program," Paper No. 340, Corrosion 98, NACE, Houston, 1998.

[6] Boschek, P. and Knotkova, D., "Derivation of Corrosivity Categories from Environmental Data. Verification of the system applied in ISO 9223," Doc. ISO/TC 156/WG 4 N 329, 1998.

[7] Boschek, P. and Knotkova, D., "Contribution to Adjustment of Corrosivity Classification," Doc. ISO/TC 156/WG 4 N 340, N 349, N 366, 1999 – 2000.

[8] Zhang, X.G., *Corrosion and Electrochemistry of Zinc*, Plenum Press, New York and London, 1996, pp. 254-255.

[9] Dean, S.W. and Reiser, D.B., "Comparison of the Atmospheric Corrosion Rates of Wires and Flat Panels," Paper 00455, Corrosion 2000, NACE, Houston, 2000.

[10] King, G.A., Pikul, S., Shermen, N., and Ganther, W.D., "Influence of Composition of Thickness on the Atmospheric Corrosion of Steel and Zinc and the Implications for Standards and Specifically ISO 9226," Doc. ISO/TC 156/WG 4 N 375, 2000.

[11] Tidblad,J., Kucera,V., Mikhailov, A.A., and Knotkova,D., "Improvement of the ISO System of Classification for the Corrosivity of Outdoor Atmospheres," *Outdoor and Indoor Atmospheric Corrosion, ASTM STP 1421*, American Society for Testing and Materials, West Conshohocken, PA, 2002.

[12] Dean, S.W. and Reiser D.B., "Analysis of Long-Term Atmospheric Corrosion Results from ISOCORRAG Program," *Outdoor and Indoor Atmospheric Corrosion, ASTM STP 1421*, American Society for Testing and Materials, West Conshohocken, PA, 2002.

[13] Johansson, E., "Corrosivity Measurements in Indoor Atmospheric Environments A Field Study", Licentiate Thesis, Royal Institute of Technology, Stockholm, Sweden, 1998.

LABORATORY TESTING AND SPECIALIZED OUTDOOR TEST METHODS

Jonas Weissenrieder[1] and Christofer Leygraf[2]

In-situ Studies of the Initial Atmospheric Corrosion of Iron

Reference: Weissenrieder, J., Leygraf, C., "***In-situ* Studies of the Initial Atmospheric Corrosion of Iron,**" *Outdoor Atmospheric Corrosion, ASTM STP 1421,* H. E. Townsend, Ed., American Society for Testing and Materials International, West Conshohocken, PA, 2002.

Abstract: A unique experimental setup has been developed with the intent to provide molecular information during initial atmospheric corrosion of the iron surface. Infrared reflection absorption spectroscopy (IRAS) and quartz crystal microbalance (QCM) have been integrated into one surface analytical system. The results show that an aqueous adlayer of constant mass was physisorbed on the surface at a given relative humidity. The aqueous adlayer was found to be thicker when compared to previous studies performed on copper. A linear relationship between mass change and the intensity of the water absorption band at 3400 cm^{-1} (IRAS) was found when altering the relative humidity. At high relative humidity a thick aqueous adlayer was formed, whereby an absorbance band at 1100 cm^{-1} was observed that disappeared when dry air was introduced. When introducing SO_2 and O_3 in the sub-ppm range, the formation of sulfate surface species could be monitored quantitatively with monolayer sensitivity and a significant increase in reaction kinetics could be discerned.

Keywords: iron, IRAS, QCM, O_3, SO_2, corrosion, *in-situ*, AFM, SEM

Introduction

Interaction between a metal and the surrounding atmosphere is of great importance in many technically important and naturally occurring processes, including atmospheric corrosion, heterogeneous catalysis and weathering of minerals [*1-5*]. Atmospheric corrosion of iron has been extensively studied, both in laboratory [*6-10*] and field exposures [*1,2,11*]. In order to understand the process involved, there is an obvious need

---

[1] Ph.D. student, Division of Corrosion Science, Royal Institute of Technology, Drottning Kristinas väg 51, SE-100 44 Stockholm, Sweden

[2] Professor, Division of Corrosion Science, Royal Institute of Technology, Drottning Kristinas väg 51, SE-100 44 Stockholm, Sweden

to characterize the metal/atmosphere interface on a molecular level. Surface characterization techniques such as x-ray photoelectron spectroscopy (XPS) [*12*], Auger electron spectroscopy (AES) [*13*], low energy electron diffraction (LEED) [*13*], secondary ion mass spectroscopy (SIMS) [*14*], scanning electron microscopy (SEM) [*15*] and other vacuum requiring techniques have all increased the knowledge of atmospheric corrosion. They can, however, not be used for *in-situ* studies of the metal/atmosphere or metal/water interface at ambient pressures. Such studies are necessary as a bridge to relate vacuum studies to field studies and can be performed with techniques like scanning probe microscopy (SPM) [*12*], Raman spectroscopy [*16*], sum frequency generation (SFG) [*17*], infrared reflection absorption spectroscopy (IRAS) [*7,8,18-21*], quartz crystal microbalance (QCM) [*9,18*] and x-ray absorption spectroscopy (XAS) [*22*].

The aim of this paper is to explore the influence of humidity, sulfur dioxide (SO_2) and ozone (O_3) on the initial atmospheric corrosion of iron by an integrated IRAS/QCM system.

Experimental

The IRAS/QCM-setup is described in detail in a previous paper by Aastrup *et al* [*18*]. It consists of a teflon coated reaction chamber, in which the sample is exposed to a well-controlled atmosphere at ambient temperature (22°C). Four external features are connected to the chamber: a Fourier transform infrared spectrometer (Bio-Rad FTS 60a), a modified commercial QCM probe (Maxtek MPS 550) with a frequency counter (Maxtek PM 740), a corrosive air generation system and gas analysis facilities. The humidity in the atmosphere can be regulated between 0% and 95% relative humidity. SO_2 is generated in a well-controlled fashion through permeation tubes. The outlet gas of the chamber is continuously analyzed by SO_2 (Thermo Environmental 43A) and O_3 (Environment 41MC) gas analyzers. The O_3 analyzer is a combined analyzer/generator and it is used for providing O_3 to the corrosive air generation system, meanwhile it is analyzing the outlet gas of the chamber. The gas flow through the chamber was 1.3 l/min, which corresponds to an airflow rate of 3.5 cm/s over the sample. All iron samples were exposed to dry air for at least 30 minutes after insertion into the reaction chamber in order to stabilize the QCM and to remove residual humidity in the atmosphere present after the insertion of the sample.

IRAS spectra were recorded in absorbance units, -log(R/R_0), where R is the reflectance of the sample during exposure at any given time and R_0 is the reflectance of the sample prior to the exposure. All background spectra were recorded after 30 minutes exposure to dry air. Each spectrum was taken as an average of 1024 scans during approximately 8 minutes with a resolution of 4 cm^{-1} and the time noted in the figures is the time after 4 minutes of scanning.

The frequency shift measured by QCM can be transformed to a mass change through the Sauerbrey equation (1) [*23*]:

$$\Delta f = -2f_0^2(\Delta m)(\rho_q v_q)^{-1} \tag{1}$$

Where Δf is the frequency shift, f_0 the frequency at the starting point, ρ_q the density of quartz, v_q the shear wave velocity of quartz and Δm the mass change. A frequency change of 1 Hz, which corresponds to a mass change of about 18 ng/cm^2 at the actual resonance frequency 5 MHz, can easily be monitored by the QCM. This corresponds to 2/3 of a monolayer of water [*24*].

A Kratos AXIS HS Spectrometer equipped with an Al-K_α x-ray source was used for complementary XPS measurements of the iron samples after completed IRAS/QCM experiments. The instrument resolution was measured to be 1.2 eV and the binding energies were calibrated relative to the C-C peak in the C 1s core level. It was set to 285.0 eV. Detailed spectra of the following core levels were recorded: C 1s, Fe 2p, O 1s and S 2p.

The atomic force microscopy (AFM) used in this study was a Digital Instruments Nanoscope IIIa. The measurements were performed in tapping mode (TM-AFM) at constant amplitude in order to minimize the influence of the tip on the sample surface. Commercial silicon cantilevers (Digital Instruments) were used.

A JEOL JSM-820 scanning electron microscope (SEM) equipped with a Tracor Series II microanalyzer was used in order to observe larger features than was possible with AFM and to perform x-ray elementary analysis of species on the sample surface.

Two different kind of samples were analyzed in this study, 750 nm thick electron beam deposited iron films (purity 99.95%) on quartz crystal substrates (Maxtec Inc.) and a 2.0 mm thick iron foil (purity 99.9985%) (Johnson & Matthey). The iron thin film samples were mechanically polished by diamond pastes down to 1 μm and rinsed in 99.5% ethanol (Kemetyl AB) before placement in the reaction chamber. The QCM was used to monitor the polishing of the iron films and the measured removed thickness was about 25 nm. The iron foil sample was initially wet-ground and then diamond polished down to 1 μm in ethanol. The mean roughness of the polished surfaces was measured with AFM and was found to be about 0.4 nm.

As an alternative preparation method the iron samples were electrochemically prepared in a sodium tetraborate ($Na_2B_4O_7 \cdot 10H_2O$) solution with a pH value of 9.22 (Merck) using an EG&G 273A potentionstat-galvanostat system and a three-electrode cell. A platinum grid served as counter electrode and Ag/AgCl was used as reference electrode. The iron samples were cathodically polarized for 120 s at −1 V relative to the reference electrode prior to passivation in order to remove the air formed oxide films. They were subsequently anodically polarized for 600-1000 s at +0.5 V relative to the reference electrode in order to create a well-defined passive film [*25*]. The electrochemically prepared passive film was found to be slightly more corrosion protective than the air formed film after the mechanical polishing procedure. Because of its enhanced chemical reactivity, the mechanically polished surface was chosen as the standard operation for preparing the iron samples.

Results

Iron exposed to humidified air

In the initial experiments the iron samples were exposed to humidified air at different relative humidities. The results obtained by IRAS and QCM reveal that an aqueous adlayer of constant mass was physisorbed on the surface at a given relative humidity. The aqueous adlayer on iron was found to be thicker when compared to previous studies performed on copper [*18*]. The QCM measurements of the physisorbed water layer indicate that the water film on iron is about two times thicker than that on copper for a given relative humidity. Even prolonged exposures at 90 % relative humidity did not show any substantial iron oxide growth in either IRAS or QCM. With IRAS no oxide absorption band, similar to the cuprous (Cu_2O) absorption band at 645 cm^{-1} in the case of copper, could be detected. This does not exclude the possibility of an iron oxide growth. The IRAS cross section of the oxide may be very low or the orientations of the dipoles could be in the plane of the surface and thereby not possible to detect with IRAS. The mass gain, measured by QCM, was typically below 200 ng/cm^2 after 24 hours exposure.

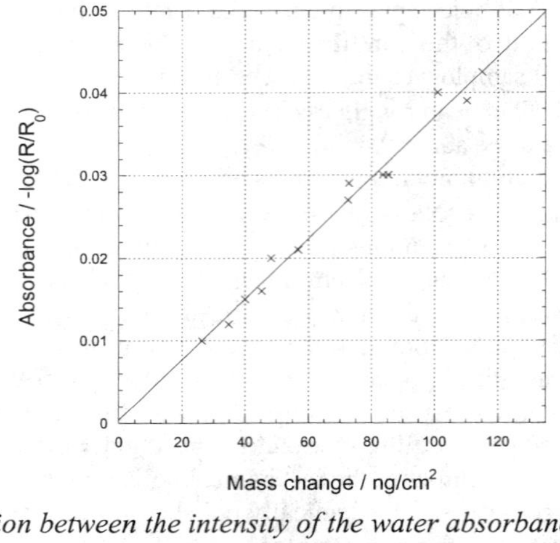

Figure 1 - *Relation between the intensity of the water absorbance band at 3400 cm^{-1}, measured by IRAS, and the equivalent water mass change measured by QCM.*

Ex-situ XPS investigations of the Fe 2p core levels did not reveal any distinct difference in the relation between the intensities of the metallic $2p_{3/2}$ peak at 707.0 eV and the iron oxide $2p_{3/2}$ peak (Fe_2O_3) at 710.9 eV after the exposure compared to before. This indicates that the iron oxide growth is very slow. When plotting the IRAS absorbance in the water band at 3400 cm^{-1} versus the mass change measured by QCM a linear relationship was found when altering the relative humidity (Figure 1). The mass changes were measured by rapidly altering the relative humidity from humidified air to dry (or *vice versa*) followed by the determination of mass change after stabilization. This procedure was assumed to give a mass response mostly from reversibly adsorbed (physisorbed) water. At high relative humidity a thick aqueous adlayer was formed, whereby an absorbance band at 1100 cm^{-1} was observed that disappeared when dry air was introduced (Figure 2). The intensity of this band strictly follows the intensity of the water band at 3400 cm^{-1} as the water content on the surface changes by changing the relative humidity in the reaction chamber. This absorbance band has not, to our knowledge, been reported in the literature before. The water molecule possesses three normal modes of vibration. These

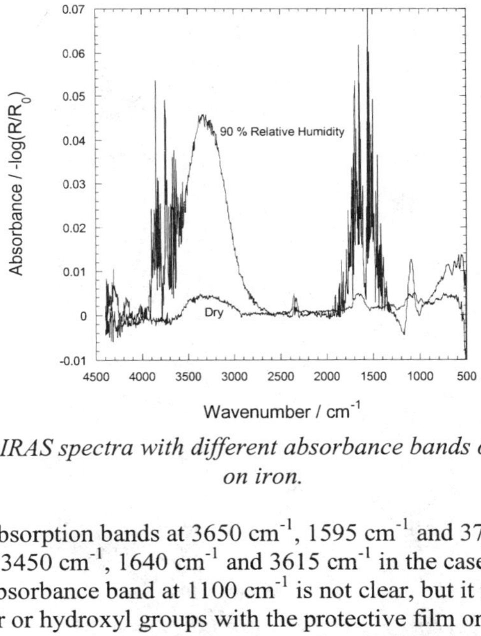

Figure 2 – *In-situ IRAS spectra with different absorbance bands originating from water on iron.*

modes have their absorption bands at 3650 cm^{-1}, 1595 cm^{-1} and 3755 cm^{-1} in the case of water vapor and at 3450 cm^{-1}, 1640 cm^{-1} and 3615 cm^{-1} in the case of liquid water [*26*]. The origin of the absorbance band at 1100 cm^{-1} is not clear, but it may arise from some interaction of water or hydroxyl groups with the protective film on the surface since adsorbed molecules tend to have their absorbance bands shifted towards lower frequencies due to the molecule substrate interaction.

Iron exposed to humidified air and SO_2

In order to examine the influence of SO_2 on the initial atmospheric corrosion of iron, SO_2 was introduced into the reaction chamber. In the experiments the iron samples were initially exposed to humidified, SO_2 free, air for time periods between 30 min and 24 h. In the IRAS spectra collected during this period the absorbance bands observed were in good agreement with previous data collected for the SO_2 free exposures. The mass change measured by QCM also followed the previously described behavior. After the initial sequence, the iron samples were exposed to 90% relative humidity and 200 ppb SO_2. No significant change compared to exposure without SO_2 could be observed. IRAS did not reveal any additional absorbance bands and the mass/frequency change of the QCM did not indicate any different reaction behavior compared to the results in the SO_2 free environment. The reaction behavior of copper in analogous exposures with relative humidity and SO_2 shows distinct differences. Sulfite and even small quantities of sulfate ions could be detected by IRAS and a considerable increase in mass change rate upon SO_2 introduction was observed with QCM. As a comparison, the mass change of copper after 500 min of exposure to relative humidity and SO_2 was about 1750 ng/cm^2 [*27*] while after the same exposure time the mass change of iron was found to be below 200 ng/cm^2. Surprisingly iron seems to form a quite corrosion protective film compared to copper under present exposure conditions [*27*]. This may possibly be due to the fact that copper is nobler and does not form a passive film as fast as iron does.

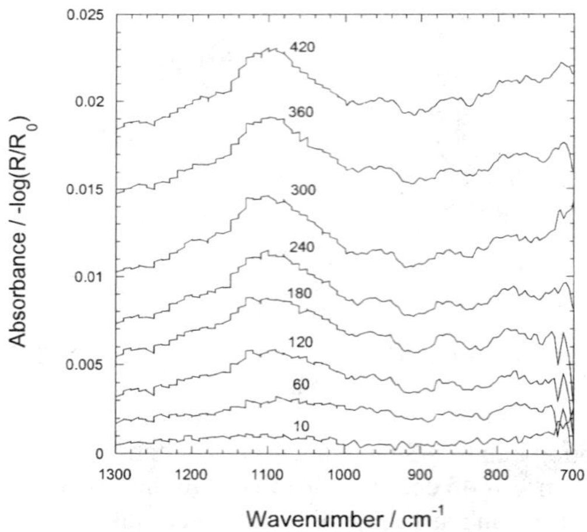

Figure 3 - *In-situ IRAS spectra of iron in 90% relative humidity, 200 ppb SO_2 and 200 ppb O_3. The reference spectra (R_0) was recorded at 90% relative humidity. The spectra originate from different exposure times, as given in minutes above each spectrum.*

Iron exposed to humidified air, SO_2 and O_3

In order to possibly increase the corrosion speed and complexity of the system, O_3 was introduced. The experiments were performed in a similar fashion as the experiments described in the section about SO_2. First the samples were exposed to pure humidified air, then 200 ppb SO_2 was added and finally, after 30 min to 24 hours in SO_2, 200 ppb of

O_3 was also introduced into the system. The exposure times were as long as 100 hours with up to 90% relative humidity, 200 ppb SO_2 and 200 ppb O_3. Higher reaction speeds compared to without O_3 were observed, but the IRAS cross section for the formed species was very low. Broad and not very intense absorbance bands centered at approximately 1100 cm^{-1} could be seen in the IRAS spectra (Figure 3). The reference spectra (R_0) was recorded immediately before O_3 exposure. This band was later determined with XPS to be sulfate. QCM showed an immediate increase in mass upon O_3 introduction. This O_3-induced mass increase continued over time with a constant mass gain of about 0.9 ng/cm^2min (Figure 4). The reaction pattern is very similar to the one observed on copper [*27*]. The reaction speed however, is significantly lower than the

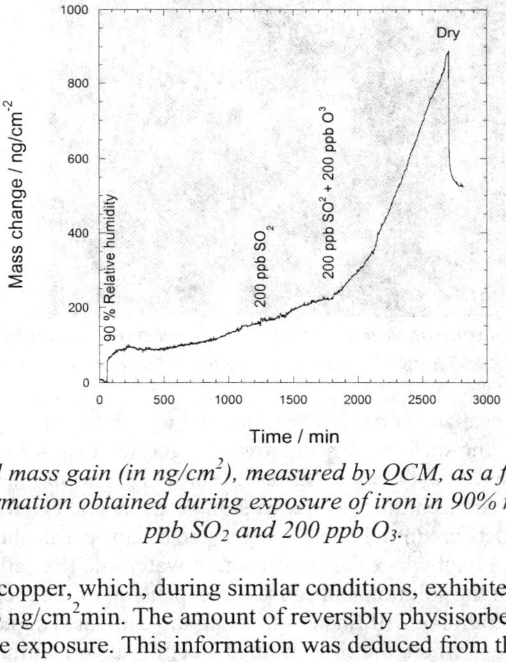

Figure 4 - *Total mass gain (in ng/cm^2), measured by QCM, as a function of exposure time. In-situ information obtained during exposure of iron in 90% relative humidity, 200 ppb SO_2 and 200 ppb O_3.*

reaction speed of copper, which, during similar conditions, exhibited a constant mass gain of about 17.6 ng/cm^2min. The amount of reversibly physisorbed water was found to increase during the exposure. This information was deduced from the mass change measured by QCM when turning on and off the relative humidity. The mass of physisorbed water after exposure could be as much as four times larger than before exposure (Figure 4). This indicates that the corrosion products formed are hygroscopic.

Ex-situ AFM and SEM measurements

Inspection of the surface with SEM after completed exposure indicates that the corrosion species are non-homogenously distributed over the surface. In fact, small localized corrosion attacks distributed over narrow areas are formed, while large areas seem unaffected (Figure 5). There seems to be quite large differences in composition of the corrosion products at different parts of the surface. According to x-ray elementary

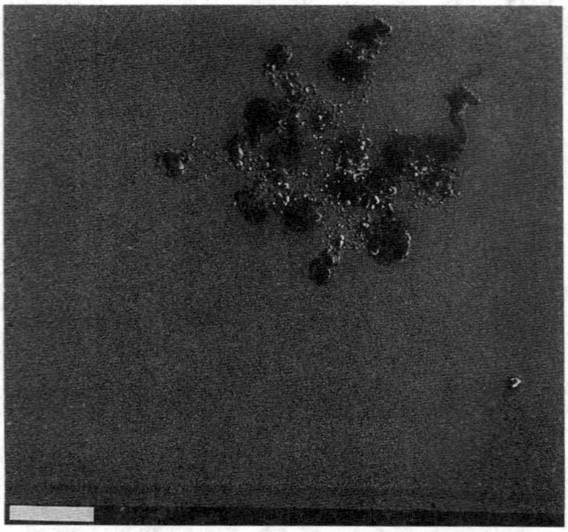

Figure 5 - *SEM image of an iron sample exposed to 90% relative humidity, 200 ppb SO_2 and O_3 for 48 h. The white bar in the bottom-left corner is 100 μm.*

analysis the areas with localized corrosion are enriched with sulfate compared to the more unaffected parts of the surface. It is tentatively suggested that the sulfate-enriched areas represent initial stages of so-called "sulfate nests" [*11*]. The sulfate nests are supposed to contain high concentrations of corrosion products and electrolyte. According to the model these products are initially occluded inside a semi-permeable membrane of precipitated and colloidal oxyhydroxides. Diffusion of water into the cell is induced by the high ionic strength of the solution and causes an expansion of the membrane. At a certain expansion the membrane will finally burst and the content is spread over the nearby surrounding surface where the concentrated electrolyte will stimulate further corrosion attacks. The model is not a complete mechanistic description since no analytical evidence of the constraining hydroxide membranes exists.

In order to obtain higher magnification and surface topology information AFM measurements were performed. The measurements were conducted in tapping mode, in

Figure 6a) – *AFM image of an iron sample exposed to 90% relative humidity for 22 h. The image size is 5x5 μm² and the black to white height range is 10 nm.*

Figure 6b) - *Iron sample exposed like 6a) followed by 7 h in 90% relative humidity and 200 ppb SO_2. The image size is 5x5 μm² and the black to white height range is 10 nm.*

Figure 6c) – *Iron sample exposed like 6b) followed by 15 h in 90% relative humidity and 200 ppb SO_2 and O_3. The image size is 5x5 μm² and the black to white height range is 10 nm.*

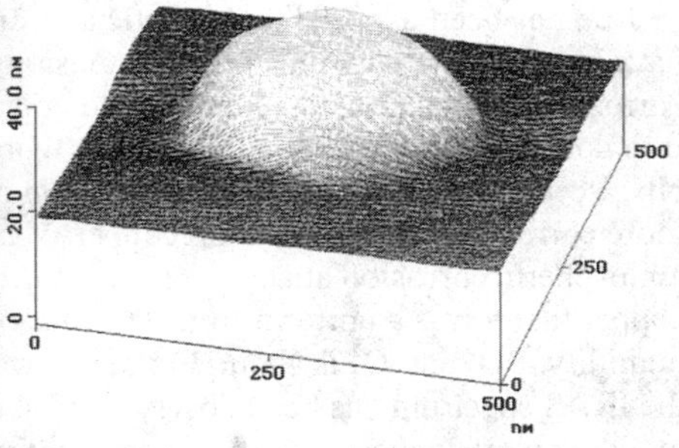

Figure 7 - *AFM image of one of the protrusions in Figure 6c.*

order to minimize the influence of the tip on the sample surface. The surface areas investigated with AFM were selected from areas that seemed flat in SEM images. In Figure 6a-c the evolution of the surface topography after different exposures measured by AFM are displayed. Figure 6a) shows a sample exposed to 90% relative humidity for 22 h, b) a sample exposed to 90% relative humidity for 22 h followed by 7 h in 90% relative humidity and 200 ppb SO_2, and c) a sample exposure just as b) but followed by 15 h in 90% relative humidity, 200 ppb SO_2 and 200 ppb O_3. The unexposed polished surface is not shown since no difference in topography between that surface and Figure 6a) could be discerned. The parallel lines present in all images are scratches from the polishing procedure. It is obvious that the sample exposed to SO_2 and O_3 is much more corroded than without O_3. Figure 7 is a magnification of one of the small protrusions in Figure 6c. The height of the protrusions is up to about 60 nm and the diameter approximately 500 nm. All these protrusions are likely to be nucleation sites for corrosion reactions where the protective oxide film has failed. After longer exposures, protrusions like this may give rise to features like those seen in Figure 5.

Within present exposure conditions, the corrosion behavior observed on iron differs greatly from the behavior observed on copper by Aastrup *et al* [*27*]. Whereas copper forms a uniform film of corrosion products across the surface, iron seems to have a more profound localized corrosion behavior. SEM images of corroded surfaces also reveal some cracks in the oxide. This may be an effect of the vacuum conditions. It may, however, be the result of variation in the relative humidity or a genuine oxide growth behavior.

Conclusions

Iron has been exposed to synthetic air containing humidity, SO_2 and O_3. The results have been compared with similar studies on copper. Great difference in atmospheric corrosion behavior of iron compared to copper has been discerned. In contrast to copper, iron does not form a homogeneous film with corrosion products and the amount of physisorbed water on iron is larger than on copper. The protective film on iron is initially more corrosion resistant than on copper. When the protective film of iron fails, atmospheric corrosion attacks occur on narrow areas of the iron surface, in contrast to copper that forms a uniform film. The mass gain of iron during exposure in relative humidity, SO_2 and O_3 is about 20 times lower than of copper. A new absorbance band in the IRAS spectrum has been observed, at about 1100 cm^{-1}. This absorbance band seems to be strongly connected to the presence of water on the surface.

Acknowledgements

Dr. Ted Aastrup is acknowledged for discussion and ideas, Dr. Saeid Zakipour for lending us the O_3 generator/analyzer, Ass. Prof. Inger Odnewall-Wallinder for XPS and SEM measurements and Mr. John Österman for assistance with AFM measurements.

Financial support from the Swedish Research Council for Engineering Sciences (TFR) is gratefully acknowledged.

References

[*1*] Leygraf, C. and Graedel, T., "Atmospheric Corrosion", John Wiley & Sons, New York, 2000

[*2*] Silver, J., "Chemistry of Iron", Blackie Academic & Professional, London, 1993

[*3*] Davenport, A.J., Oblonsky, L.J., Ryan, M.P., Toney, M.F., "The structure of the Passive Film That Forms on Iron in Aqueous Environments", *Journal of The Electrochemical Society*, 147, 2000, pp. 2162

[*4*] Oblonsky, L.J., Devine, T.M., "Corrosion of Carbon Steels in CO_2-Saturated Brine", *Journal of The Electrochemical Society*, 144, 1997, pp. 1252

[*5*] Arroyave, C., Lopez, F.A., Morcillo, M., "The Early Atmospheric Corrosion Stages of Carbon Steel in Acidic Fogs", *Corrosion Science*, 37, 1995, pp. 1751

[*6*] Johansson, L-G, "SO2 – induced corrosion of carbon steel in various atmospheres and dew point corrosion in stack gases", Thesis, Chalmers University of Technology, Sweden, 1982

[*7*] Poling, G.W., "Infrared Reflection Studies of the Oxidation of Copper and Iron", *Journal of The Electrochemical Society*, 116, 7, 1969

[*8*] Street, S.C., Xu, C., Goodman, D.W., " The Physical and Chemical Properties of Ultrathin Oxide Films", *Annual Review of Physical Chemistry*, 48, 1997, pp. 43

[*9*] Lee, S., Staehle, R.W., "Adsorption of Water on Copper, Nickel and Iron", *Corrosion Science*, 53, 1997, pp. 33

[*10*] Duncan, J.R., Spedding, D.J., "Discrete Areas of SO_2 Adsorption From the Atmosphere onto Iron", *Corrosion Science*, 13, 1973, pp. 69

[*11*] Graedel, T.E., Frankenthal, R.P., "Corrosion Mechanisms for Iron and Low Alloy Steels Exposed to the Atmosphere", *Journal of The Electrochemical Society*, 137, 1990, pp. 2385

[*12*] Maurice, V., Yang, W.P., Marcus, P., "X-ray Photoelectron Spectroscopy and Scanning Tunneling Microscopy Study of Passive Films Formed on (100) Fe-18Cr-13Ni Single Crystal Surfaces", *Journal of The Electrochemical Society*, 145, 1998, pp. 909

[*13*] Leygraf, C., Ekelund, S., "A LEED-AES study of the oxidation of Fe(100) and Fe(110)", *Journal of Vacuum Science and Technology*, 11, 1974, pp.189

[*14*] Graham, M.J., "The application of surface techniques in understanding corrosion phenomena and mechanisms", *Corrosion Science*, 37, 1995, pp. 1377

[*15*] Odnevall, I., "Atmospheric Corrosion of Field Exposed Zinc", Thesis, Royal Institute of Technology, Sweden, 1994

[*16*] Maslar, J.E., Hurst, W.S., Bowers, W.J., Hendricks, J.H., Aquino, M.I, "*In Situ* Raman Spectroscopic Investigation of Aqueous Iron Corrosion at Elevated Temperatures and Pressures", *Journal of The Electrochemical Society*, 147, 2000, pp. 2532

[*17*] Baldelli, S., Schnitzer, C., Shultz, M.J., Campbell, D.J., "Sum frequency generation investigation of water at the surface of H_2O/H_2SO_4 and H_2O/Cs_2SO_4 binary systems", *Chemical Physics Letters*, 287, 1998, pp.143

[*18*] Aastrup, T. and Leygraf, C., "Simultaneous Infrared Reflection Absorption Spectroscopy and Quartz Crystal Microbalance Measurements for *In Situ* Studies of the Metal/Atmosphere Interface", *Journal of The Electrochemical Society*, 144, 9, 1997, pp. 2986

[*19*] Itoh, J., Sasaki T., Seo M., Ishikawa T., "In situ simultaneous measurement with IR-RAS and QCM for investigation of corrosion of copper in a gaseous environment", *Corrosion Science*, 39, 1997, pp.193

[*20*] Jasinski, R., Iob, A., "FTIR Measurements of Iron Oxides on Low Alloy Steel", *Journal of The Electrochemical Society*, 135, 1988

[*21*] McDevitt, N.T., Baun, W.L., "Infrared absorption study of metal oxides in the low frequency region (700-240 cm^{-1})", *Spectrochemica Acta*, 20, 1964, pp. 799

[*22*] Kim, H.-J., Park, J.-H., Vescovo, E., "Oxidation of the Fe(110) surface: An Fe_3O_4(111)/Fe(110) bilayer", *Physical Review B.*, 61, 22, 15284, 2000

[*23*] Sauerbrey, G., Z. Phys., 155, 1959, pp. 266

[*24*] Rice, D.W., Phipps, P.B.P., Tremoureux, R., "Atmospheric corrosion of nickel", *Journal of The Electrochemical Society*, 127, 1980, pp. 563

[*25*] Büchler, M., Schmuki, P., Böhni, H., "Iron Passivity in Borate Buffer", *Journal of The Electrochemical Society*, 145, 1998, pp. 609

[*26*] Darling, B.T., Dennison, D.M., "The Water Vapor Molecule", *Physical Review*, 57, 1940, pp. 128-139

[*27*] Aastrup, T., Wadsak, M., Leygraf, C., Schreiner, M., "In Situ Studies of the Initial Atmospheric corrosion of copper", *Journal of The Electrochemical Society*, 147, 2000, pp. 2543

J.Y. Yoo,[1] W.Y. Choo,[1] and M. Yamashita[2]

Effect of Ca and S on the Simulated Seaside Corrosion Resistance of 1.0Ni-0.4Cu-Ca-S Steel

---

Reference: Yoo, J.Y., Choo, W.Y., and Yamashita, M., "**Effect of Ca and S on the Simulated Seaside Corrosion Resistance of 1.0Ni-0.4Cu-Ca-S Steel**," *Outdoor Atmospheric Corrosion, ASTM STP 1421*, H. E. Townsend, Ed., American Society for Testing and Materials International, West Conshohocken, PA, 2002.

Abstract: Atmospheric corrosion resistance of 1.0Ni-0.4Cu-Ca-S steels was investigated. The effects of Ca and S on the simulated seaside corrosion resistance of 0.1%C-1.0%Mn-0.4%Cu-1.0%Ni-0.01P-Ca-S steels were examined by CCT (cyclic corrosion test), SVET (scanning vibrating electrode technique) and Mossbauer experiments. It was found that the simulated seaside corrosion resistance of steel could be increased through Ca and S additions. Sulfur induced the rapid formation of the stable rust film and Ca was found to increase highly the corrosion resistance in the later stage. Especially, the rust layer formed in CCT tested 0.007%Ca-0.008%S steel was fully protective in 3.5%NaCl solution and thus this composition steel showed the possibility as a seaside corrosion resistant steel. The rust structures of CCT tested steels were analyzed by Mossbauer experiment and it was found that the amount of fine α-FeOOH with a diameter below 13nm remarkably increased with the increase of Ca and S contents. Ca contributed to the suppression of iron dissolution reaction by increasing the pH near the steel surface and also by enhancing the formation of fine α-FeOOH of a diameter below 13nm. Probably, Ca was assumed to activate the formation of α-FeOOH through increasing pH near the steel surface. Sulfur was thought to increase the simulated corrosion resistance by enhancing the formation of α-FeOOH under the seaside atmosphere by a similar mechanism in the case of the industrial atmosphere.

Keywords: steel, corrosion resistance, seaside atmosphere, calcium, sulfur, □-FeOOH

Introduction

Conventional Cu-Cr-P weathering steel possesses high corrosion resistance and therefore has been widely used without painting in rural and industrial environments *[1,2]*. However, conventional weathering steel cannot be used under the seaside

---

[1]Senior researcher and manager, Plate, Rod and Welding Research Group, Technical Research Laboratories, Pohang Iron and Steel Co., 1 Goedong-dong, Pohang, 790-785, Korea.

[2]Professor, Department of Mechanical Engineering, Himeji Institute of Technology, Hyogo, 671-2201, Japan.

atmosphere because the stable anti-corrosion rust is not formed under the chloride ion atmosphere. With the increase of the infrastructure in the seaside area, seaside corrosion resistant steel is required.

Under these backgrounds, seaside corrosion resistant steels like 3%Ni-0.4%Cu steel*[3]*, 0.6Ni-0.004Ca steel*[3,4]*, 1.2Ni-0.3Mo-0.06P steel*[5,6]* and 1.0Cu-1.0Ni-0.05Ti steel*[7]*, were actively developed, especially in Japan. Among these steels, Ca added Ca-Ni steel is known as the most cost-effective. In this steel, it is known that CaO inclusion dissolves in the corrosion media and thus increases the pH near the steel surface, suppressing the corrosion. However, the corrosion resistance of Ca-Ni steel is not sufficient and should be more increased.

As a purpose of searching for the cost-effective seaside corrosion resistant steel, it was tried to more increase the corrosion resistance than before researched Ca-Ni steel. As one approach to increase the seaside corrosion resistance, the effects of Ca and S on the simulated seaside corrosion resistance of Ni-Cu-Ca-S steel were investigated.

Experimental Procedure

Ni-Cu-Ca-S steels with the composition shown in Table 1 were melted in the laboratory vacuum induction furnace. Ni and Cu were fixed as 1 and 0.4wt%, respectively. Ca and S contents were varied. Ca contents of S1, S2 and S3 steels were below 10ppm. S content was changed from 60ppm to 130ppm. S4 and S5 steel had the higher Ca content. S4 steel has 30ppm Ca and 60 ppm S. In case of S5 steel, Ca and S were 70ppm and 80ppm, respectively. W was the conventional weathering and G was the general structure steels.

The melted ingot was reheated at 1250℃ for 2 hours and rolled to 5 mm thick plate with an average reduction ratio of 17%. The corrosion resistance of the rolled plate was investigated by CCT (cyclic corrosion test), SEM (scanning electron microscope), Mossbauer spectroscopy and SVET (scanning vibrating electrode technique) experiments.

Table 1-*Chemical compositions of Ni-Cu-Ca-S steels (wt%).*

No.	C	Si	Mn	P	S	Cu	Al	Ni	Nb	Ca
S1	0.097	-	1.01	0.02	0.006	0.4	0.04	0.980	-	0.0003
S2	0.101	-	1.01	0.02	0.010	0.4	0.03	0.995	-	0.0004
S3	0.095	-	1.01	0.02	0.013	0.4	0.06	1.0	-	0.0008
S4	0.100	-	1.21	0.02	0.006	0.39	0.03	0.984	-	0.003
S5	0.094	-	0.98	0.02	0.008	0.39	0.03	0.970	-	0.007
W	0.096	0.26	1.05	0.02	0.005	0.35	0.04	0.170	0.03	-
G	0.13	0.55	1.46	0.02	0.005	-	0.05	-	-	-

The CCT experiment consisted of five basic stages. : (1) salt spray of 5%NaCl - 30°C and 10min, (2) humid stage - 30°C and RH80%, 60 min, (3) dry stage - 50°C and RH50%, 30 min, (4) washing stage – 10 min, (5) dry stage - 40°C and RH35% , 80 min. This CCT experiment is to simulate the corrosion behavior in which the air-borne salt content is about 0.7 mdd (mg/dm^2/day).*[8]* In this CCT experiment, 240 cycles were thought to correspond to the exposure for two years and 480 cycles for four years. For the each steel, 3 samples were exposed for 240 cycles and the other 3 samples were exposed for 480 cycles in the cyclic corrosion test chamber. The corroded depth was estimated by measuring the corrosion weight losses of the three samples. Upon measuring the corrosion weight loss, the rust was removed by immersing the sample into the di-ammonium citrate solution (200g per purified water 1000ml) at 70℃, in accordance with ASTM Test Method for Chemical Cleaning Procedure for Removal of Corrosion Products (G1-88).

The rust phases in the steel were quantitatively analyzed through Mossbauer spectroscopy and SEM. The protectiveness of the rust formed by the CCT experiment was evaluated by the SVET experiment. In the SVET experiment, the gap between the specimen and platinum probe was 40 μm and the used electrolyte was 3.5% sodium chloride solution. The frequency of vibration was 130Hz and the vibration amplitude of electrode was 10 μm. Scanning area was 40X 40mm^2 and step distance was 0.2mm.

The inclusions in Ca added steel were analyzed by SEM and EDS (energy dispersed spectroscopy) and the pH of a thin water film covering the surface of the steel was measured with a micro-pH electrode having a tip diameter of 1.3mm. The water film was made of distilled water and 0.1M NaCl solution. The pH measurement was carried out every 10 minutes over 2 hours of exposure time. About 1cm^2 of the specimen was covered with the water film, and the thickness of water film was controlled at about 2mm.

Results and Discussion

Changes of Corroded Depth with Ca and S Contents

Figure 1 plots the changes of corroded depth of CCT tested steels with Ca and S contents. With the increase of sulfur content, corrosion depth increased in the initial stage from 0 to 240 cycles, but decreased after 480 cycles. This implies that sulfur induced the rapid formation of stable rust film and thus increased the corrosion resistance of the steel with the exposure time. Ca addition did not change the initial corrosion phenomenon from 0 to 240 cycles but was found to increase highly the corrosion resistance in the later stage. Especially, in case of S5 steel with 70ppm Ca and 80ppm S, the difference of corrosion depth between the two exposure stages of 240 and 480 cycles was nearly zero. From these results, it was well understood that seaside corrosion resistance of steel could be very much increased by Ca and S additions and 1.0Ni-0.4Cu-0.007Ca-0.008S steel had a possibility of application as a seaside corrosion resistant steel.

Protection Ability of the Rust Formed in the CCT Tested Steel

The protectiveness of the rust in the steels tested for 480 cycles by CCT machine was evaluated by SVET experiment in 3.5%NaCl solution.

Corrosion current distributions in the S1 steel (0.0003%Ca-0.006%S) and S3 steel (0.0008%Ca-0.013S) are shown in Figures 2(a) and (b). In Figures 2(a) and (b), the upper region corresponds to the anode current and the lower part corresponds to the cathode current. In case of S1 steel (0.0003%Ca–0.006%S), considerable anode and cathode current peaks were detected at the exposure stage of 480 cycles. But, in S3 steel (0.0008%Ca-0.013%S) as shown in Figure 2(b), the number of anode and cathode corrosion current peaks considerably decreased. This means that the rust of high S steel was more immune to the corrosion attack of the sodium chloride solution and thus the rust became protective with the increase of the S content.

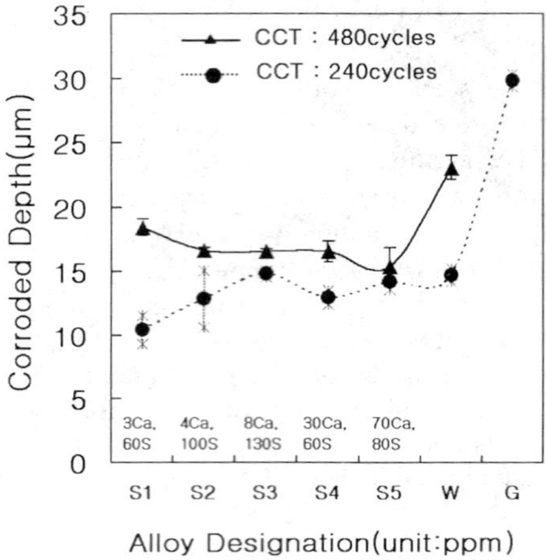

Figure 1 - *Changes of corroded depth with Ca and S contents in CCT (cyclic corrosion Test) tested Ni-Cu-Ca-S steels.*

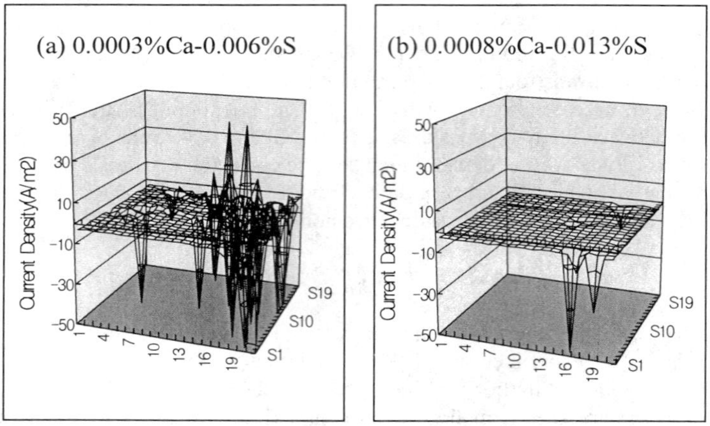

Figure 2 - *Corrosion current distributions over S1 steel (0.0003Ca-0.006S) (a) and S3 steel (0.0008Ca-0.013S) (b) covered with the rust layers formed in the cyclic corrosion test for 480 cycles.*

The protection ability of rust much highly increased with Ca content. Figures 3(a) and (b) show the corrosion current distributions in the steels including a Ca content of more than 30ppm. From Figure 2(b) and Figure 3(a), it was known that the number and height of corrosion current peak of S4 steel (0.003%Ca-0.006%S) was at a similar level with those of S3 steel (0.0008%Ca-0.013%S). But, in S5 steel (0.007%Ca-0.008%S) as shown in Figure 3(b), there was no corrosion current peak after exposure for 480 cycles. Therefore, it is easily conjectured that the fully protective rust to the 3.5%NaCl was formed in S5 steel (0.007%Ca-0.008%S) during the exposure period and thus this composition steel has a possibility as a seaside corrosion resistant steel.

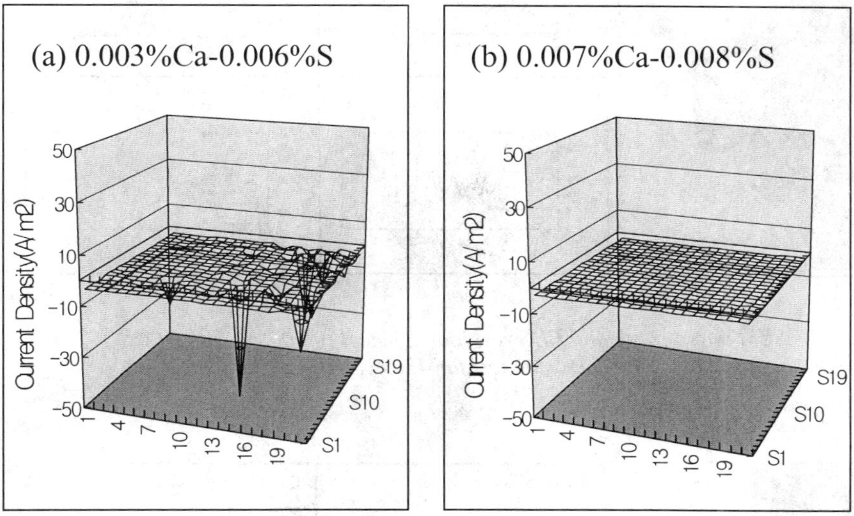

Figure 3 - *Corrosion current distributions over S4 steel (0.003Ca-0.006S) (a) and S5 steel (0.007Ca-0.008S) (b) covered with the rust layers formed in the cyclic corrosion test for 480 cycles.*

Rust Structure of Ca and S Added Steel

To examine why the simulated seaside corrosion resistance of Ni-Cu-Ca-S steel greatly increased with Ca and S addition, the rust layers formed in the CCT tested steel for 480 cycles were investigated through SEM, EDS and Mossbauer experiments.

Figures 4 and 5 show the SEM observation and EDS result on the rust layer in the S3 steel (0.00008%Ca-0.013%S) and S5 steel (0.007%Ca-0.008%S), respectively. As a whole, S5 steel, with better corrosion resistance, had the densest rust layer. Cu and Ni content was less in the rust layer than in matrix steel, which is thought because these noble elements remained in the matrix steel in the rust forming process. However, it was found that Ca and S contents was higher in rust layer than in matrix steel. Up to now, it is not clear why Ca and S enriched in the rust layer. But it is conjectured that CaS inclusion in the matrix steel dissolved in the corrosive media and then are entrapped into the rust layer. At any rate, it is thought that the enriched Ca and S modified the physical and chemical properties of the rust layer and thus enhanced the corrosion resistance.

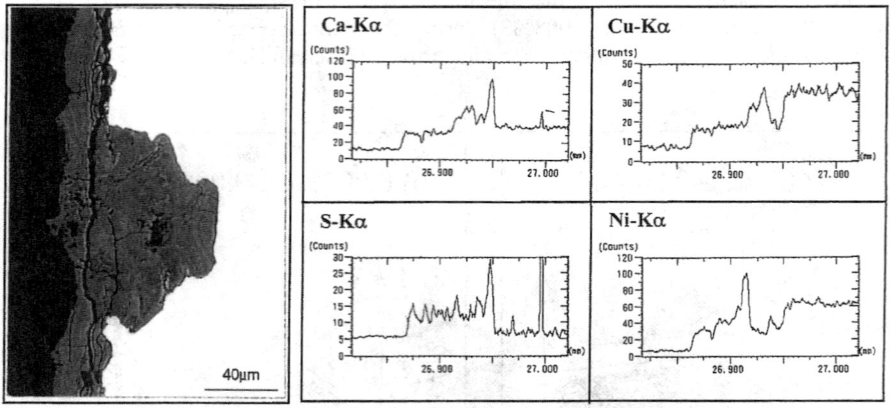

Figure 4 - *SEM micrograph and EDS result on S3 steel (0.0008%Ca-0.013%S) Exposed by cyclic corrosion test for 480 cycles.*

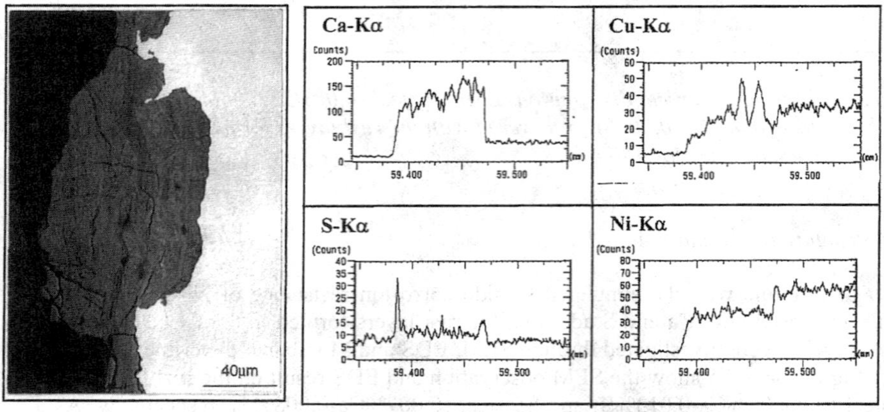

Figure 5 - *SEM micrograph and EDS result on S5 steel (0.007%Ca-0.008%S) exposed by cyclic corrosion test for 480 cycles.*

Figure 6 shows the rust phase fraction as a function of Ca and S contents in the CCT tested steels for 480 cycles. In Figure 6, the rust fraction was quantitatively analyzed by Mossbauer spectroscopy. The rust phases of the experimental steels were composed of γ-FeOOH, β-FeOOH and α-FeOOH. As a whole, 1.0%Ni-0.4%Cu-Ca-S steel had more γ-FeOOH and less β-FeOOH phases than W and G steels. This means that Ni contributes in enhancing the corrosion resistance by suppressing the formation of β-FeOOH.

With the increase of Ca and S contents, the amount of β-FeOOH increased but that of γ-FeOOH decreased. The amount of α-FeOOH slightly increased with additions of Ca and S. However it was found that the fine α-FeOOH with below 13nm in diameter drastically increased with Ca and S contents. In general, the increase of β-FeOOH is known to be not so good to the corrosion resistance. In this research, even though the amount of β-FeOOH increased with Ca and S addition, it was noticed that the simulated seaside corrosion resistance increased very much with Ca and S contents. Considering that fine α-FeOOH is a protective rust against atmospheric corrosion *[9]*, it is deduced that the increase of the simulated seaside corrosion resistance with Ca and S additions was attributed that the formation of fine α-FeOOH of diameter below 13 nm was enhanced.

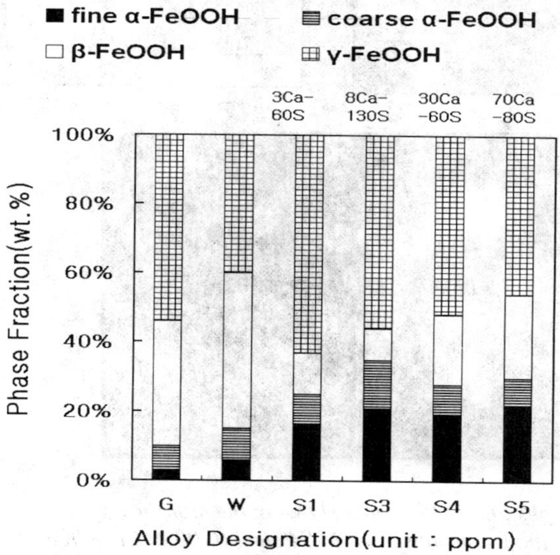

Figure 6 - *Mossbauer analysis result showing the change of the rust phases with Ca and S contents in Ni-Cu-Ca-S steels exposed by cyclic corrosion test for 480 cycles.*

Effect of Ca and S Additions on the pH Change in the Thin Electrolyte Film

In Ca added steel, it is well-known that pH near the corroded steel surface depends on the nonmetallic inclusion and thus the corrosion resistance can be affected by the species and morphology of the nonmetallic inclusion. The morphology and composition of the inclusions were observed with SEM and EDS.

Figure 7 shows the morphology and constitution of nonmetallic inclusions observed in S4 steel (0.003%Ca-0.006%S steel). As shown in Figure 7, the inclusions were round in shape with a diameter of 2-8 μm and were mainly composed of Ca-Al-Mn-O-S. When a small amount of Ca was added to the steel, it was thought to react preferentially with oxygen in the molten steel, leading to the formation of CaO-Al_2O_3 in the inner part of the inclusions, and later react with S by forming CaS-MnS in the outer part.

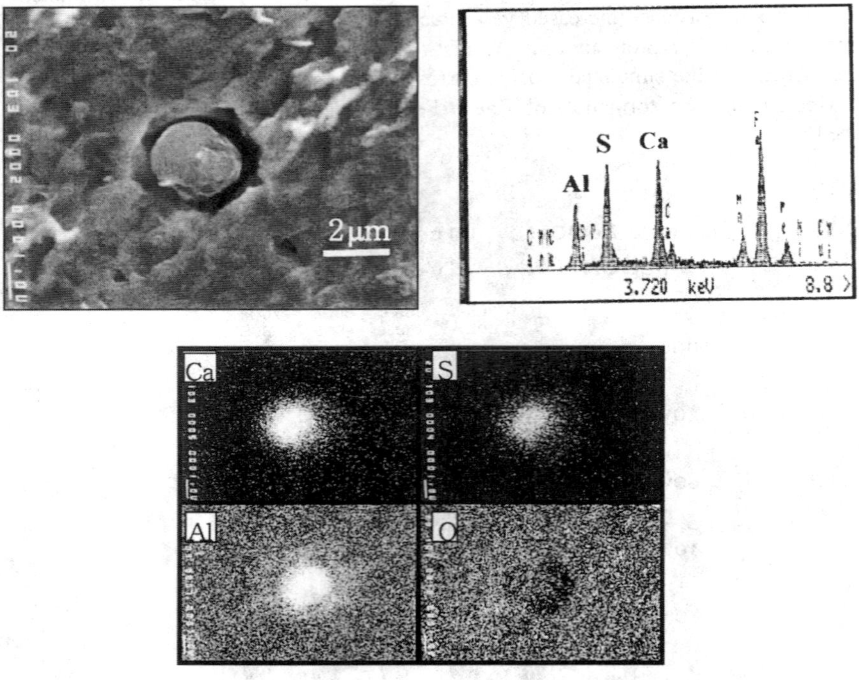

Figure 7 - *SEM micrographs and EDS results showing the morphology and compositions of non-metallic inclusions observed in S4 steel (0.003%Ca-0.006%S).*

Figure 8 shows changes in pH value of thin water films of the distilled water and 0.1M NaCl solution covering the steel. In general, the specimens containing Ca showed higher pH value than those containing no Ca. In distilled water, the specimen containing

0.007%Ca showed an increasing trend in pH value. However, in 0.1M NaCl solution, all the specimens showed a decreasing trend. This means that the positive effect of Ca on the pH increase in steel surface diminished with the increase of the chloride ion content. Therefore, it is assumed that Ca addition is useful in increasing the corrosion resistance in the region under not so severe air borne salt.

From the above results, it is concluded that Ca existed as CaO or CaS in the steel matrix and had an important role in increasing the pH near the steel surface contacting corrosion media, and thus increased the corrosion resistance of the steel.

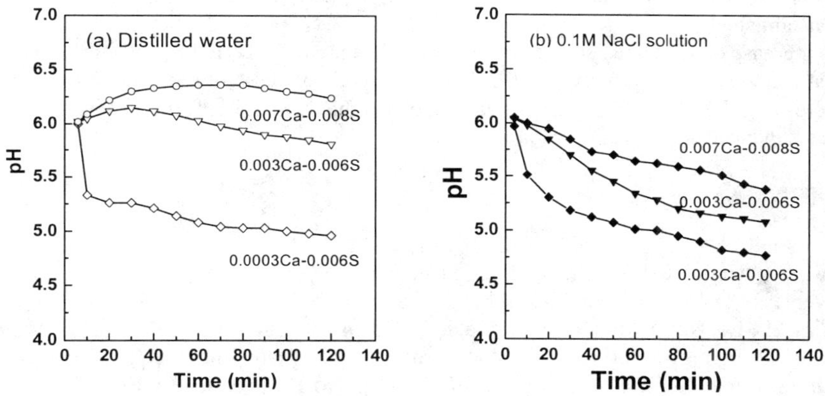

Fig.8 Changes of pH with time near the Ni-Cu-Ca-S steel surface in thin distilled water and thin film solution of 0.1M NaCl.

Corrosion Resistance Mechanism of Ca and S Added Steel

Considered from the above experimental results, the following two factors are thought to be the causes of the good corrosion resistance of Ca and S added steel. First is because the added Ca contributed to the suppression of iron dissolution reaction by increasing the pH near steel surface. In this steel, CaO and CaS inclusions were thought to dissolve in the corrosive media and thus increase the pH near the steel surface. Probably, CaO and CaS were thought to increase the pH by $CaO + H_2O \rightarrow Ca^{2+} + 2OH^-$ and $CaS + H_2O \rightarrow Ca^{2+} + 2OH^- + HS^-$ reactions, respectively. Second is because the formation of fine α-FeOOH was accelerated with the increase of Ca and S contents.

The reason why the formation of fine α-FeOOH was enhanced through Ca and S additions, is not clear. But that is supposed due to that pH increase near the steel surface by CaS inclusion accelerated the formation of fine α-FeOOH. According to the several researchers, the formation of fine α-FeOOH is accelerated under the alkaline atmosphere. Moreover, S was thought to enhance the formation of α-FeOOH like the case of industrial atmosphere. According to many researchers, the amount of α-FeOOH increases with the increase of SO_2 content in case of industrial and rural atmosphere *[10]*.

From these experimental results, it was understood that a cheaper seaside corrosion resistant steel than the currently supplied steel could be developed.

Conclusions

(1) Simulated seaside corrosion resistance of Ni-Cu-Ca-S steel was very much increased by the Ca and S additions.

(2) The rust layer in 1.0%Ni-0.4%Cu-0.007%Ca-0.008%S exposed by cyclic corrosion test, was fully protective to the 3.5%NaCl solution. From this, it was deduced that 1.0%Ni-0.4%Cu-0.007%Ca-0.008%S steel has a possibility as a seaside corrosion resistant steel.

(3) The increase of simulated seaside corrosion resistance with Ca and S was attributed that the added Ca contributed to the suppression of iron dissolution reaction by increasing the pH near the steel surface and also the formation of fine α-FeOOH was enhanced.

References

[1] *Weathering Steel,* Japan Steel Materials Club, 1977.

[2] Shoji, Y., "Unpainted Weathering Steel Structures," *Nippon Kokan Technical Report* , Overseas No.45, 1985, PP.44-50.

[3] Kihira, H., Ito, S., Mizoguchi, S., Murata, T., Usami, A. and Tanabe, K., "Creation of Alloy Design Concept for Anti Air-Born Salinity Weathering Steel," *Zairyo-to-Kankyo*, Japan Society of Corrosion Engineering, No.1, 2000, pp. 30-40.

[4] Yamamoto, M., Kihira, H., Usami A., Tanabe, K., Masuda, K. and Tsuauki, T., "Corrosion Resistance of Ca-Ni Added Weathering Steel in Marine Environment," *Tetsu-to-Hagane*, The Iron and Steel Institute of Japan, Vol. 84, No.3, 1998, pp. 36-41.

[5] Takaku, T., Mihara, Y., Fujita, S., Matsui, K., Kato, M. and Honda, M., "NKK's Minimum Maintenance Bridges," *NKK Technical Report*, No. 165, 1999, pp. 17-21.

[6] Tanaka, K., Nishimura T., and Suzuki S., "A Newly Developed Weathering Low Alloy Steel for Coastal Use," *CAMP-ISIJ*, The Iron and Steel Institute of Japan, Vol. 10, 1997, pp. 1239.

[7] Nakayama, T., Yuse F., and Kan T., "Effect of Alloying Elements on Corrosion Resistance of Painted Steels in Chloride Environments," *CAMP-ISIJ*, The Iron and Steel Institute of Japan, Vol. 11, 1998, pp. 454.

[8] Yoshii, T., Ito, K. and Nogami, M., "Development of Accelerated Atmospheric Corrosion Test of Stainless Steels," *Nisshin Steel Technical Report*, No. 59, 1988, pp. 1-15.

[9] Yamashita, M., Nagano, H., Misawa, T., Townsend H.E., "Structure of Protective Rust Layers Formed on Weathering Steels by Long-term Exposure in the Industrial Atmospheres of Japan and North America," *ISIJ International*, The Iron and Steel Institute of Japan, Vol. 38, No.3, 1998, pp. 285-290.

[10] Misawa, T., "Corrosion Science in Rusting of Iron and Weathering Steel," *Boshoku Gijutsu*, Japan Society of Corrosion Engineering, Vol. 37, 1988, pp. 501-506.

Masato Yamashita,[1] Hitoshi Uchida,[1] and Desmond C. Cook[2]

Effect of Cr^{3+} and SO_4^{2-} on the Structure of Rust Layer Formed on Steels by Atmospheric Corrosion

Reference: Yamashita, M., Uchida, H., and Cook, D. C., "**Effect of Cr^{3+} and SO_4^{2-} on the Structure of Rust Layer Formed on Steels by Atmospheric Corrosion,**" *Outdoor Atmospheric Corrosion, ASTM STP 1421,* H. E. Townsend, Ed., American Society for Testing and Materials International, West Conshohocken, PA, 2002.

Abstract: The effect of chromium ion (Cr^{3+}) and sulfate ion (SO_4^{2-}) on the structure of the rust layer formed by thin electrolyte-film corrosion of low-alloy steels has been examined. By using X-ray diffraction spectroscopy, it was found that coexistence of Cr^{3+} and SO_4^{2-} in the electrolyte film covering the steel surface quickly forms the Cr-goethite (α-$(Fe_{1-x},Cr_x)OOH$) layer which has been known as the final protective rust layer. Scanning vibrating electrode measurements showed that the rust layer formed under the electrolyte film containing Cr^{3+} and SO_4^{2-} possesses higher protective ability against the aggressive chloride environment. Mössbauer spectroscopy revealed that most of the Cr-goethite formed by corrosion of the Fe-5at%Cr alloy under thin electrolyte-film containing SO_4^{2-} was the superparamagnetic ultra-fine Cr-goethite. It can be said that the Cr-goethites possess the high protective ability against aggressive corrosives.

Keywords: atmospheric corrosion, electrolyte film, ion species, rust layer, goethite

Introduction

Atmospheric corrosion of steels strongly depends on the nature of oxide or rust layer on the steel surface [*1*]. Thus, we expect the rust layer to give certain beneficial effects for prevention of degradation of the steels. It is well-known that the so called "protective" rust layer formed on a weathering steels which contain a small amount of Cr, P, and Cu has strong protective ability for atmospheric corrosion of the steel. Thus, weathering steels possess high corrosion resistance, approximately twice as good as carbon steels, and therefore has been widely used as a structural material in an atmospheric corrosion environment.

Because formation of the protective rust layer usually takes time, the initial corrosion rate of the weathering steel is not substantially low and the initial unprotective

[1] Associate Professor and Professor, respectively, Faculty of Engineering, Himeji Institute of Technology, 2167 Shosha, Himeji, Hyogo 671-2201, Japan.

[2] Professor, Department of Physics, Old Dominion University, 4600 Elkhorn Ave, Norfolk, VA, 23529.

rust contaminates the steel surfaces and other environing materials. In addition, the protective rust layer cannot form in the coastal environment where the amount of air-borne salts is relatively high. These are significant problems to reduce maintenance tasks for structures made of weathering steel.

According to the recent understanding of the protective rust layer, the main constituent of the rust layers is changed with the exposure period from lepidocrocite (γ-FeOOH) (less than a few years), via amorphous substance (several years), to goethite (α-FeOOH) type phase (decades). The corrosion rate of the weathering steel decreases with these phase transformations. The protective rust layer possesses the structure of Cr-goethite (α-$(Fe_{1-x},Cr_x)OOH$) [*2*], which might be formed due to concentration of Cr in the rust layer after long-term exposure. By this Cr concentration in the goethite structure, the Cr-goethite possesses a densely packed structure that provides higher protective ability and prevents penetration of aggressive corrosive anions in the atmosphere. It is also pointed out that the Cr-goethite possesses cation selective ability. Therefore, the protective rust layer of the Cr-goethite impedes the penetration of aggressive corrosive anions such as Cl^-, as well as its densely aggregated structure, which provides the resistance against a corrosive penetration.

Based on the above basic knowledge, the Cr-goethite is a highly protective rust that is useful for obtaining corrosion-resistant steels. It has been found that chromium sulfate ($Cr_2(SO_4)_3$) is effective for obtaining the protective Cr-goethite rust layer in a short period [*3*]. Because of the quick formation of the protective layer, this effect of $Cr_2(SO_4)_3$ can be a key technology which prevents the damage by the briny air. It is now very important to know how Cr^{3+} and SO_4^{2-} contribute the quick formation of the Cr-goethite. In this paper, we try to understand the effect of Cr^{3+} and SO_4^{2-} on the detailed rust structure of steel.

Experimental Method

Specimen

Specimens employed were 0.5%Cr-0.3%Cu-0.06%P-0.5%Ni steel disks 15mm diameter and 5mm thick. The surfaces of the specimens were polished until #1500 emery paper.

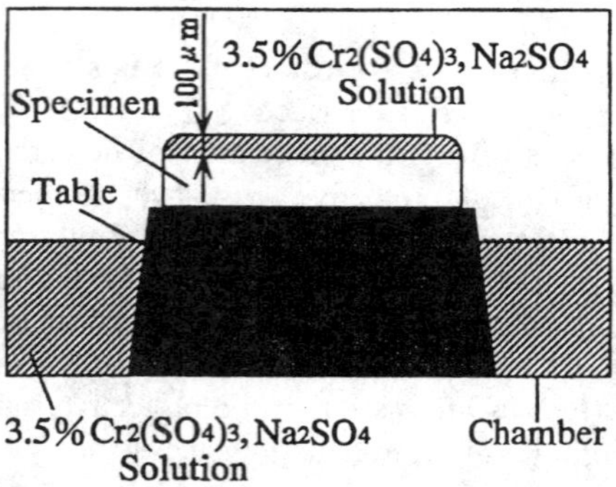

Figure 1 - *Schematic illustration of electrolyte film corrosion chamber.*

Corrosion under Electrolyte Film

In order to obtain atmospheric rust layers on the steel surfaces, the steel disk specimens were corroded under thin layer of electrolytes of $Cr_2(SO_4)_3$ or sodium sulfate (Na_2SO_4) at concentration of 3.5% each for 24 hours. The thickness of the electrolytes, h, was supposed to be 1-100

μm by adjusting the amount of the electrolytes and relative humidity of the surrounding air in an electrolyte-film corrosion chamber, as shown in Figure 1. The rust layers formed under the $Cr_2(SO_4)_3$ and Na_2SO_4 electrolyte-films will be referred to in the remainder of this report as the CS and SS layers, respectively.

Measurement of Corrosion Current Distribution

The specimens with the rust layers formed under the 100 μm thick electrolyte film on the steel disks were immersed in 3.5%NaCl solution at ambient temperature, and then the corrosion currents were measured using the scanning vibrating electrode technique (SVET). The probe of the SVET was Pt wire with diameter of 0.05mm. The measurements were done after certain immersion periods, t_i, of 0, 3, and 5 hours in the 3.5%NaCl solution. The amplitude and frequency of probe vibration were controlled to be 10 μm and 130Hz, respectively.

Measurement of Corrosion Rate

Specimens with the CS layer formed on the steel disks were corroded under 3.5%NaCl solution film with the same thickness of the electrolyte film, h, employed to obtain the rust layers. The corrosion rate during this NaCl solution film corrosion was measured by oxygen consumption measurement [*4*], that measures the oxygen partial pressure decreased by oxygen consumption as a cathodic reaction in the tightly sealed corrosion chamber. The purpose of this measurement is to know the effect of the electrolyte-film thickness on the protective ability of the rust layer.

Metallographic Cross Sections

Metallographic cross sections were prepared by shearing cross sections and mounting in two-part, room temperature-curing epoxy. The mounts were ground repeatedly with successively finer silicon carbide, and finished with alumina powder. The cross sections of the rust layers were characterized by optical microscopy and by electron probe microanalyzer to determine the distribution of some important elements.

Structural Analysis of Rust Layer

Rust layers formed on the steel were removed by a razor until the steel surface appeared, and then ground into powder. The powdered rust samples were desiccated for a week in advance of the analysis. The powdered samples were characterized by means of X-ray diffraction spectroscopy employing a Co target. The quantitative determination of rust constituents comprising goethite, lepidocrocite, akaganeite, magnetite, and the X-ray amorphous substances was carried out by measuring the diffraction intensities and comparing that of KCl powder that had been mixed with rust samples as an internal standard. The remainder other than the above crystalline constituents was considered as the X-ray amorphous substances in this quantifying estimation because the amounts of other crystalline materials and air-borne dusts might be very small.

Fe-Cr Alloy Film

Since the role of Cr as an alloying element on the formation of rust layer is very important to discuss the protective ability of the rust layer, the rust layer formed from Fe-5at%Cr alloy was also examined using Mössbauer spectroscopy. Fe-5at%Cr alloy film with a thickness of 100nm was vapor-deposited on a Si(100) single crystal substrate. The surface of the alloy film was covered with 100 μm-thick (0.1mol/l Na_2SO_4 + 0.005 mol/l H_2SO_4) film for 200h at ambient laboratory temperature, so that all the alloy film was corroded and finally the uniform rust layer was formed. Because the alloy film was corroded in a short period, the rust layer formed is categorized as an initially formed rust layer.

In order to identify separately the iron oxides that were either superparamagnetic at 300K or magnetically ordered between 300 and 77K, the transmission Mössbauer spectroscopy was used. The powdered sample of the rust layer on the Si substrate was mixed with BN powder and pressed into a 1cm diameter tablet. The Mössbauer spectra were recorded using a 20mCi ^{57}Co in rhodium source at room temperature 300K and 77K in a liquid nitrogen cooled cryostat. A commercial proportional counter filled with the mixture of xenon and carbon dioxide gases was used.

Results and Discussions

Corrosion Current Distribution

A large number of cracks were observed in the loosely packed SS layer by the optical and scanning electron microscopies. On the contrary, densely aggregated structures were found in the CS layer. The corrosion current distributions are shown in Figure 2. Many corrosion current peaks were observed over the SS layer. However, little corrosion current was measured in the CS layer; this indicates that the CS layer possesses high protective ability against corrosives even in the aggressive chloride environment.

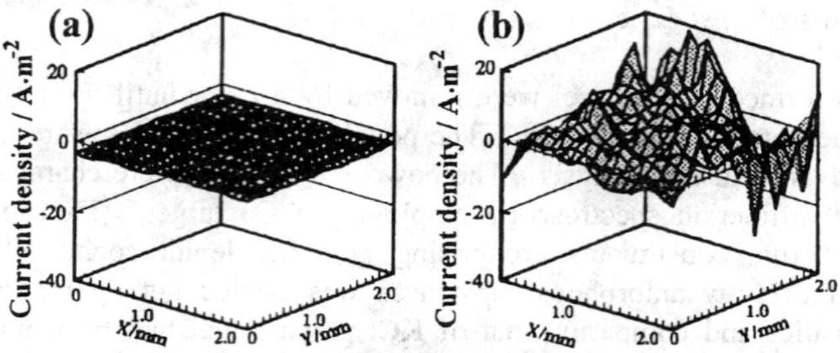

Figure 2 - *Corrosion current distributions over the steel specimens covered with the rust layers formed under the $Cr_2(SO_4)_3$ (a) and Na_2SO_4 (b) electrolytes, in the NaCl solution (t_i=0h).*

The X-ray diffraction spectroscopy revealed that main phase of rust constituents is goethite, as shown in Figure 3. It has been pointed out that lepidocrocite is an initial unprotective rust on the weathering steel, as was mentioned above. The large amount of goethite in the CS and SS layers, therefore, implies that SO_4^{2-} accelerates the rust formation and leads the formation of goethite-phase rust. It was found using electron probe microanalysis that the CS layer contains Cr in its rust. It can be said that coexistence of Cr^{3+} and SO_4^{2-} preferentially forms Cr-goethite which has been known as the final protective rust layer [*2,3*]; probably Cr^{3+} positions Fe^{3+} lattice site of goethite structure as a substitutional ion and forms stable goethite structure. It was pointed out that Cr^{3+} precipitates as hydroxide much faster than Fe^{2+} during air oxidation [*5*]. Thus, a large number of nuclei of hydroxide might be formed at the Cr^{3+} precipitation sites, which may result in formation of very fine particles of the Cr-goethites. Aggregation of this ultra-fine Cr-goethite causes higher protective performance of the rust layer.

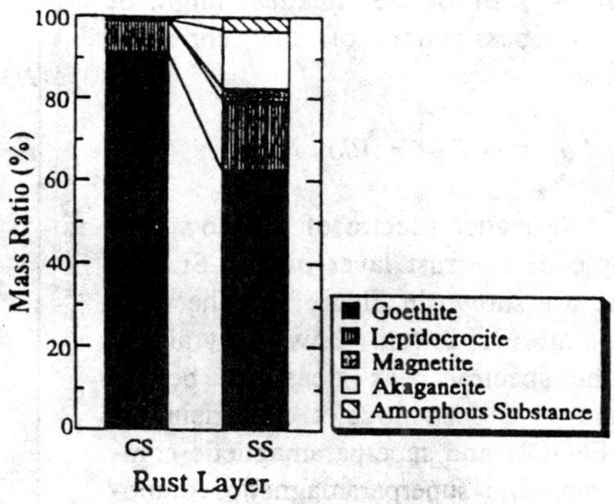

Figure 3 - *Mass ratio of rust constituents evaluated after the SVET measurement by X-ray diffraction intensities of the rust layers formed under the $Cr_2(SO_4)_3$ (CS) and Na_2SO_4 (SS) electrolytes (t_i=5h).*

Corrosion Rate

Figure 4 shows the relationship between the electrolyte-film thickness, *h*, to form the rust layer and the corrosion rate of the steel with the CS layer under the NaCl solution film with the same thickness. Even though the corrosion rate of the steel with the CS layer was very low when *h* is large, the corrosion rate peak in its electrolyte-film thickness dependence was observed at *h* of about 4 μm. This indicates that the protective ability of the CS layer might be low when *h* is very small. The existence of the corrosion rate peak was due to increasing in oxygen diffusion rate to the surface resulting in acceleration of the cathodic reaction. The decrease in the corrosion rate

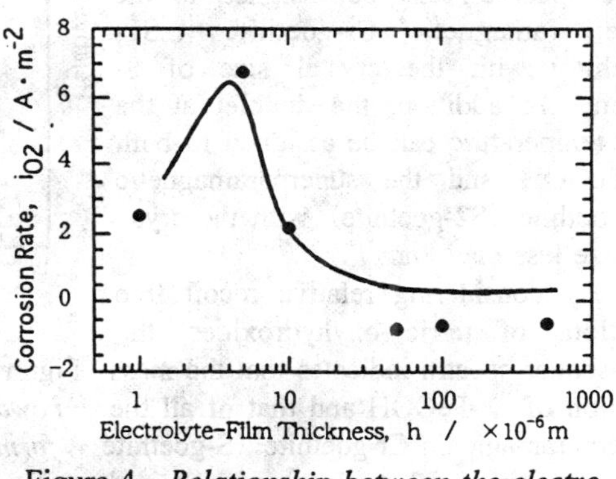

Figure 4 - *Relationship between the electrolyte-film thickness, h, to form the rust layer and the corrosion rate of the steel with the CS layer under the NaCl solution film with the same thickness (t_i=0h).*

below 4 μm of the thickness might be due to passivation of the corroding surface.

Rust Layer of Fe-Cr Alloy Film

Mössbauer spectra of the powdered sample of the rust layer on the Si substrate are shown in Figure 5. The room temperature spectrum shows that almost all the spectral peak areas can be assigned to a deep doublet comprising of γ - FeOOH and superparamagnetic components. The superparamagnetic components would correspond to ultra-fine crystals that might show no clear X-ray diffraction peaks because they have a small number of crystal planes in the crystal grain.

From the low temperature spectrum, it is found that the quadruple splitting showing the doublet at room temperature was split out to the sextet of the magnetic splitting. By fitting the hyperfine parameters of iron oxides and oxy-hydroxides, the sextet in the low temperature spectrum can be assigned to the superparamagnetic Cr-goethite (S1-goethite) with the crystal size of 8-15nm. In addition, the doublet at the low temperature can be assigned to both γ-FeOOH and the superparamagnetic Cr-goethite (S2-goethite) with the crystal size less than 8nm.

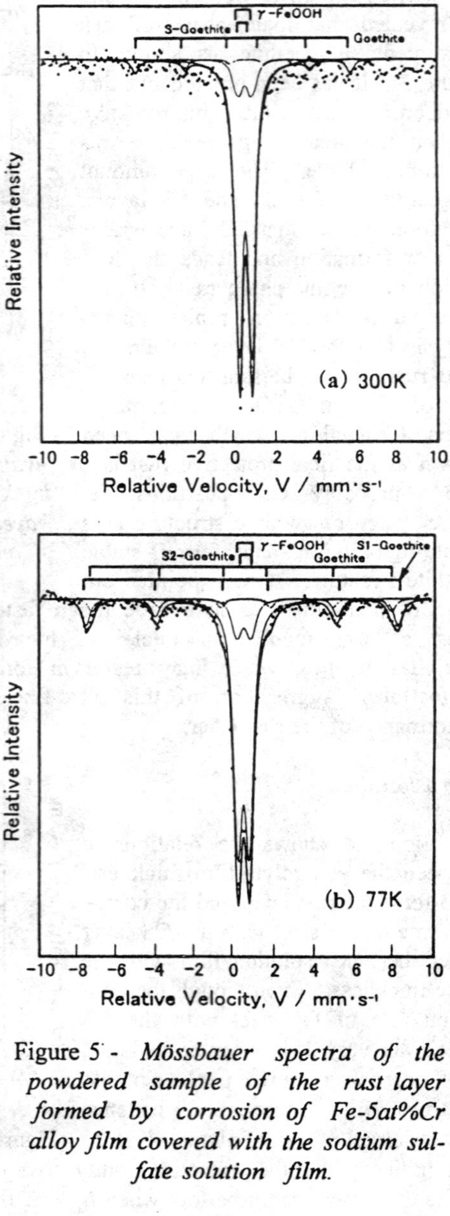

Figure 5 - *Mössbauer spectra of the powdered sample of the rust layer formed by corrosion of Fe-5at%Cr alloy film covered with the sodium sulfate solution film.*

By considering relative recoil free fractions of ferric oxyhydroxides, the Mössbauer spectra indicates that the mass fraction of γ-FeOOH and that of all the superparamagnetic Cr-goethite (S-goethite = S1-goethite + S2-goethite) are 12% and 88%, respectively. Therefore, it is said that the rust layer contains a large amount of the X-ray undetectable superparamagnetic Cr-goethites corresponding to the X-ray amorphous substance.

As was mentioned above, the main constituent of the initial rust layer within a few year exposure on a weathering steel is γ-FeOOH and the protective rust layer obtained after long-term exposure, depending on the atmosphere and steel composition, has the Cr-goethite structure. However, most of rust constituents on the present Si substrate was the superparamagnetic ultra-fine Cr-goethite. This quick formation of the untra-fine Cr-goethite might be due to the high Cr concentration in the Fe-Cr alloy film. Usually, the weathering steel contains 0.3-0.5 mass% Cr, which concentrates in the rust layer during long-term corrosion and helps formation of the Cr-goethite [*6,7*]. On the other hand, the Cr concentration in the Fe-Cr alloy was much higher than that in the weathering steel. This higher concentration of Cr in the Fe-Cr alloy, inevitably in its rust layer too, would result in the quick formation of the ultra-fine protective Cr-goethite phase.

In the case that the Cr concentration in the Cr-goethite is more than 5mass%, the crystal size becomes less than 10nm [*8*]. This agrees well to the present result that shows high concentration of Cr results in the formation of the ultra-fine Cr-goethite. In addition, it has been pointed out that the cation selective property is appeared when the Cr content exceeds approximately 3 mass% [*9*]; i.e. the Cr-goethite impedes the penetration of aggressive corrosive anions such as Cl^-. Thus, the ultra-fine Cr-goethite possesses the high protective ability against aggressive corrosives. Finally, it is said that the results obtained in this research can be directed toward development of new technologies which enable us to obtain the protective rust layer on the weathering steel very quickly.

Conclusions

The effect of Cr^{3+} and SO_4^{2-} on the structure of rust layer on steels by atmospheric corrosion under thin electrolyte film has been examined using the scanning vibrating electrode method, X-ray diffraction and Mössbauer spectroscopy. The rust layer formed under the electrolyte film containing Cr^{3+} and SO_4^{2-} possesses higher protective ability against corrosives even in the aggressive chloride environment. Coexistence of Cr^{3+} and SO_4^{2-} in the electrolyte film covering the steel surface quickly forms the Cr-goethite, α-$(Fe_{1-x},Cr_x)OOH$ layer which has been known as the final protective rust layer. Most of the Cr-goethite formed by corrosion of the Fe-Cr alloy under thin electrolyte film containing SO_4^{2-} was the superparamagnetic ultra-fine Cr-goethite. It can be said that the Cr-goethites possess the high protective ability against aggressive corrosives.

Acknowledgments

One of the authors (M.Y.) would like to thank the Ministry of Education, Science and Culture of Japan, as a Grant-in-Aid for Scientific Research on the Research No. 12750625. The support by the Sumitomo Metal industries, Ltd is also greatly appreciated.

References

[1] Leygraf,C. and Graedel, T.E., "Atmospheric Corrosion", Wiley-Interscience, New York, 2000.

[2] Yamashita,M., Miyuki,H., Matsuda,Y., Nagano,H. and Misawa,T., "Long-Term Growth of Protective Rust Layer Formed on Weathering Steel", Corrosion

Science, Vol.36, 1994, pp.283-299.

[3] Yamashita,M., Miyuki,H. and Nagano,H., "Corrosion Resistance of Weathering Steel and Its Application", The Sumitomo Search, No.57, 1995, pp.12-17.

[4] Yamashita,M., Nagano,H. and Oriani,R.A., "Dependence of Corrosion Potential and Corrosion Rate of a Low-Alloy Steel upon Depth of Aqueous Solution", Corrosion Science, Vol.40, 1998, pp.1447-1453.

[5] Misawa,T., "Corrosion Morphologies and Products of Iron and Steels in Wet Corrosion", Bulletin of the Japan Institute of Metals, Vol.24, 1985, pp.201-210.

[6] Yamashita,M., Nagano,H., Misawa,T. and Townsend,H.E., "Structure of Protective Rust Layer formed on Weathering Steels by Long-Term Exposure in the Industrial Atmospheres of Japan and North America", ISIJ International, Vol.38, 1998, pp.285-290.

[7] Yamashita,M. and Uchida,H., "Effect of Ion Species in the Electrolyte Environment on the Structure and Protective Ability of the Rust Layer on Steel", *Proc. NACE 55th Annual Conference* (*Corrosion 2000*), NACE, Orlando, 2000, pp.450/1-3.

[8] Cook,D.C., Balasubramanian,R., Oh,S.J. and Yamashita,M., "The Role of Goethite in the Formation of Protective Corrosion Layers on Steels", Hyperfine Interactions, Vol.122, 1999, pp.59-70.

[9] Yamashita,M. and Misawa,T., "Recent Progress in the Study of Protective Rust-Layer Formation on Weathering Steel", *Proc. NACE 53rd Annual Conference* (*Corrosion 98*), San Diego, 1998, pp.357/1-9.

Erin T. McDevitt[1] and Fritz J. Friedersdorf[1]

Analysis of the Sources of Variation in the Measurement of Paint Creep

Reference: McDevitt, E. T. and Friedersdorf, F. J., "**Analysis of the Sources of Variation in the Measurement of Paint Creep,**" *Outdoor Atmospheric Corrosion, ASTM STP 1421*, H. E. Townsend, Ed., American Society for Testing and Materials International, West Conshohocken, PA, 2002.

Abstract: Performance testing of coil coated steel sheet products usually involves the measurement of paint creep back from a sheared edge or scribe (ASTM D 1654-92) on samples subjected to accelerated and atmospheric exposure tests. Measurements of paint creep are used for screening paints, pretreatments, and inorganic coatings for use on steel sheets. Because these tests are used for product development and selection, it is important to have a clear understanding of the limitations and sensitivity of these destructive test methods. A measurement system analysis has been undertaken to determine the sources of variation in edge creep measurements. Coil coated Galvalume®[2] sheet material tested in salt spray is the subject of this investigation. A method for measuring edge creep that will provide satisfactory repeatability and reproducibility in the measurement system is described. It was found that without rigorous attention to the measurement method, the measurement system could contribute to as much as 25% of the variation in the measurement of edge creep. It has been shown that the operator preparing the samples is not a significant factor influencing the extent of edge creep. However, edge creep can vary by up to 1 mm simply because the test panels were cut from different positions across the width and along the length of a coil.

Keywords: corrosion, paint, edge creep, scribe creep, steel, galvanized, Al-Zn coated steel

Introduction

Assessing the quality of coil coated steel sheet products usually involves the measurement of paint creep back from a sheared edge or a scribe on samples that have been subjected to accelerated and atmospheric exposure tests. Measurements of paint creep are used for quality assurance purposes as well as for screening paints, pretreatments, and inorganic coatings for use on steel sheets. At Bethlehem Steel, we routinely use edge creep measurements to qualify new primers and to monitor the performance of coil painting facilities. In new product development, measurement of edge creep is used to assess the viability of new metallic coatings or paint systems.

---

[1] Research Engineers, Bethlehem Steel, Homer Research Laboratories, Bldg. G, Bethlehem, PA 18016

[2] Galvalume® is a registered trademark of BIEC International, Vancouver, WA, for steel sheet with a 55% Al-Zn coating.

Because of the widespread use of these edge creep measurements, it is important to have a clear understanding of the limitations and sensitivity of these destructive evaluation methods. Regardless of the validity of using accelerated and atmospheric exposure tests to judge product performance, the sensitivity of the measurement technique will determine the usefulness of the method in discriminating between products.

The standard method for evaluating the corrosion on painted sheet steel, ASTM Test Method for Evaluation of Painted or Coated Specimens Subjected to Corrosive Environments (D 1654-92), provides guidelines for preparing the samples for measurement and conducting the corrosion tests, but is less diligent in prescribing a standard method for evaluating the extent of edge creep. There is not a procedure describing the number of measurements that need to be made to arrive at the "representative mean creepage" for a test panel and the measurement of the edge creep is left to the user.

At Bethlehem Steel, the standard practice for measuring edge creep after performing outdoor exposure tests or accelerated corrosion tests such as ASTM Standard Method for Operating Salt Spray (Fog) Apparatus (B 117-97) or SAE J2334 has been to make 19 measurements of the extent of creepback on both long (non-drip) edges of the panel for a total of 38 measurements. These 38 measurements provide an average creepback for a single test panel. Typically, three panels of a single material are tested simultaneously and the grand average of the creepback on all three panels is reported as the average creepback for the material. Similar methods are practiced throughout the steel and coil coating industries [*1,2,3*] with differences occurring in the number of replicate panels, number of replicate measurements, method of measuring, etc. The performance of paint systems or paint lines are often ranked by the results of these tests. It has been assumed that the variation in the measurement system is small enough so that the value reported as the grand average for a test material does not depend on the person who prepares the panels or performs the measurements. However, prior to the present investigation, a detailed measurement system analysis had not been performed.

Sources of Variation

In the measurement of the average creepback for a given material, there are several potential sources of variation: test panel-to-test panel variation, operator-to-operator variation (where an operator is the person performing the measurement and/or preparing the panels), test cabinet-to-test cabinet variation, measurement-to-measurement variation, and within panel variation. Generally, one is interested in variation within a panel or the panel-to-panel variation in order to make statistical comparisons of the edge creep on different test materials, and needs the other sources of variation (the noise) to be small enough so that the variation of interest can be determined.

The variance in the grand average is simply the algebraic sum of the individual components of variance in the test method and measurement system

$$\sigma^2_{total} = \sigma^2_{preparation} + \sigma^2_{panel\text{-}to\text{-}panel} + \sigma^2_{within\ panel} + \sigma^2_{reproducibility} + \sigma^2_{repeatability} \quad (1)$$

where σ^2_{total} is the total variance in the calculation of the average edge creep for a material type, $\sigma^2_{preparation}$ is the variance that can be attributed to the preparation of the panels, $\sigma^2_{panel\text{-}to\text{-}panel}$ is the variance in the average edge creep from one panel to the next panel within a given material, $\sigma^2_{within\ panel}$ is the variance due to differences in the extent of edge creep at different locations on an individual panel, $\sigma^2_{reproducibility}$ is the variance arising from the measurements performed by different operators, and $\sigma^2_{repeatability}$ is the variance that arises from errors in making repeated measurements of the same sample. Generally, a measurement system analysis is intended to determine the reproducibility and the repeatability of the measurement system and ensuring that those components of the total variation are a small percentage of the total variation that is measured. In this investigation, the variance due to dissimilarities in panel preparation was also of interest.

The goals of this study were to develop and verify a protocol for determining the sensitivity of edge creep measurements, quantify the variances and sources of variance in edge creep measurements, and provide guidance in using edge creep measurements for comparative performance testing of painted sheet products. One standard for the performance of a measurement system is that used by the Automotive Industry Action Group. That group states that a good measurement system will contribute less than 2% to the total variance for a process, and an acceptable measurement system will contribute less than 8% to the total variance[*4*]. General Electric uses 2% and 7.7% to define a good and an acceptable measurement system, respectively[*5*]. Our goal was to define a measurement system that would at least meet the criteria for an acceptable method (7.7% contribution to the total variance).

Experimental Method

Materials

The test material used in this investigation was prepainted Al-Zn alloy coated (Galvalume) steel sheet produced according to ASTM Specification for Steel Sheet, 55% Aluminum-Zinc Alloy-Coated by the Hot-Dip Process (A 792M - 97a). The metallic coating thickness was nominally 21.2 μm on the top side of the sheet and 22.7 μm on the bottom side. The coil was pretreated, primed with a urethane primer, and painted with a fluorocarbon topcoat on a commercial coil coating line. This material is used as control material in corrosion tests at Bethlehem Steel; therefore, the performance has been well documented.

Six pieces of this steel measuring approximately 38 cm x 104 cm were sequentially cut from the coil stock and labeled A - F. From each mill edge, 4.75 cm were trimmed. The remaining material was cut into seven 13.5 cm x 38 cm panels for a total of 42 panels. Each panel was labeled according to where the panel originated (A1 - F7). The 42 panels were randomized and three sets of five panels were selected. Each set was distributed to one of three operators who subsequently prepared the final panels for testing. Each panel was further trimmed to a final size of 10.8 cm x 30.5 cm. The edges were sheared burr down. A V-bend, 80 inch-lb reverse impact dimple (ASTM Test Method for Resistance of Organic Coatings to the Effects of Rapid Deformation (Impact), D 2794 - 93(1999)e1), and scribe were placed in each panel. The 15 prepared panels were again randomized and placed into the salt spray cabinet and subjected to 750 hours of testing (B117 - 97)[3].

At the completion of the testing, the operator who prepared a given panel was responsible for cleaning the panel and preparing it for measurement of the edge creep according to ASTM D 1654 - 92. Each panel was rinsed and soaked in water for approximately one hour before being scraped. Then any remaining loose material was blown off using an air blast.

Measurement Methods

Initially, each of the three operators measured the edge creep on all 15 panels using the standard Bethlehem Steel procedure. Measurements were made by placing a template onto the panel which indicated the position where each measurement should be taken (Figure 1). The template consisted of 19 hash marks separated by 1.3 cm. The

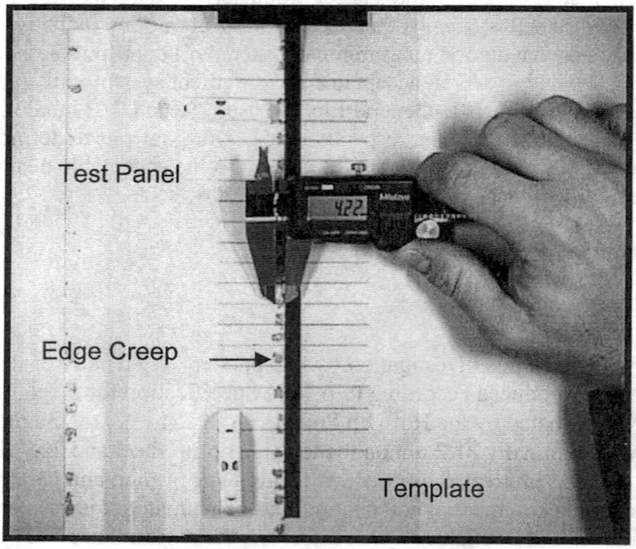

Figure 1-*Representative photograph showing the method used to measure edge creep on a panel.*

center point of each edge of the panel was aligned with the center point on the measurement template. Nineteen measurements were made using a digital micrometer, then the procedure was repeated on the second edge of the panel. If the line on the

[3] Salt Spray testing was chosen because it is a test routinely performed at Bethlehem Steel and throughout the industry and provides samples with measurable edge corrosion in an appropriate time frame. The authors are not making a judgement regarding the validity of Salt Spray testing or any corrosion test in reproducing service performance. This investigation is a study of the manner in which paint creep is measured, regardless of the test method employed.

template fell on an area of the panel where no edge creep had occurred, the measurement was recorded as a zero.

In addition to these initial measurements, two types of reproducibility and repeatability studies were performed. The first assessed the repeatability and reproducibility of using the digital calipers to measure the extent of edge creep. The second looked at the entire measurement system and its ability to provide reproducible and repeatable measurements of the average edge creep on a panel.

The repeatability and reproducibility study of the caliper measurements was performed using three operators measuring the extent of creepback at three designated areas on the panel. On a single panel, three discrete locations were designated for measurement. Each operator measured the extent of creepback at each location five times for a total of 15 measurements per operator. The operators were instructed to stop and leave the room between making repeated measurements. The results were analyzed using the "Gauge R&R" macro in the MINITAB™[4] software package, which is simply analysis of variance using operator, replicate, and measurement position as the factors in a three level, three factor, full factorial design.

The repeatability and reproducibility study of the entire measurement system was conducted in a similar manner. Three operators performed three replicate measurements of the average edge creep on three individual panels. On each panel 38 individual measurements are made using the procedure described above. Again, the operators left the room between making repeated measurements. The 27 values for the average edge creep on a panel were analyzed using the Gauge R&R macro in the MINITAB software.

Results and Discussion

Measurement of Edge Creep Using Digital Calipers

The first analysis was a gauge repeatability and reproducibility (R&R) study of the effectiveness of using the digital calipers to measure the extent of creepback at a specified location (Table 1). Each operator made five measurements at each of three locations (Figure 2). The total variance in the measurements was $\sigma^2 = 3.34$ mm^2. Of that

Table 1-*Results of gauge R&R study for the digital calipers.*

Source of Variation	*Variance (mm^2)*	*Standard Deviation (mm)*	*Contribution to Total Variance (%)*
Total Gauge R&R	0.014	0.119	0.42
Repeatability	0.008	0.090	0.24
Reproducibility	0.006	0.077	0.18
Position-to-Position	3.325	1.823	99.58
Total Variation	3.339	1.827	100.00

[4] MINITAB is a registered trademark of Minitab Inc., 3081 Enterprise Dr., State College, PA, 16801, USA.

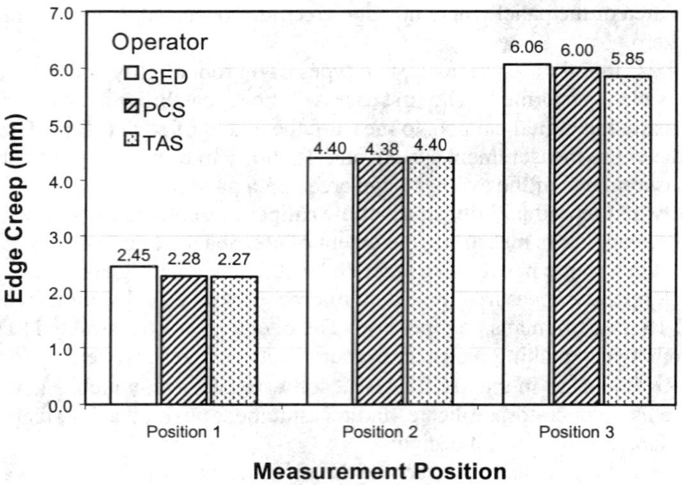

Figure 2-*Edge creep measurements at three specified positions on a single panel used to evaluate the suitability of the digital calipers to effectively measure edge creep. The average of five measurements at each position is shown.*

total variance, the position-to-position variance was 3.32 mm^2, the variance due to repeatability was 0.008 mm^2 and the variance due to reproducibility was 0.006 mm^2. The variance due to the measurement technique was only 0.42% of the total variance, giving a discrimination ratio of 22. These data indicate that the use of digital calipers is a very good method for measuring the edge creep at a designated position on the panel because the variation in the edge creep measurements that are the result of the measurement technique are very small compared to the variation in the amount of edge creep we typically measure.

If the total variance of the edge creep measurements on a panel is less than approximately 0.2 mm^2, then the variance due to the use of the calipers to measure the edge creep will be approximately 7.7% of the total variance. At that point, one should begin to use caution in interpreting the results as the contribution of the measurement technique is becoming a significant contributor to the total variance in the results.

First Measurement System Analysis

The second study was to determine the repeatability and reproducibility of edge creep measurements using the existing measurement technique described in the previous section. Each of the three operators made 38 measurements of the edge creep on each of three panels (A3, C5, D4). They repeated the procedure three times and the average edge creep for each panel was calculated (Figure 3). Again, the contribution to the variance in the measurement of the average edge creep by repeatability and reproducibility was determined (Table 2). The total variance in the measurements was $\sigma^2 = 0.30$ mm^2. Of that total variance, the panel-to-panel variance was 0.23 mm^2, the variance due to repeatability was 0.042 mm^2 and the variance due to reproducibility was 0.032 mm^2. Therefore, the variance due to the measurement technique was 24.8% of the total variance, giving a discrimination ratio of 2. These results indicate that the amount of variation due to the measurement technique is too high to make this a satisfactory measurement to reliably differentiate between materials. A discrimination ratio of 2 indicates that using this measurement technique, the amount of edge creep on a panel can be divided only into two categories, high or low.

Table 2-*Results of the first measurement system analysis.*

Source of Variation	*Variance (mm^2)*	*Standard Deviation (mm)*	*Contribution to Total Variance (%)*
Total Gauge R&R	0.075	0.273	24.75
Repeatability	0.042	0.204	13.85
Reproducibility	0.033	0.181	10.90
Panel-to-Panel	0.227	0.476	75.25
Total Variation	0.301	0.549	100.00

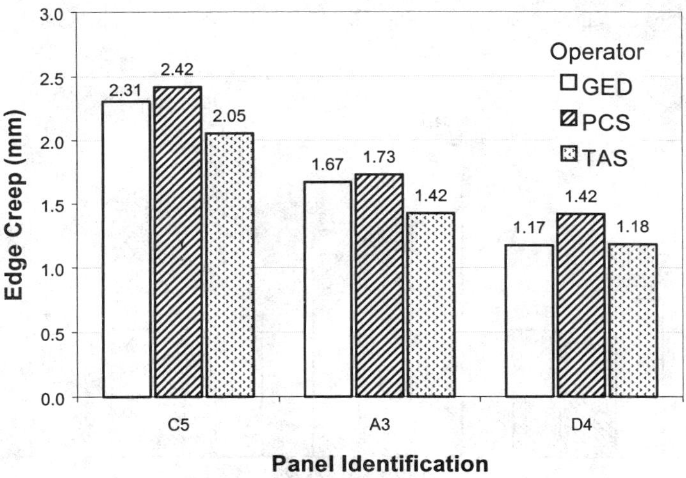

Figure 3-*The average edge creep measurements by three different operators. The values shown are the average of three panels per operator.*

This result is particularly troublesome because this method has routinely been used to make decisions regarding the acceptability of a particular paint system or to opt for one paint system over another.

It was noticed that, of the 38 edge creep measurements on a given panel, a number of the measurements were recorded as zero millimeters (0 mm) of creep back, indicating that the location of the hash mark intersected a portion of the panel where no edge creep had occurred. Furthermore, the number of zero measurements on a given panel varied significantly from operator to operator and between replicate measurements by the same operator. The number of zero measurements would have a large influence on the average edge creep measured for a given panel. Therefore, the analysis was repeated, this time by computing the average edge creep for each panel excluding the measurements of zero edge creep (Figure 4). Instead of 38 measurements per panel, now each average was computed using fewer individual measurements. Each panel had an average of 20 non-zero measurements with the number of measurements per panel ranging from 14 to 32.

Using the data with the zero values excluded from the calculation of the average edge creep, the repeatability and reproducibility analysis was repeated (Table 3). The contribution to the total variance by the measurement system was reduced from 24.8% to 13.0%, but is still unacceptable. The total variance was similar, $\sigma^2 = 0.33$ mm^2, the panel-to-panel variance was 0.29 mm^2, and the variance due to repeatability decreased to 0.013 mm^2. The contribution by the variance due to reproducibility was nearly unchanged at 0.030 mm^2.

These results show that the variation in the extent of the edge creep within a given panel strongly impacts the ability to adequately measure the average edge creep. The greater the variation within a given panel, the greater the repeatability and reproducibility contribution to the total variance will be unless the number of measurements on each panel or the number of panels tested is increased.

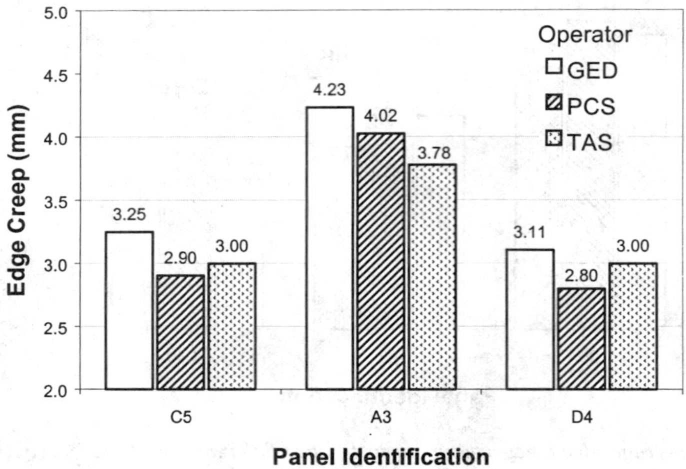

Figure 4-*The average edge creep measurements by three different operators when analyzing the data with the zero edge creep values removed from the data set. The values shown are the average of three panels per operator.*

Table 3-*Results of the first measurement system analysis when data are analyzed after removing values of zero edge creep.*

Source of Variation	*Variance (mm^2)*	*Standard Deviation (mm)*	*Contribution to Total Variance (%)*
Total Gauge R&R	0.043	0.208	13.02
Repeatability	0.013	0.114	3.91
Reproducibility	0.030	0.174	9.11
Panel-to-Panel	0.290	0.539	86.98
Total Variation	0.334	0.578	100.00

The number of panels required to achieve a given contribution by repeatability and reproducibility to the total measurement variation can be calculated. If one uses a general industry standard of a maximum of 7.7% contribution by the repeatability and reproducibility to the total variance as acceptable (a maximum 2% contribution indicates a good measurement system), then the following relationship holds

$$\frac{s_{\bar{X}}^2}{s_{total}^2} \leq 0.077 \qquad (2)$$

where $s_{\bar{X}}^2$ is the variance due to the measurement technique and s_{total}^2 is the total variance in the measurements. Further, $s_{\bar{X}}^2$ approximates the variance in the population, σ_{pop}^2, by

$$s_{\bar{X}}^2 = \frac{\sigma_{pop}^2}{n} \qquad (3)$$

where n is the number of samples used to compute the average. Combining equations (2) and (3) yields

$$\frac{\left(\frac{\sigma_{pop}^2}{n}\right)}{s_{total}^2} \leq 0.077 \qquad (4)$$

By performing a gauge R&R study, $s_{\bar{X}}^2$ can be computed and one can back calculate the value for σ_{pop}^2 using equation (3) (with $n = 3$ in this particular study). Then, using the value for σ_{pop}^2 in equation (4), the number of samples required can be calculated.

For example, when analyzing the data and including all 38 measurements from each panel (including zeros), the variance due to the repeatability and reproducibility, $s_{\bar{X}}^2$, was computed to be 0.075 mm^2 and $s_{total}^2 = 0.301$ mm^2. Using $n = 3$ in equation (3), σ_{pop}^2 can be calculated to be 0.224 mm^2. Then, using that value for σ_{pop}^2 in equation (4), n is calculated to be 9.6. This result means that a minimum of 10 replicate panels need to be included in the corrosion testing to make the repeatability and reproducibility contribution to the total variance in the measurements less than 7.7%. To reduce their contribution to less than 2%, 37 panels would have to be included in a test.

If only non-zero measurements are included in the analysis of the same data, five panels are required to obtain an acceptable measurement method with a 7.7% contribution to variance from the gauge R&R. Twenty panels are required for 2% contribution to the total variance.

Clearly, it is impractical to begin testing a significantly greater number of panels for each test that is conducted. Thirty-seven panels per material is unreasonable for research purposes, and could never be used for quality control. Therefore, the measurement system needed to be improved.

Second Measurement System Analysis After Improving the Measurement System

The approach to improving the measurement method was to incorporate what was learned in the preliminary gauge R&R study with the comments and experiences of the operators who performed the study. It was noticed that a major source of variability was the number of measurements of zero edge creep in a given data set. By analyzing the same data sets, but excluding measurements of zero edge creep, the contribution to the total variation by the measurement system was reduced by 48% even though there were fewer measurements in the data set. In addition, the operators noted the difficulty in repeatedly aligning the template with the center point on the panel. They felt that the difference in the number of zeros was caused by the lack of a good reference point for aligning the template and the panel. They also noted that the lines on the template were too thick and allowed the operator significant discretion on where to take the measurement of the creep back. As a result, a new template was constructed that had finer lines to indicate the position where the measurement should be made and had a reference mark that was to be aligned with the top of the panel.

The measurement system analysis was repeated using the improved template (Figure 5). The result was a significant reduction in the contribution to the total variation made by the measurement system (Table 4). The contribution of the measurement system was reduced from 24.8% of the total variation to 3.6%. The discrimination ratio was increased to 7 by using the improved template. This improvement makes the measurement system satisfactory for use with only three panels per material being tested. To achieve a 2% contribution to the total variation, six panels per material are required.

The individual components of the total variation are further broken down into repeatability ($\sigma^2_{repeatability}$ = 0.82% of total variance) and reproducibility ($\sigma^2_{reproducibility}$ = 2.78% of the total variance). Reproducibility is the highest contributor to measurement system variation. In fact, these data from the measurement system analysis show that there is a systematic operator bias where one operator consistently measured the edge creep to be approximately 0.2 mm less than the other two operators. If the data from the third operator are excluded from the analysis, the percent contribution is reduced further to 2.2% with a discrimination ratio of 9. The reproducibility component is reduced to 1.63% of total variation. This level of contribution from the measurement system is what should be achieved using the described methodology provided that the operators are equally skilled in performing the measurements. Further improvement in the instructions for performing the measurements needs to be implemented to eliminate the operator bias and reduce variation due to reproducibility. In order to reach a measurement system contribution to the variance of 2%, 42 measurements per panel need to be made along with the improved instructions on how to perform the measurements.

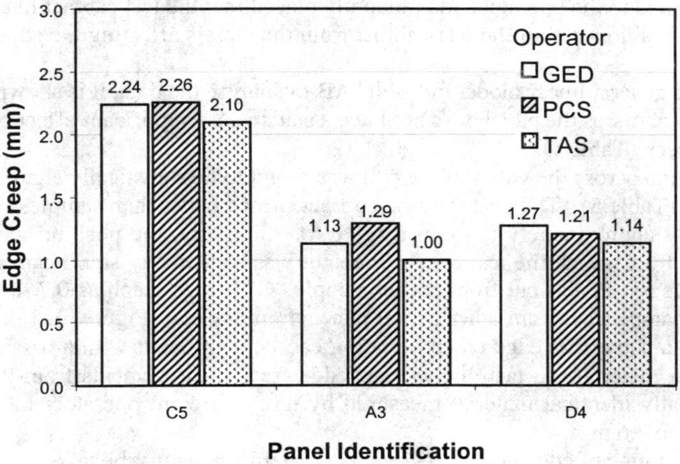

Figure 5-*The average edge creep measurements by three different operators using the improved measurement template. The values shown are the average of three panels per operator.*

Table 4-*Results of the Second measurement system analysis after improving the measurement method.*

Source of Variation	*Variance (mm^2)*	*Standard Deviation (mm)*	*Contribution to Total Variance (%)*
Total Gauge R&R	0.013	0.114	3.60
Repeatability	0.003	0.054	0.82
Reproducibility	0.010	0.100	2.78
Panel-to-Panel	0.349	0.590	96.40
Total Variation	0.362	0601	100.00

Variation Due to Panel Preparation and Panel Location within the Coil

The three operators each prepared and cleaned five panels. These panels were then evaluated for edge creep in an effort to quantify the variation due to the individual who prepared the panels and the variation due to where the sample was cut from the sheet. All of the test panels were nominally identical material. The materials were randomly distributed to the operators and randomly placed into the test cabinet in an effort to remove position-within-the-test-cabinet from the factors affecting the edge creep.

Using the general linear model in MINITAB to analyze the data, it is shown that the operator who prepares the panel does not have a statistically significant effect on the extent of edge creep (Table 5).

The position across the width of the coil was found to be statistically significant at 95% confidence (Table 5). The center of the coil performed better than samples selected from the edge by approximately 0.3 mm (Figure 6). Similarly, the position along the rolling direction from which the test panels were cut was statistically significant at 91% confidence (Table 5). Panels cut from master sample "C" had as much as 0.7 mm more edge creep than samples cut from other master panels from the coil (Figure 6).

These data show that edge creep variation can be significant within a sample of nominally similar material. In fact, the average edge creep measurements from these 15 panels of nominally identical material measured by three different operators fall in the range of 0.7 mm to 2.6 mm.

This large range of edge creep suggests that extreme care must be taken when making decisions based on edge creep data alone. It has been demonstrated that the measurement of edge creep is subject to extensive variation if proper measures are not taken to insure good repeatability and reproducibility in the measurements. It has been further shown that real differences in the extent of edge creep from panel to panel can be seen both across the width of the coil and along the length of the coil.

Table 5. *Results of general linear model analysis for effect of operator preparing the panel, position of panel across coil width, and position of panel along coil length.*

Factor	*F-statistic*	*p-value*
Operator preparing panel	0.12	0.89
Position across width of coil	2.92	0.05
Master sample in the coil	2.28	0.09

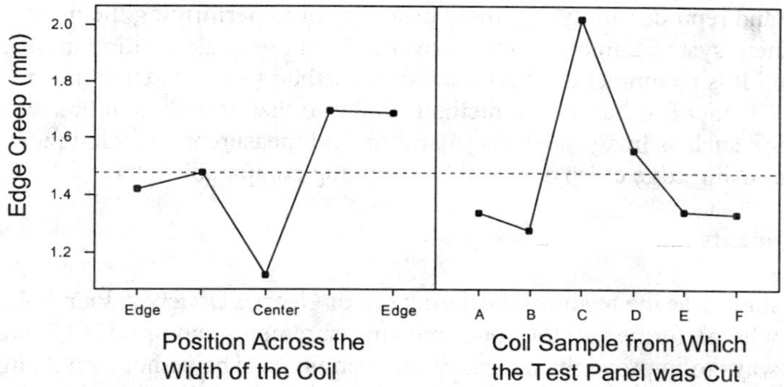

Figure 6-*Main effects plot for the position of the test panel from across the width of the coil, and the master sample within the coil from which the test panel was cut.*

These measurements were performed prior to the improvements in the measurement system. Because of the variation in the measurement system, it is possible that small differences from operator to operator preparing the panels were masked due to variation induced by the measurement system. However, the variation due to position across the width of the coil and along the length of the coil were found to be statistically significant even though the measurement system was less than adequate. A poor measurement system reduces the ability to discriminate between values. Improving the measurement system improves one's ability to discriminate and can only increase the confidence level at which an affect is significant.

Conclusions

The measurement of edge creep is subject to several sources of variation, both controllable and uncontrollable. In order to make valid comparisons between edge creep measurements on different materials, it is important to sample the material and measure the edge creep in a way such that the variation due to the measurement does not mask the real differences in performance between panels.

It has been shown that using three test panels to evaluate edge creep and making 38 measurements per panel is satisfactory if special practices are taken to ensure repeatability and reproducibility of the measurements. However, it was also shown that without a careful measurement practice, at least 10 panels may be necessary to have a satisfactory measurement system. A good edge creep measurement system will use a minimum of three panels per material per test.

It has also been shown that the panel preparation methods described in ASTM D 1654 - 92 adequately remove variation from operator to operator preparing the panels. This standard provides a robust method for panel preparation.

The present investigation indicates that the measurement of "representative mean creepage" as required in ASTM D1654 - 92 is subject to extreme variation from both repeatability and reproducibility. Without proper rigor in performing the measurements, the measurement system can contribute to over 25% of the total variation in measurement of edge creep. It is recommended that a standard method to measure the mean creep back be defined. Whether it is based on a method similar to that described in this work, or another method such as image analysis [*6*], a standard measurement technique will ensure that all parties using edge creep data will be receiving comparable data.

Acknowledgments

This study was the result of the hard work of George Donchez, Pam Salgado, and Tom Suchy, who performed all the edge creep measurements and provided valuable feedback on ways to improve the measurement technique. The authors would also like to thank Bethlehem Steel Corporation for permission to publish this paper.

References

[1] Walter, G. W., "Paint Undercutting Accelerated Test (PUCAT): Test Development and Mechanistic Information," *Corrosion Science*, Vol. 35, 1993, pp. 1391-1404.

[2] Oka, J. and Takasugi, M., "Edge Creep of Prepainted Zn--55% Al Coated Steel Sheet," *Tetsu-to-Hagane*, Vol. 77, No. 7, July 1991, pp. 1130-1137.

[3] Howard, R.L., Lyon, S.B., and Scantlebury, J.D., "Accelerated Tests for Prediction of Cut Edge Corrosion of Coil-Coated Architectural Cladding. Part II: Cyclic Immersion," *Progress in Organic Coatings*, Vol. 37, No. 1, 1999, pp. 99-105.

[4] Automotive Industry Action Group, *Statistical Process Control (SPC) Reference Manual,* A.I.A.G., Detroit, MI, 1991.

[5] General Electric, "Six Sigma Black Belt Training Manual," GE Industrial Systems, 1998.

[6] Reddy, C. M., Yasuda, H. K., Moffitt, C. E., Wieliczka, D. M., and Deffeyes, J., "Quantifying Corrosion Test Results," Plating and Surface Finishing, October 1999, pp. 77-79.

Hideki Katayama,[1] *Masahiro Yamamoto*[2] and *Toshiaki Kodama*[1]

Atmospheric Corrosion Monitoring Sensor in Outdoor Environment Using AC Impedance Technique

REFERENCE: Katayama, H., Yamamoto, M., and Kodama, T., **"Atmospheric Corrosion Monitoring Sensor in Outdoor Environment Using AC Impedance Technique,"** *Outdoor Atmospheric Corrosion, ASTM STP 1421,* H. E. Townsend, Ed., American Society for Testing and Materials International, West Conshohocken, PA, 2002.

Abstract: An alternating current (AC) impedance technique has been applied to the monitoring of corrosion rate in outdoor environments and humidity chambers. A corrosion sensor used for corrosion rate monitoring consisted of a pair of ring and rod of steel electrodes embedded co-axially in epoxy resin. The instantaneous corrosion rates have been monitored by the continuous measurement of impedance at 10mHz and 10kHz. In comparing corrosion loss obtained from impedance measurement with actual corrosion loss by laser microscope, it was found that the present electrochemical technique could be applied for the atmospheric corrosion monitoring. Furthermore, from the thermodynamic analysis for drying and water absorption process of the artificial seawater, the atmospheric corrosion rate of steels showed its a maximum at an average water film thickness of 100 μm from 10 μm.

Keywords: AC impedance technique, steel, atmospheric corrosion, corrosion monitoring, relative humidity

Atmospheric corrosion is greatly influenced by various environmental factors. In order to clarify the mechanisms, it is necessary to examine in detail the relationship between environmental factors and corrosion behavior. Most of the research for atmospheric corrosion has been carried out by field tests. Although long-term corrosion loss can be easily estimated by outdoor exposure, precise daily change of corrosion behavior is hard to obtain. Accordingly, the technique that can continuously monitor the change in corrosion rate is necessary to clarify the atmospheric corrosion mechanism.

Motoda et al. analyzed the time of rainfall, wetness and drying in actual environment by in-situ measurement using an atmospheric corrosion monitoring (ACM) sensor consisting of iron and silver electrodes [*1*]. Though the galvanic couple is useful for the estimation of the time of wetness, it does not give quantitative information on the

[1] Researcher and Supervising Researcher, respectively, Corrosion Resistant Materials R. G., National Institute for Materials Science (NIMS), 1-2-1 Sengen, Tsukuba-shi Ibaraki, 305-0047 Japan.

[2] Senior Researcher, Surface Treatment Laboratory, Steel Research Laboratories, Nippon Steel Corp. (NSC.), 20-1 Shintomi, Futtsu-shi Chiba, 293-8511 Japan.

corrosion rate in atmosphere. Nishikata et al. monitored the atmospheric corrosion rate of steels by the AC impedance method using the probe electrode embedded a pair of plate electrodes in epoxy resin [2]. They continuously measured impedance both at 10kHz and 10mHz, which gave more accurate estimation of atmospheric corrosion [2, 3].

In order to analyze the atmospheric corrosion process of steels in more detail, we developed a concentric-ring corrosion sensor. The corrosion rates of steels were monitored in outdoor environment by AC impedance method using this sensor. Furthermore, corrosion monitoring was carried out in a humidity chamber in order to investigate the effect of water film thickness on atmospheric corrosion rate.

Experiment

Concentric-Ring Corrosion Sensor

The schematic diagram of the concentric-ring atmospheric corrosion monitoring sensor (hereafter, corrosion sensor) is shown in Figure 1. The corrosion sensor consisted of ring and rod electrodes of SS400 steel, which were embedded concentrically in epoxy resin with a narrow gap. Since the two electrodes are placed concentrically, there is no anisotropy existed in electrochemical measurement. The areas of the exposed surface of ring and rod electrodes are the same.

The gap is obtained by wrapping the rod electrode with Teflon tape before inserting the ring electrode. And it was determined at 100 μm, which is limit on preparation of the sensor. As the result of calculating the effect of the current distribution for this sensor under a thin electrolyte layer [3], it was confirmed that the accurate impedance could be measured at the thickness of the electrolyte layer down to a few micrometers without effect of the current distribution.

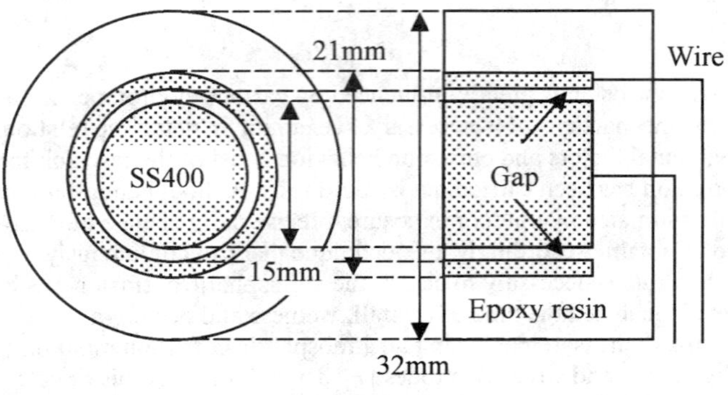

Figure 1 - *Scheme of concentric ring type atmospheric corrosion monitoring sensor.*

After the insulation between two electrodes and the concentrical arrangement of two electrodes were confirmed, this sensor was used for the corrosion monitoring measurements.

Outdoor Corrosion Monitoring

The corrosion sensor was installed outdoors, as shown in Figure 2. The corrosion sensors are placed under a shielded condition with a concrete roof to prevent the washing of salt deposition on steel surface by rainfall. The impedance analyzer was placed in a shelter 15m apart from the sensor. The diluted artificial seawater 1600 μ L was dropped on the corrosion sensor installed horizontally. The chemical compositions of artificial seawater are listed in Table 1. Seawater levels were adjusted to 10, 1.0, and 0.1mg/cm^2 NaCl conversion. The exposure test was carried out at Tsukuba located 50km northeast of Tokyo. The exposure lasted for 1 month. Temperature and relative humidity as environmental factors were simultaneously measured at intervals of five minutes.

The impedance was measured by an AC impedance corrosion monitor (Riken Denshi CT-5) with a multiplexer controlled by a computer thorough the RS-232C interface. The bias ac potential was 10mV. The system allowed the continuous measurement of the impedance values at two different frequencies of 10kHz and 10mHz, Z_{10kHz} and Z_{10mHz}. It was confirmed that the sensor shown in Figure 1 represented the typical impedance characteristic of electrode and solution interface in aqueous solution. Thus, the values of Z_{10kHz} and Z_{10mHz} correspond to the solution resistance (R_s) and the sum of the solution resistance and the polarization resistance (R_s +R_p), respectively. Thus, the polarization resistance R_p is determined by subtracting Z_{10kHz} from Z_{10mHz}.

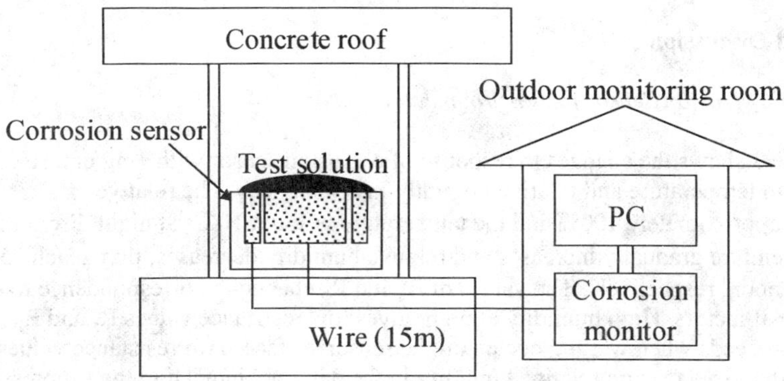

Figure 2 - *Schematic diagram of the outdoor monitoring system.*

Table 1 - *Chemical compositions of artificial seawater for corrosion test of metals (x $10^3 mol/m^3$).*

$MgCl_2$ ($6H_2O$)	$CaCl_2$ ($2H_2O$)	$SrCl_2$ ($6H_2O$)	KCl	$NaHCO_3$	KBr	H_3BO_3	NaF	NaCl	Na_2SO_4
54.6	10.4	0.159	9.32	2.39	0.844	0.435	0.071	419.7	28.8

The actual corrosion loss of the sensor was estimated from the change in roughness and surface level change using a laser microscope. Before microscopic observation, the rust layer was removed by immersing a sulfuric acid solution containing an inhibitor. The surface level change of steel portion was measured with relative to that of the resin. By assuming that the surface level of the resin was unchanged even after the corrosion of metal, corrosion loss of the metal was estimated from the average surface level difference between resin and metal multiplied by the area of metal.

Corrosion Monitoring in Humidity Chamber

Outdoor simulation tests were carried out using the corrosion sensor in a humidity chamber, for calibrating the relationship between relative humidity and corrosion rate. The test solution is similar to the experiment in outdoor environment. The humidity was changed 60 to 95%RH with a cycle of 12 hours in the chamber while the temperature was kept constant at 298K. The impedance measurement was conducted under the same condition as that experiment in atmosphere.

Results and Discussion

Response of Corrosion Sensor for Environmental Factors

Figure 3 shows the changes in response of corrosion sensor with $1 mg/cm^2$ sea salt deposition, in temperature and relative humidity in a sunny day. The relative humidity is approximately 100% and the temperature is about 10℃ at night. From the dawn, temperature gradually increases and relative humidity decreases; they reach 25℃ and 40% at noon, respectively. The values of R_s and R_p change in correspondence to the environmental factors. High humidity at night gives low resistance values R_s and R_p: Corrosion proceeds when wetting occurs on the sensor surface. Two resistance values increase with the decreasing relative humidity in the daytime, indicating the suppression of corrosion.

The results on a rainy day are shown in Figure 4. The vertical bars indicate precipitation. Because of the continuous rainfall, the temperature retained a constant level about 15℃, and relative humidity in most cases exceeded 80%. R_s was maintained at a low value (about 100 Ω), and R_p also showed constant value at several hundred Ω. This demonstrates that the corrosion progresses owing to the wetness of corrosion sensor surface.

Figure 5 shows the changes in the reciprocal of polarization resistance $1/R_p$ for 1 month. It is known that $1/R_p$ is proportional to corrosion rate. The corrosion sensor with10mg/cm^2 sea salt deposition showed relatively constant values ranging 10^{-3} from 10^{-4} Ω^{-1}. The $1/R_p$ for the sensor with 1mg/cm^2 deposition showed higher fluctuation and its value is often larger than that with 10mg/cm^2 deposition. By comparison with the meteorological data, this occurs when the RH exceeds 70%.

From the above results, it is concluded that the atmospheric corrosion is not a steady process, but it proceeds intermittently under humid conditions particularly at higher RH values than 70% RH in the night and in the rainy day.

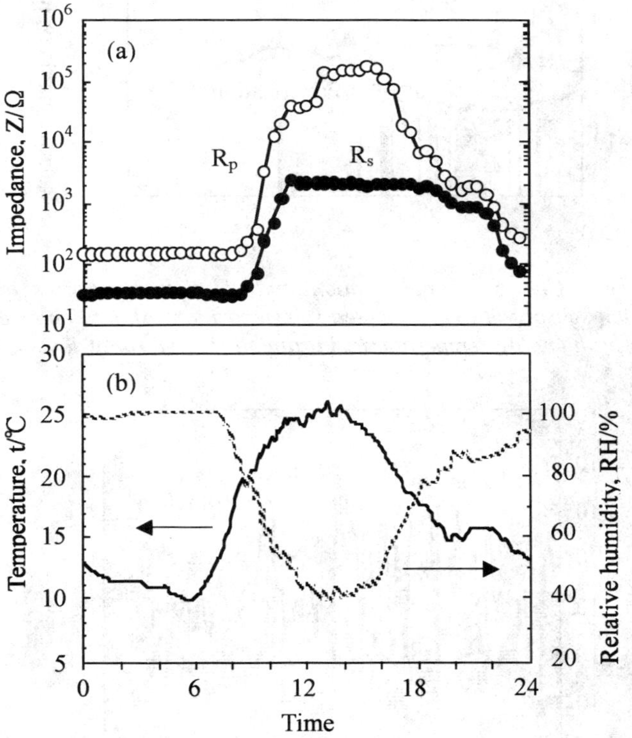

Figure 3 - *Change in polarization and solution resistances of corrosion sensor with 1mg/cm^2 sea salt deposition (a), and change in relative humidity and temperature (b) in a sunny day.*

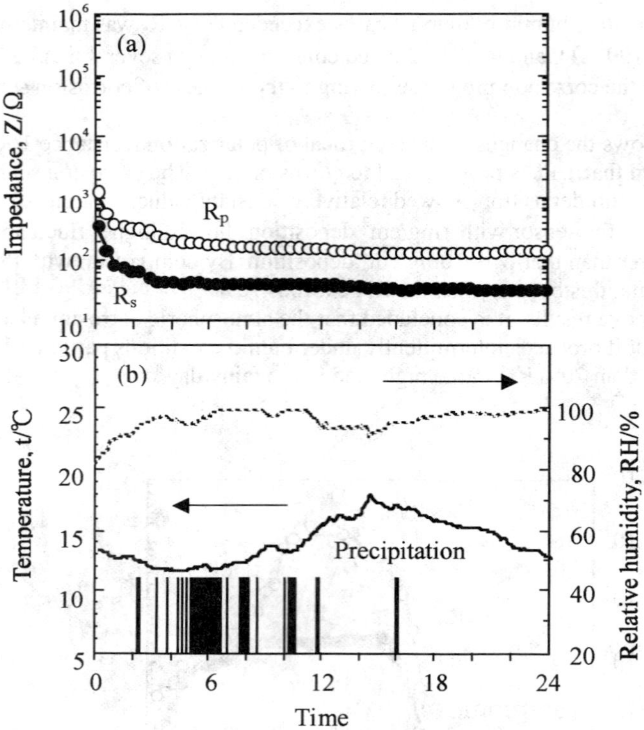

Figure 4 - *Change in polarization and solution resistances of corrosion sensor with 1mg/cm² sea salt deposition (a), and change in relative humidity, temperature and rainfall (b) in a rainy day.*

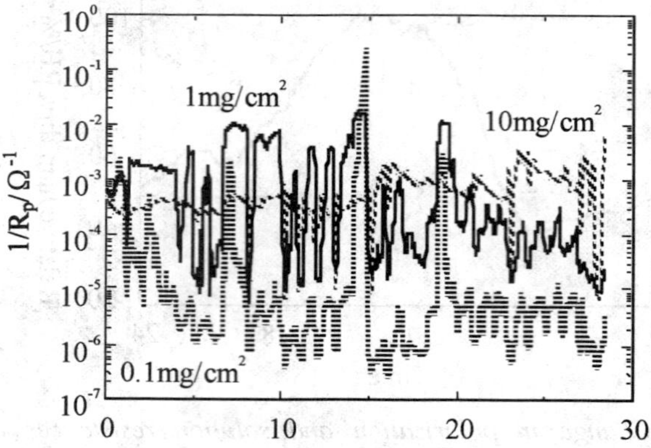

Figure 5 - *Change in 1/Rp of corrosion sensor with sea salt deposition for a month in outdoor environment.*

Corrosion Loss Obtained from Response of Corrosion Sensor

The corrosion rate was calculated from the data of $1/R_p$ shown in Figure 5 in order to compare with actual corrosion loss. It is known that the relation between $1/R_p$ and corrosion rate i_{corr} is represented in the following equation.

$$i_{corr} = k/R_p$$

where k is 25mV on the assumption that the anodic reaction is dissolution of iron and the cathodic reaction is reduction of oxygen [*4*]. The measurement data, however, include the reaction of rust reduction and the other reaction, and the water film also affects the measuring area. In this paper, the amount of electricity was obtained by integrating the estimated i_{corr} with exposure time. The average corrosion loss was calculated by assuming that all the current is consumed for the dissolution of iron to Fe^{2+}.

The surface roughness of the corrosion sensor after removing the rust layer was analyzed by laser microscope in order to obtain the actual corrosion loss. The result is shown in Figure 6. The white color part corresponds to the resin, and the other part corresponds to the electrodes. The corrosion loss was obtained from the difference of height in resin and electrode parts. Figure 7 shows the comparison of the corrosion loss calculated from the response of the corrosion sensor with the actual corrosion loss. The dashed line with a unit slope in the figure shows the case of agreement between mass loss estimated from electrochemical measurement and that estimated by laser microscope. Fairly good agreement was seen between values of different evaluation methods, indicating that the present electrochemical technique can be applied for the atmospheric corrosion monitoring.

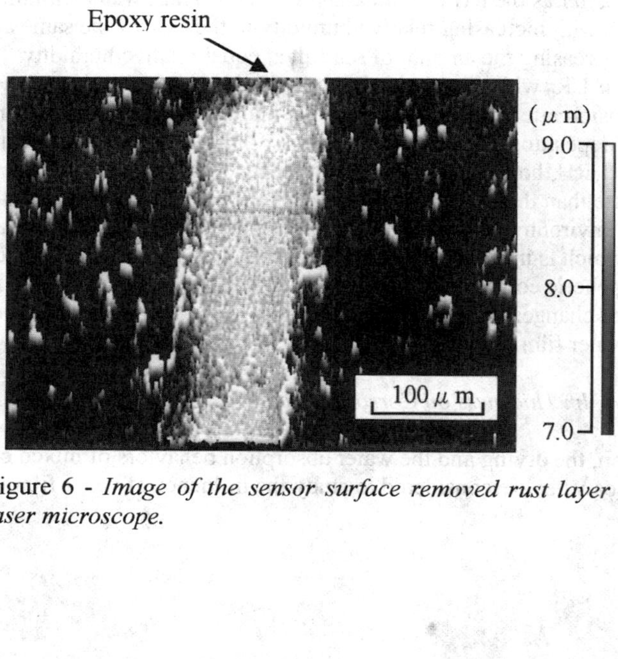

Figure 6 - *Image of the sensor surface removed rust layer by laser microscope.*

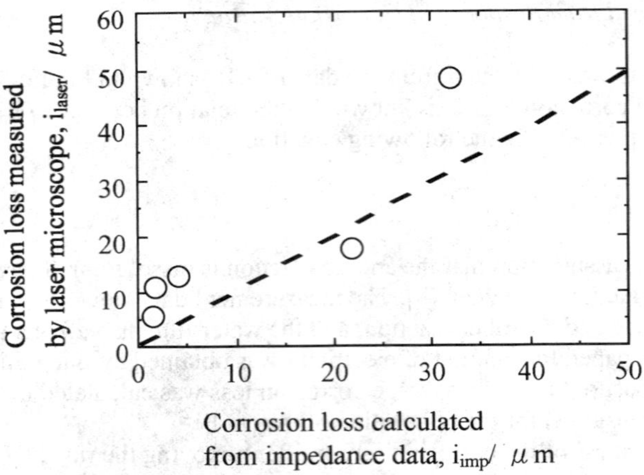

Figure 7 - *Comparison of corrosion loss calculated from impedance data with measured by laser microscope.*

Influence of Relative Humidity on Corrosion Rate

Assuming that equilibrium exists between water in gas phase and water in liquid phase (chloride solution film), the RH value should be equivalent to the activity of water in liquid phase [*5*]. The chloride concentration should be the same regardless of the amount of sea salt as far as the RH is maintained constant. Thus, water film thickness on the surface thickens with increasing relative humidity in the case of the same amount of sea salt, and with increasing the amount of sea salt at equal relative humidity.

The change in $1/R_p$ with the relative humidity in humidity chamber is shown in Figure 8. The corrosion rate increases with relative humidity. The corrosion sensor with 10mg/cm^2 sea salt deposition showed a higher corrosion rate than that with 1mg/cm^2 at 60%RH. Above 70%RH, however, the corrosion sensor with 1mg/cm^2 sea salt gave higher corrosion rate than that with 10mg/cm^2. The reverse of the corrosion rate is also observed in actual environment shown in Figure 5. Apparently reverse occurs at RH higher than 70%, which is in agreement with the fact that this value is the critical humidity for atmospheric corrosion. Considering that the water film thickness formed on the corrosion sensor changes with relative humidity, it is presumed that the reverse is greatly related to water film thickness of the chloride aqueous solution.

Influence of Water Film Thickness on Corrosion Rate

In this section, the drying and the water absorption behaviors of mixed chloride solutions such as seawater are considered to study the influence of water film thickness of

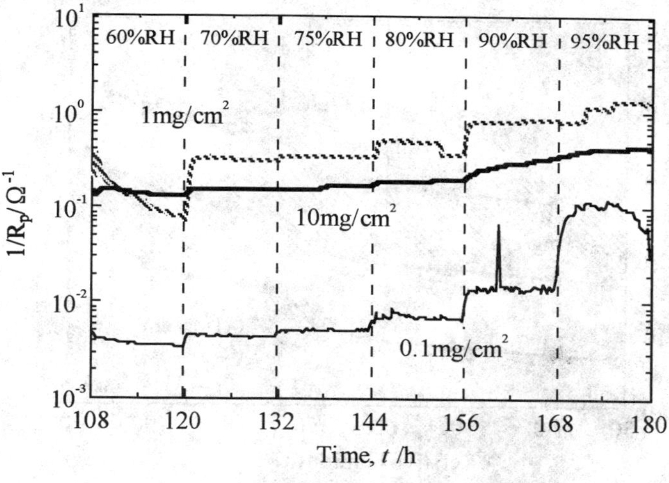

Figure 8 - *Change in $1/R_p$ measured by corrosion sensor with sea salt deposition at different RH values in humidity chamber.*

chloride aqueous solution on corrosion rate of steels. However, it is difficult to calculate the water film thickness thermodynamically because of the nonuniformity of size and thickness of the droplet. It is assumed that the water film is formed uniformly on the sensor surface.

It is known that deposited sea salt absorbs and desorbs water from air to balance the activities of water in atmosphere and in water film. Tang [*6*] and Muto et al. [*7*] reported the thermodynamical analyses of H_2O-NaCl-$MgCl_2$ system and presented a solubility diagram for the mixed-chlorides. Using the diagram the chloride ion concentration can be estimated at an arbitrary value of relative humidity [*8*, *9*, *10*]. Furthermore, the mass of the absorbed water at each relative humidity can be estimated. The results for the known amount of deposited sea salt are shown in Figure 9, where the ordinate represents the average thickness of water film. In Figure 8, the corrosion sensor with 10mg/cm^2 sea salt deposition showed higher corrosion rate than that with 1mg/cm^2 at 60%RH, and the latter gave a higher corrosion rate than the former at above 70%RH. Therefore, the corrosion rate of steel showed a maximum in the hatched region in Figure 9, i.e., at an average thickness of several tens to about a hundred μ m of water film.

There are some reports researched on the dependence of corrosion rate on the water film thickness. Tomashov conceptually explained that the largest corrosion rate appeared at water film thickness of about 1 μ m [*11*]. Nishikata et al. estimated the corrosion rate of metals covered with a thin electrolyte layer by an electrochemical impedance spectroscopy (EIS): They reported that the corrosion rate of copper covered with acidic solutions of pH 3 gave a maximum at a thickness of several tens of micrometers [*12*]. In recent years, Stratmann et al. demonstrated that the corrosion rate showed the highest at

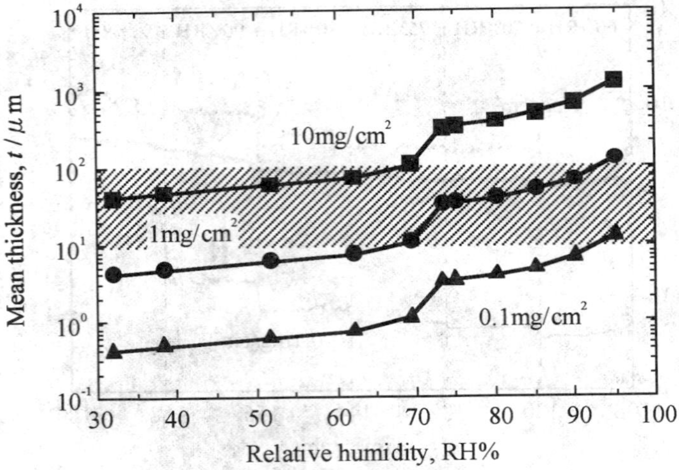

Figure 9 - *Relationship between the relative humidity and the mean thickness of liquid phase formed by the water absorption of the deposited artificial sea salt.*

about 10 μm by the polarization curve under thin water film using the Kelvin probe [*13*]. Yamashita et al. also reported the similar results in which they estimated corrosion rate from the consumption rate of oxygen [*14*].

As for the experimental method, Nishikata et al. [*12*] and Stratmann et al. [*13*] directly measured the water film thickness, while Yamashita et al. calculated the water film thickness experimentally by the concentration of the liquid phase in equilibrium gas phase [*14*]. The technique of the latter case is similar to that of present work with the difference being salt concentration estimation. In the present work salt concentration was estimated by thermodynamically. Even with the differences in experimental procedures, fairly good agreement was seen regarding the relationship between corrosion rate of steel and water film thickness in the initial stage of atmospheric corrosion.

Conclusions

From the corrosion monitoring of steels in outdoor environment and in humidity chamber, the following conclusions were drawn.

1. The atmospheric corrosion rate is greatly influenced by relative humidity, and the rate shows large value at high relative humidity in the night and on the rainy day.
2. The monitoring technique using concentric ring type corrosion sensor by AC impedance method is useful for investigating the atmospheric corrosion behavior.
3. Though the corrosion sensor with 10mg/cm^2 sea salt deposition gave a higher corrosion rate than that with 1mg/cm^2 at 60%RH, the corrosion rate of the latter is larger

than that of the former above 70%RH.

4. From the thermodynamic analysis for drying and water absorption process of the artificial seawater, it was found that the atmospheric corrosion rate of steels showed a maximum at an average water film thickness of 100 μ m from 10 μ m.

References

[1] Motoda, S., Suzuki, Y., Shinohara, T., Kojima, Y., Tsujikawa, S., Oshikawa, W., Itomura, S., Fukushima, T., and Izumo, S., "ACM (Atmospheric Corrosion Monitor) Type Corrosion Sensor to Evaluate Corrosivity of Marine Atmosphere," *Zairyo-to-Kankyo*, vol.43, No.10, 1994, pp. 550-556.

[2] Nishikata, A., Takahashi, T., Bao-Rong , H., and Tsuru, T., "Monitoring of Corrosion Rate of Carbon Steel under Wet/Dry Cycle Conditions and Its Corrosion Mechanism," *Zairyo-to-Kankyo*, vol.43, No.4, 1994, pp. 188-193.

[3] Nishikata, A., Kumagai, S., and Tsuru, T., "The Application of AC Impedance Technique to Atmosphere Corrosion Study - Impedance Characteristics of Metal/Thin Electrolyte Layer Interface - ," *Zairyo-to-Kankyo*, vol.43, No.2, 1994, pp. 82-88.

[4] M., Stern, and L., Geary, "Electrochemical Polarization I. A Theoretical Analysis of the Shape of Polarization Curves," *J. Electrochem. Soc.*, vol.104, No.1, 1957, pp. 56-63.

[5] Majima, H, and Awakura, Y., "Activity of Water in Concentrated Aqueous Solutions," *Bull. Japan Inst. Metals*, vol.29, No.6, 1990, pp. 449-452.

[6] I.N., Tang, "Phase Transformation and Growth of Aerosol Particles Composed of Mixed Salts," *J. Aerosol Sci.*, vol.7, 1976, pp. 361-371.

[7] Muto, I., and Sugimoto, K., "Modeling of Atmospheric Corrosion Environments and Its Application to Constant Dew-point Corrosion Test," *Zairyo-to-Kankyo*, vol.47, No.8, 1998, pp. 519-527.

[8] W.F., Linke, and A., Seidell, *Solubilities of Inorganic and Metal Organic Compounds (4th Edition)*, Am. Chem. Soc., Washington, 1965, pp.489-490.

[9] R.A., Robinson, and R.H., Stokes, *ELECTROLYTE SOLUTIONS (2nd Edition)*, Butterworths, London, 1959, pp.476-478.

[10] Landolt-Bornstein New Series Group IV, *Macroscopic and Technical Properties of Matter*, **1b**, Springer-Verlag, Berlin, Heidelberg, New York, 1977, pp.64-80.

[11] N.D., Tomashov, "Development of the Electrochemical Theory of Metallic Corrosion," *Corrosion*, vol.20, No.1, 1964, pp. 7t-14t.

[12] Nishikata, A., Ichihara, Y., and Tsuru, T., "An Application of Electrochemical Impedance Spectroscopy to atmospheric Corrosion Study," *Corr. Sci.*, vol.37, No.6, 1995, pp. 897-911.

[13] M., Stratmann, and H., Streckel, "The Investigation of the Corrosion of Metal Surfaces, Covered with Thin Electrolyte Layers - A New Experimental Technique," *Ber.Bunsenges.Phys. Chem.*, vol.92, No.11, 1988, pp. 1244-1250.

[14] Yamashita, M., and Nagano, H., "Corrosion Potential and Corrosion Rate of Low-Alloy Steel under Thin Layer of Solution," *J. Japan Inst. Metals*, vol.61, No.8, 1997, pp. 721-726.

EFFECTS OF CORROSION PRODUCTS ON THE ENVIRONMENT

Inger Odnevall Wallinder[1] and Christofer Leygraf[1]

Environmental Effects of Metals Induced by Atmospheric Corrosion

---

Reference: Odnevall Wallinder, I., and Leygraf, C., **"Environmental Effects of Metals Induced by Atmospheric Corrosion,"** *Outdoor Atmospheric Corrosion, ASTM STP 1421*, H. E. Townsend, Ed., American Society for Testing and Materials International, West Conshohocken, PA, 2002.

Abstract: A major on-going research project is described that aims at generating relevant data for future risk assessments of environmental effects caused by copper, zinc, chromium and nickel release from roof materials. Cross-disciplinary activities have been implemented that range from the metal release situation through transport in soil and aquatic environments to the metal receptor situation. Key concepts include atmospheric corrosion, metal runoff, metal chemical speciation, bioavailability, changes in metal speciation and bioavailability during transport, metal retainment, and ecotoxicity effects. In all, tools are available to describe fully the situation from the roof to the recipient. These tools will be described briefly, together with selected results. The results show consistency with respect to chemical speciation, bioavailability and ecotoxicity, at least when describing the metal effect immediately after release from the roof. However, much work remains in order to characterize more complex situations closer to the recipient.

Keywords: atmospheric corrosion, metal runoff, chemical speciation, bioavailability, ecotoxicity, copper, zinc, chromium, nickel, stainless steel

Background

During the 1990s the hazards posed by the emission of metals has been shifted from trace metals, with primary sources within the metal extraction and manufacturing sectors, to a number of dispersed sources with emission rates and environmental

---

[1]Assistant Professor and Professor, respectively, Department of Materials Science and Engineering, Division of Corrosion Science, Royal Institute of Technology, Dr. Kristinas v. 51, SE-100 44 Stockholm, Sweden.

effects that have not yet been well studied. Such dispersed sources include the corrosion-induced runoff from buildings and other parts of our infrastructure. Transportation, agriculture, and landfills are other sectors with significant dispersed metal sources.

Frequently used metals in building applications, such as building shells, roofing materials and fencing, are copper (in pure form or as a copper-based alloy), zinc (mostly as a plating on steel- galvanized steel), and chromium and nickel (as alloying elements of stainless steel). Their dispersion is caused by atmospheric corrosion and the subsequent precipitation-induced release ("runoff") process.

As a result of increased hazards, the use of, e.g., copper, zinc, chromium and nickel has been questioned. Restricted and regulated use of selected metals has been introduced in several countries within Europe [*1*]. These regulations, however, have been based on the principle of caution and not of a full risk assessment, mainly because of missing adequate information on runoff rates and environmental effects. Hence, whereas long-term atmospheric corrosion rates of, e.g., copper and zinc are well-established from several national and international field- exposure programs, the dispersion of these metals and their potential environmental impact are relatively unknown. Forced by the need to fill the most important gaps of knowledge and financed by Swedish environmental funding organizations and European industries, the Division of Corrosion Science at the Royal Institute of Technology (KTH) in Stockholm, Sweden, has initiated a major research effort directed towards environmental effects of metals induced by atmospheric corrosion. The aim is to generate data that can be used for future risk assessments.

A *risk assessment* has to be preceded by an *exposure assessment* and an *effect assessment*. The exposure assessment includes generation of relevant data on metal release rates and on atmospheric exposure parameters. The effect assessment includes determination of the chemical form of the metal, and hence its bioavailability, during interaction with different reacting species in, e.g., soils or aquatic systems. The effect assessment also includes ecotoxicity effects when the metal finally is exposed to organisms in the biosphere. Not until such data are available is there a possibility to perform a risk assessment, which then also considers the overall budgets of the actual metal as it is used in the technological society. Such budgets have also to consider opportunities to replace the metal by alternative materials. To date, however, very few metal budgets of this type have been realized.

This presentation is an over-view of current research activities at KTH to provide data within the exposure assessment and the effect assessment domains, respectively. It includes short descriptions of exposure procedures and analytical tools used and also selected key results obtained from copper, zinc and stainless steel as roofing materials exposed to outdoor atmospheric environments. By necessity the information gained reaches far outside the field of atmospheric corrosion and emphasizes the cross-disciplinary character of the studies performed. Three investigations, forming parts of the KTH-effort, are presented as separate papers from this ASTM symposium [*2-4*].

Key Concepts

The debate on dispersion effects of metals has sometimes been confused because of lack of stringent definitions of the most important concepts. Hence, a few key concepts are shortly described below.

The *corrosion rate* is the amount of a metal (g) that corrodes per surface area unit (m^2) and time unit (yr). The rate is then given in $gm^{-2}yr^{-1}$. An alternative expression is the average thickness (μm) of the removed metal layer per time unit (yr). In this case, the rate is given in μmyr^{-1}.

Part of the corroded metal will precipitate and retain on the metal as a corrosion product, whereas another part will run off during precipitation events. The *runoff rate* is the amount of metal (g) that is released from the corrosion product per surface area unit (m^2) and time unit (yr). It is usually given in $gm^{-2}yr^{-1}$.

Corrosion and runoff are generally ruled by completely different physical, chemical and electrochemical processes, and also governed by different atmospheric exposure parameters. From this follows that corrosion and runoff proceed independently of each other and with rates that not necessarily equal each other [*5*]. The difference between corrosion rate and runoff rate is mainly attributed to successive accumulation, or reduction, of metal in the corrosion product.

Upon runoff the metal is released as a chemical species with a particular *chemical form or speciation*, depending on its ambient chemical environment. The chemical speciation is what determines its *bioavailability*, which is a measure of the rate and extent of absorption of that chemical species into the biosphere. The *biosphere* encompasses all forms of life on Earth, and extends from the ocean depths, through the surface of landmasses to a few thousand meters of altitude in the atmosphere.

Ecotoxicity, finally, is the damaging action of any chemical species upon a particular environmental system in the biosphere.

Exposure Assessment

Exposure Conditions and Samples

The exposure of samples for runoff measurements is based on the same procedure as for corrosion rate measurements according to ASTM and ISO [*6-7*], and with slight modifications to collect the necessary runoff sampling. The exposure site is on the roof of a seven-storey building located within the KTH campus area. It has been selected because of its location near the laboratory and the absence of nearby point sources of emission of corrosive substances. The environment is urban with annual average concentrations of SO_2 and NO_2 around 3 and 50 $\mu g/m^3$, respectively, and the annual precipitation approximately 500 mm.

The panels, sized 30 by 10 cm^2, are mounted on a rack made of aluminum. They are exposed facing south with an inclination of 45° from the horizontal, and well separated from each other. In order to avoid metal release from the panel back, each specimen is mounted single-sided by covering the back with a well-adhering tape. Experience has shown that single panels are sufficient for runoff rate measurements, as opposed to corrosion rate measurements that usually require triplicates.

Prior to exposure the panels are cleaned through degrease in acetone and isopropylic alcohol to minimize surface contamination and to ensure surface uniformity. All panels are usually mounted on the same day to avoid the influence of initial exposure conditions. Each panel is individually mounted on a fixture with an inclined gutter into which the runoff water is collected. From the gutter the runoff solution is transported to a container (Figure 1).

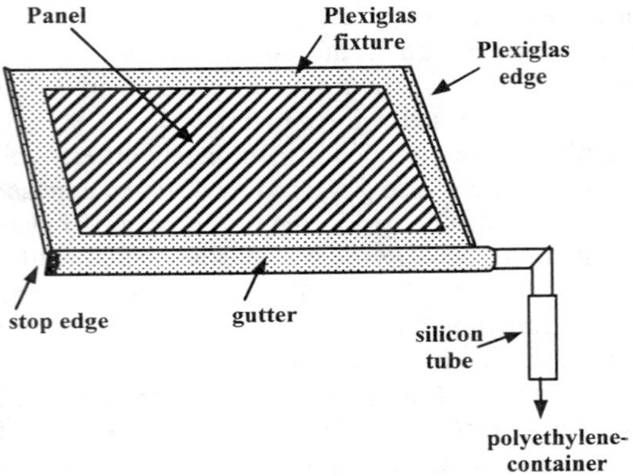

Figure 1 - *Experimental setup for runoff rate measurements.*

The fixture and gutter are made of Plexiglas, whereas the containers are acid-cleaned (10% HNO_3) polyethylene. The runoff water in the container is collected typically within two weeks or so, depending on the amount of precipitation. A blank specimen, consisting of a Plexiglas fixture with no panel inserted, is included to correct the metal runoff rate with the actual background level of dry- or wet-deposited metal.

Analysis of Metal Runoff

Before analysis of total metal content, the runoff water is acidified to a pH less than 3 by means of additions of conc. HNO_3. Depending on metal and concentration, one of the following analytical methods are regularly used:

* Ion chromatography: with a detection limit for copper, zinc and nickel of 10 ppb (ppb= molar parts per billion 10^{-9});
* Polarography with a detection limit for copper of 10 ppb, for zinc of 5 ppb and nickel of 2 ppb; and
* Flame atomic absorption spectroscopy with a detection limit for copper of 200 ppb and for zinc of 40 ppb.

The metal content in the runoff from a given sampling period is corrected by the metal content from the collected blank panel. The total amount of metal in the collected runoff is the sum of measured metal amounts obtained from each individual sampling event.

Key Runoff Results

The first runoff studies at KTH were implemented in 1995 so that experience has been gained from more than five years. In the following paragraphs, a list of key results will be shortly presented. The reader is referred to more detailed information in the listed references.

While the corrosion rate for both copper and zinc exhibits a clear time-dependence, characterized by initially a high corrosion rate that successively levels off, the runoff rate as a function of time exhibits a much more time-independent value, at least on a time-scale of the order of years (Figure 2). This behavior has been frequently observed on both copper and zinc, in several different atmospheric environments, and during extended periods of up to five years. During the same period of time (*i.e.*, several years) the corrosion products formed on both metals undergo significant changes in chemical composition and atomic structure. Yet, the runoff rate is relatively independent of time. Together with other observations, this suggests that precipitation parameters rather than corrosion product properties govern the runoff rate of copper and zinc.

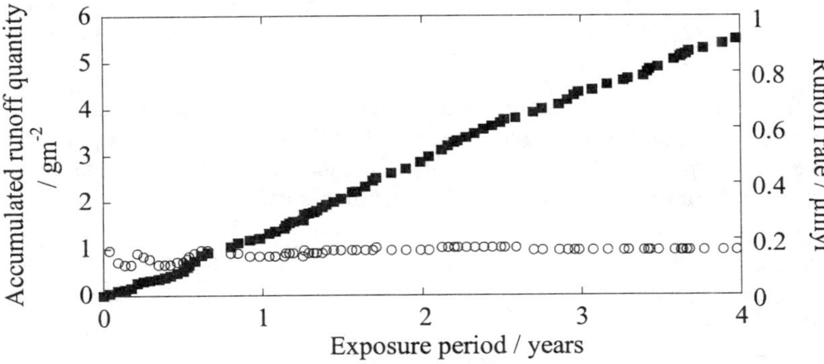

Figure 2 - *The accumulated copper runoff (filled squares) as function of time during an extended period of time, and the corresponding copper runoff rate (unfilled circles). Data are based on a four-year exposure in the Stockholm area.*

The runoff rate of both copper and zinc is initially lower than the corresponding corrosion rate. The difference between the results is an accumulation of metal in the corrosion products. As long as the average corrosion product increases in mass, the runoff rate will be lower than the corrosion rate. Eventually, the runoff rate will equal the corrosion rate, which corresponds to a situation when the corrosion product does not increase in mass any longer. This happens on zinc after approximately a few years of Stockholm exposure, and probably after a few decades on copper in the same environment. On a short time-scale, corresponding to a single precipitation event, the runoff rate exhibits a clear time-dependence characterized by an initially high rate, the so-called "first-flush", which gradually levels off to a more constant runoff rate at the later stage of the precipitation event [*8-9*] (Figure 3). The first-flush is mainly caused by dry-deposited gases and aerosol particles, which create a more concentrated aqueous adlayer atop the corrosion products and aid in dissolving more easily soluble components of the corrosion products. During later stages of the precipitation event the concentration decreases, and so does the runoff rate. The first-flush increases with the length of the dry period preceding the precipitation event.

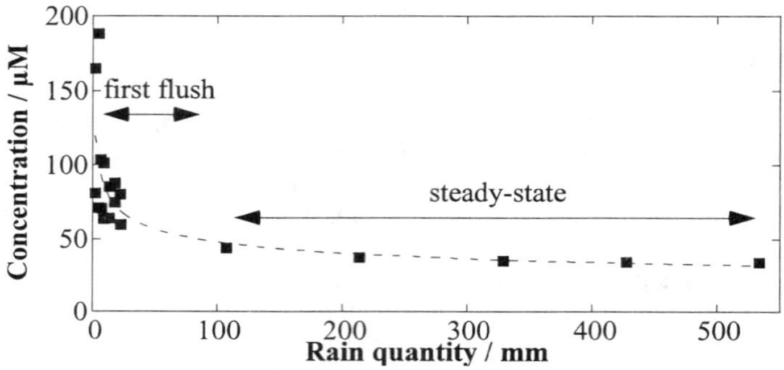

Figure 3 - *The copper concentration in runoff during a single rain event, illustrating the first-flush effect characterized by initially increased copper concentrations.*

The runoff rates of nickel and chromium from stainless steel of type 304 and 316 are approximately three orders of magnitude lower than the runoff rates of copper and zinc from the same exposure site [*10*] (Figure 4). The runoff rates for both metals are slightly higher for the more highly alloyed 316 steel than for the 304 steel.

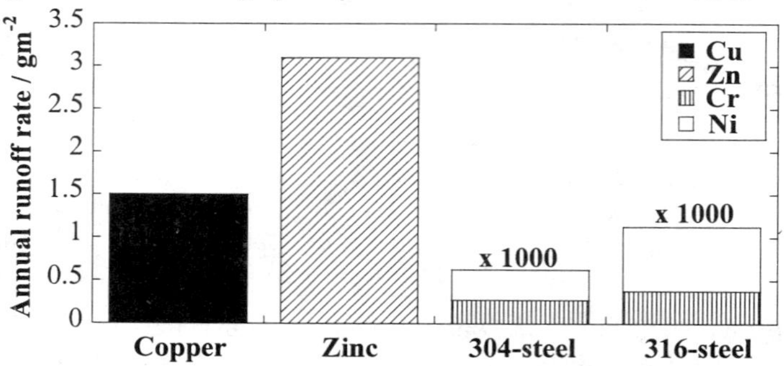

Figure 4 - *Annual runoff rates from pure copper, zinc and chromium and nickel from 304 and 316 type stainless steel. The runoff rate data for copper and zinc represent average values during four-year exposures whereas the chromium and nickel runoff rate data are based on a one-year exposure.*

The age of corrosion products on copper or zinc has no major influence on runoff rate, at least up to 140 years old copper [*3*] and 40 years old zinc [*2*].

The total amount of copper or zinc runoff increases with increasing precipitation volume. When the precipitation volume and composition are given, the runoff rate decreases with increasing inclination [*11*] from horizon and with increasing precipitation intensity [*2*].

The total runoff rate of metal from any real building is estimated by adding the runoff rates from each individual surface. It turns out that the building itself has a shielding effect. Based on an attempt to compare the weighed average zinc runoff rate from a roof with many surface inclinations and a certain degree of sheltering it could be shown that the weighed runoff rate was at least 25% lower than the zinc runoff rate from the exposure situation described above (exposure on a rack of a roof, facing south and with 45° inclination), see further in Reference 11.

Based on zinc runoff rate data from exposure sites with comparable annual precipitation volumes (500 to 1000 mm/year) and with different SO_2-concentration levels, a correlation was found between SO_2 –concentration ($[SO_2]$, in μgm^{-3}) and zinc runoff rate (in $gm^{-2}yr^{-1}$), see Reference 12:

$$\textit{Zinc runoff rate} = 1.36 + 0.16[SO_2]$$

For copper and stainless steel no similar expression has yet been established.

Effect Assessment

Chemical Speciation and Bioavailability

The runoff rates discussed so far have not considered the chemical form of the metal released. The values presented refer to the total metal content or concentration in the runoff solution, irrespective of its chemical form (speciation). As emphasized before, it is necessary to establish the chemical speciation of the metal since this determines the bioavailability of the metal in the runoff. This task must also consider the changes of metal chemical speciation during transport from the building through water systems of various kinds to the recipient. In a more general perspective one can illustrate the cycle of a metal from its most stable state, the mineral, over extraction processes to the metal, which thermodynamically is an unstable state, over various deterioration processes (including corrosion and wear) back to the mineral (see Figure 5). During its way from the metallic state to the mineral the metal can exist in various chemical forms, among which the free hydrated metal ion (e.g., Zn^{2+} or Cu^{2+}) usually is the most bioavailable form. When forming part of a chemical compound, the metal is much less available to interact with organisms in the biosphere.

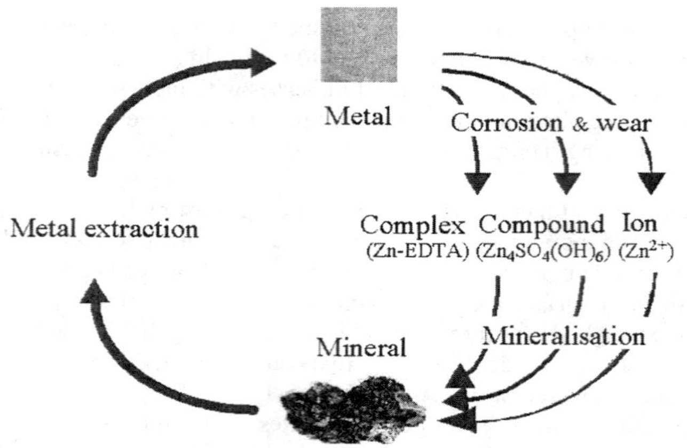

Figure 5 – *Natural cycle of zinc.*

Various tools exist to determine the chemical forms of the metal or its bioavailability after runoff. Within the KTH-effort the following have been used, largely through collaboration with the Laboratory for Environmental Toxicology and Aquatic Ecology, University of Ghent and the Department of Soil Sciences, Swedish University of Agricultural Sciences.

Computer Modeling

Metal speciation can be modeled under different chemical conditions of the environment in which the metal exists. Parameters used for the input in the model include pH, hardness, alkalinity, and concentration of complexing species. Two models have been used, MINTEQA2 [*13*] and WHAM [*14*]. One difference between the models is that the latter also considers metal-bonding to dissolved organic compounds, such as humic and fulvic acids. The output of such calculations is the equilibrium content of various metal-containing chemical compounds at given environmental conditions. As an example, Figure 6 shows how the percentage of zinc species varies as a function of pH under conditions representing the runoff solution immediately after leaving the roof. The modeling with either MINTEQA2 or WHAM predicts that a majority of released zinc, between 95 and 99.9%, is present as the hydrated free Zn^{2+}-ion. Minor concentrations of $Zn(OH)^+$ and $ZnSO_4$ (not shown in the figure) are expected at pH higher than 6. This value can be compared with the pH of the runoff immediately after release from the patina, usually in the range between 4.5 and 7 as a result of the buffering capacity of the corrosion products on zinc. In all, the results suggest that most of the zinc in the runoff immediately after leaving the roof surface is present as hydrated Zn^{2+}, the most bioavailable form of zinc.

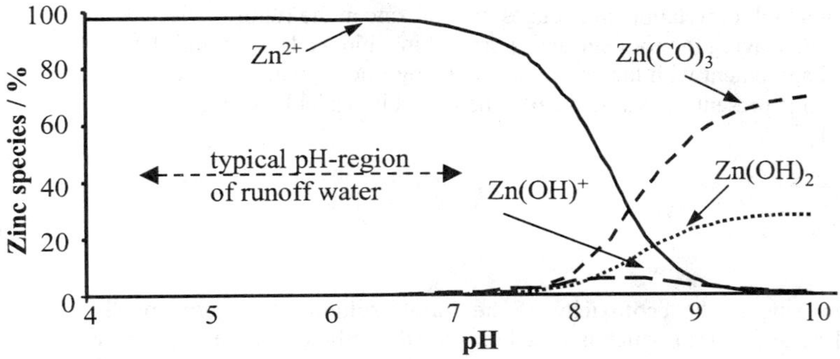

Figure 6 - *Chemical speciation of zinc as a function pH predicted by the computer models MINTEQA2 and WHAM. The data are representative of the situation immediately after release from zinc samples exposed in the Stockholm area [15].*

Bioavailability Testing

Organisms that react with the bioavailable part of a metal can be used for this purpose. As part of the collaboration with the University Ghent, Belgium, a metal-specific assay (BiometR) is applied that is based on the bacteria *Alcaligenes eutrophus* [*16*]. The bacteria respond by producing light with an intensity that increases linearly with the bioavailable fraction of the metal to which it is exposed. To illustrate the capability of the biosensor, Figure 7 is a summary of all results in which collected runoff samples from different zinc samples were analyzed with respect to total zinc content (in mg total Zn per liter, by means of atomic absorption spectroscopy) and bioavailability (in mg free Zn per liter, by means of the biosensor).

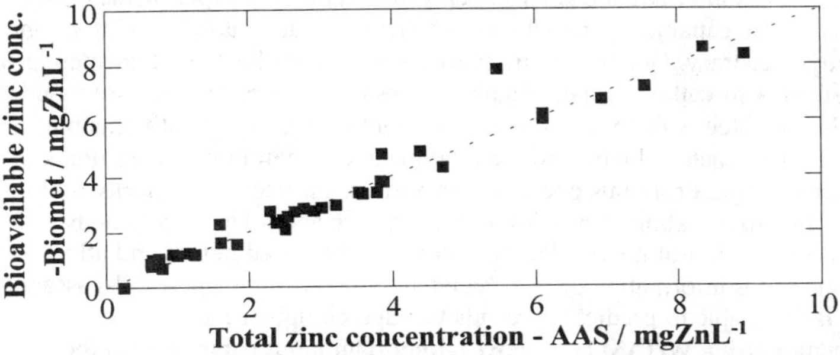

Figure 7 - *Bioavailability of released zinc, as measured by the biosensor BiometR, and the total zinc content in runoff, as measured by atomic absorption spectroscopy. Data are based on runoff solutions from various zinc-containing samples exposed in the Stockholm area. The runoff solutions were collected during the same two-week exposure period [15].*

The results exhibit a high correlation and suggest that all zinc in the runoff collected immediately after the leaving the roof surface is present in a bioavailable form. The results are in good agreement with the computer modeling finding that mostly all zinc under this condition is present as hydrated Zn^{2+}, the most bioavailable form (see preceding section).

Ecotoxicity Testing

In order to investigate the ecotoxicity of the runoff solution an internationally recognized standard test is used which is based on growth inhibition of the micro-alga *Raphidocelis subcapitata* (previously named *Selenastrum capricornutum*) during a 72 hour exposure to the runoff solution. The alga occurs in fresh water, is highly sensitive to metals and generally accepted in international legislative work on metal risk assessment [*17*]. When exposed to a nutrition solution without any toxic influence the number of species grows exponentially. By exposing to a series of metal containing solutions with increased dilution, the growth inhibition of the alga in each solution can be studied by counting the density of species every 24 hours during a 72 hours period. The metal concentration during which the growth is reduced by 50%, compared to the growth without any metal influence, is called the EC50 value (ml/l). The toxicity of the solution is expressed in Toxic Units (TU), defined as the inverse of the EC50-value multiplied by 1000. The higher the toxicity of the runoff solution, the higher the number of toxic units.

Soil Percolation Studies

An important task is to be able to monitor the change in metal chemical speciation and, hence, in bioavailability, during transport of the metal from the roof to the recipient in the biosphere. By necessity this cannot be accomplished through any precise means in real soil or aquatic environments. Instead model experiments have to be designed which are capable of predicting the behavior under natural conditions and with high enough accuracy. One type of measurements that has been implemented in our research effort is so-called soil percolation studies (see Figure 8). They are based on a set of columns filled with soils of different characteristics [*4*]. Metal-containing runoff passes through each soil core and the total metal concentration is measured before and after passage. From this one can obtain the metal retention capacity of the soil. Similarly, the bioavailability (through biosensors, see above) and ecotoxicity (through growth inhibition of micro-alga, see above) is measured before and after passage. This provides information on the change in bioavailability upon soil-passage. The intention is to be able to predict the results through computer modeling of chemical speciation using WHAM (see above). Important input parameters in the computer modeling include the total metal concentration of the soil, total organic content, pH, and concentrations of selected anions and cations.

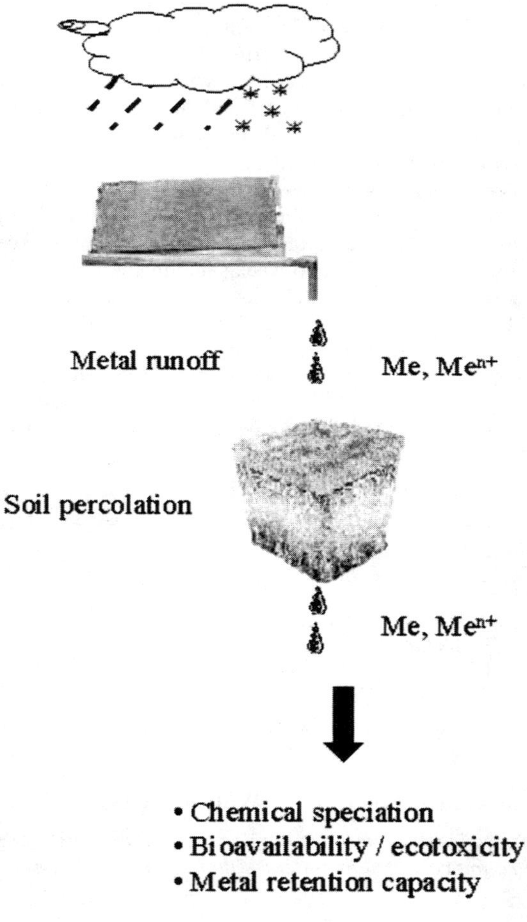

Figure 8 - *Experimental setup for soil percolation studies.*

The measurements so far have been performed during 250 hours at a constant temperature of 8 °C and a light rain intensity of 1.6 mm/h. The total rain amount corresponds to about one year of precipitation in Stockholm and preliminary results obtained on zinc indicate a very high zinc retention capacity for most soils studied. The measurements have to be extended to other metals in order to find the break-through point for each soil, corresponding to the time when the soil cannot accommodate any more metal.

Metal Retention Through Surface Adsorption

Through its passage from the roof to the environment the metal can be retained through adsorption or absorption on a variety of different inorganic and organic surfaces. As an example, a set of experiments is described below in which runoff solutions from a copper roof passed through gutters filled with or without limestone. The studies show that a significant fraction (between 5 and 50%) of the total copper in the runoff is retained in the limestone. The number depends on e.g. the rain intensity, the length of dry periods between rain events, and the concentration of copper in the runoff. The ability to retain copper increases with decreasing flow of runoff water through the limestone. To illustrate the effect of limestone, the open bars of Figure 9 display how the total concentration of copper in the runoff water is being reduced with about 50% in contact with limestone during a two-week sampling period. The filled bars of Figure 9 show the corresponding concentrations of Cu^{2+}-ions, as measured by an ion-selective electrode. The results from this particular sampling period suggest that the fraction of the bioavailable Cu^{2+}-ions that have not been in contact with limestone is about 50% of the total copper concentration released from the roof. However, in contact with limestone, the concentration of Cu^{2+}-ions becomes less than 50% of the total concentration and less than 25% of the total copper concentration that has been released from the copper roof.

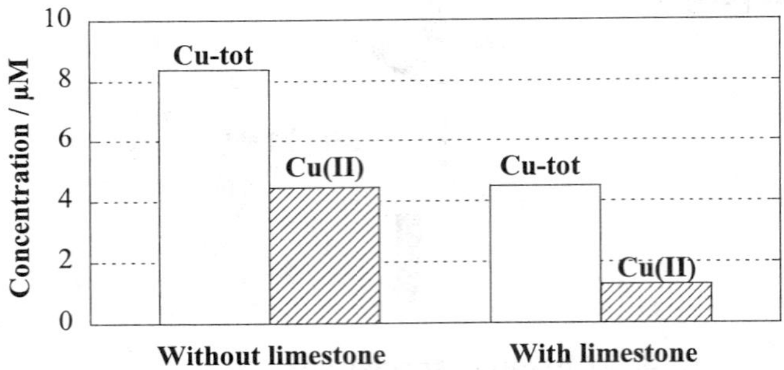

Figure 9 - *Total copper concentration (open bars) and Cu^{2+}-ion (filled bars) concentration collected in gutters without and with limestone.*

Although these specific results cannot be generalized to any wider range of exposure conditions, they nevertheless illustrate that the reduction of the bioavailable form of copper undergoes a rapid reduction already in the immediate vicinity of the roof.

Concluding Remarks

Triggered by the need to fulfill crucial gaps of knowledge, this unique research effort implemented by KTH intends to cover the whole competence area from the metal-release situation to the metal-reception situation. The aim is to generate relevant data for a full risk assessment of environmental effects of roofing materials. The risk assessment has to be preceded by an exposure assessment and an effect assessment.

The exposure assessment presented here includes research activities to generate data on runoff rates from roofing materials made from copper, zinc, and stainless steel. The rates are representative of the Stockholm urban area, characterized by relatively low pollutant concentrations and annual precipitation rates. As opposed by corrosion rates, runoff rates as a function of time exhibit a relative constant value for a given atmospheric exposure situation, at least on a time-scale of the order years. When inspecting a single rain-event, however, the runoff rate is not constant but characterized by an initial so-called first-flush event during which the metal concentration in the runoff exhibits increased values. Through gradual accumulation of more runoff data combined with atmospheric and precipitation parameters, the intention is to transform present runoff data to more general exposure situations, at least representative of the tempered climatic temperature zone.

The effect assessment aims at generating data on chemical speciation, and on concomitant bioavailability, of the metal in the runoff solution. It includes, furthermore, to establish the degree of change of chemical speciation during transport to, and toxic effect on, the recipient. Results presented show that data on chemical speciation, bioavailability and ecotoxicity are consistent with each other, at least during the situation immediately after metal release from the roof. Hence, experimental tools seem to be at hand for describing more complex situations, for instance during transport through soil or aquatic systems.

Acknowledgments

We acknowledge the initiatives and fruitful discussion over the years we have had with Rolf Sundberg, Outokumpu Copper Partner AB and Scandinavian Copper Development Association, Sweden, Hans Klang, SSAB, Sweden and Pascal Verbiest, Union Minière, Belgium. We are also grateful to our colleagues Wenle He, Camilla Karlén and Sofia Bertling, KTH, Sweden, and Prof. Colin Janssen and Dagoberth Heijerick, Laboratory for Environmental Toxicology and Aquatic Ecology, University of Ghent, Belgium, for valuable insight and skilled experimental work.

References

[1] Korenromp, R.H.J., and Hollander, J.C. T., "Diffusive Emissions of Zinc due to Atmospheric Corrosion of Zinc and Zinc Coated (galvanized) Materials," TNO Institute of Environmental Sciences, Energy Research and Processs Innovation, TNO-report, TNO-MEP-R 99/441, NL-7300 Apeldoom, The Netherlands, 1999.

[2] He, W., Odnevall Wallinder, I., and Leygraf, C., "Runoff Rates of Zinc – A Four-Year Field and Laboratory Study", *Outdoor Atmospheric Corrosion," ASTM STP 1421*, H.E. Townsend, Ed., American Society for Testing and Materials, West Conshohocken, PA, 2002.

[3] Odnevall Wallinder, I., Korpinen, T., Sundberg, R., and Leygraf, C.,"Atmospheric Corrosion of Naturally and Pre-Patinated Copper Roofs in Singapore and Stockholm – Runoff Rates and Corrosion Product Formation", *Outdoor Atmospheric Corrosion, ASTM STP 1421*, H.E. Townsend, Ed., American Society for Testing and Materials, West Conshohocken, PA, 2002.

[4] Bertling, S., Odnevall Wallinder, I., Leygraf, C., and Berggren, D.,"Environmental Effects of Zinc Runoff from Roofing Materials – A New Multidisciplinary Approach," *Outdoor Atmospheric Corrosion, ASTM STP 1421*, H.E. Townsend, Ed., American Society for Testing and Materials, West Conshohocken, PA, 2002.

[5] Odnevall Wallinder, I., and Leygraf, C., "A Study of Copper Runoff in an Urban Atmosphere," *Corrosion Science*, Vol. 39, No. 12, 1997, p. 2039.

[6] ISO 8565, "Corrosion of Metals and Alloys – Atmospheric Corrosion Testing – General Requirements for Field Tests," International Organization for Standardization, Geneva, Switzerland, 1992.

[7] Lawson, H.H., "Atmospheric Corrosion Test Methods," Houston, TX, NACE International, 1995.

[8] He, W., Odnevall Wallinder, I. and Leygraf, C., "A Laboratory Study of Copper and Zinc Runoff During First Flush and Steady State Conditions," *Corrosion Science,* Vol. 43, No. 1, 2000, p.127.

[9] Cramer, S. D., McDonald, L.G. and Spence, J.W., "Effect of Acidic Deposition on the Corrosion of Zinc and Copper," Proceedings 12th International Corrosion Congress, Houston, USA 1993, p. 722.

[10] Odnevall Wallinder, I., Lu, J., Bertling, S., and Leygraf, C., "Release Rates of Chromium and Nickel from 304 and 316 Stainless Steel During Urban Exposure – A Combined Field and Laboratory Study", *Corrosion Science*, submitted

[11] Odnevall Wallinder, I., Verbiest, P., He, W., and Leygraf, C., "Effects of Exposure Direction and Inclination on the Runoff Rates of Zinc and Copper Roofs," *Corrosion Science*, Vol. 42, No. 8, 2000, p. 1471.

[12] Odnevall Wallinder, I., Verbiest, P., He, W., and Leygraf, C., "The Influence of Patina Age and Pollutant Levels on the Runoff Rate of Zinc from Roofing Materials," *Corrosion Science*, Vol. 40, No. 11, 1998, p.1977.

[13] Allison, J.D., Brown, D.S., Novo-Gradic, A., "MINTEQA2, A Geochemical Model for Environmental Systems," U.S. Environmental Protection Agency, Athens, GA, EPA/600 3 91 021, 1991.

[14] Tipping, E., and Hurley, M.A., "A Unifying Model of Cation Binding by Humic Substances," *Geochimica et Cosmochimica Acta*, Vol. 56, No. 10, 1992, p. 3627.

[15] Karlén, C., Odnevall Wallinder, I., Heijerick, D., Leygraf, C., and Janssen, C.R., "Runoff Rates and Ecotoxicity of Zinc Induced by Atmospheric Corrosion," *The Science of the Total Environment*, in press

[16] Courbisier, P., Thiry, E., Masolijin, A., and Diesl, L., "Construction and Development of Metal Biosensors," In Bioluminescence and chemoluminescence: fundamentals and applied aspects, A.K. Campbell, L.J. Kricka, and P.E. Stanley, eds, John Wiley & Sons, Chichester, 1994, pp. 150-155.

[17] ISO 8692, "Water Quality – Fresh Water Algal Growth Inhibition Test with *Scenedesmus subsicatus* and *Raphidocelis subcapitata*", International Organization for Standardization, Geneva, Switzerland, 1989.

Sofia Bertling,[1] Inger Odnevall Wallinder,[1] Christofer Leygraf,[1] and Dan Berggren[2]

Environmental Effects of Zinc Runoff from Roofing Materials – A New Multidisciplinary Approach

Reference: Bertling, S., Odnevall Wallinder, I., Leygraf, C., and Berggren, D., "**Environmental Effects of Zinc Runoff from Roofing Materials – A New Multidisciplinary Approach**," *Outdoor Atmospheric Corrosion, ASTM STP 1421*, H. E. Townsend, Ed., American Society for Testing and Materials International, West Conshohocken, PA, 2002.

Abstract: The objective of this work is to study changes in concentration and bio-availability of zinc-containing runoff water, released from roofing materials, upon passage through soil. The experimental approach is based on simulating the interaction between zinc in artificial runoff water and soil in a column system.

The total zinc concentration of runoff was substantially reduced when passing through the soil and suggests marked zinc retention. During a constant flow and supply of zinc into the soil, equivalent to three and a half years of precipitation in Stockholm, a zinc retention capacity of approximately 99% was recorded. Not only the total concentration, but also the bio-available portion of the total zinc concentration was reduced after passage through soil. Most of the retained zinc was located in top 3 cm of the soil core and suggests the total capacity for zinc retention of the investigated soil to be about 140 years per kilogram soil in an isolated system. In real systems, changes of temperature, pH, microbial activity, weathering of minerals and deposition of new organic material must be considered.

The results form part of the effect assessment, preceding future risk assessment of the environmental effects of dispersed zinc.

Keywords: zinc, atmospheric corrosion, runoff rate, soil, zinc ion, metal retention

[1] Ph. D. student, Assistant Professor and Professor, respectively, Department of Materials Science and Engineering, Division of Corrosion Science, Royal Institute of Technology, Dr. Kristinas v. 51, SE-100 44 Stockholm, Sweden.

[2] Assistant Professor, Department of Soil Sciences, Swedish University of Agricultural Sciences (SLU), Ulvsv. 17, SE-750 07 Uppsala, Sweden.

Background

Extensive knowledge of metal dispersion rates, chemical speciation of released metals, environmental interactions, and finally, changes of the retained metal in the environment are crucial parameters to consider when evaluating the potential environmental effects of metal dispersion. The effect of various environmental parameters on the metal concentration in runoff water has recently been studied through extensive field and laboratory investigations [*1-2*]. In addition, the chemical speciation and the bioavailability of metal runoff have been investigated when the runoff water leaves the metal surface [*3*]. However, information on the interaction between metal in the runoff water and the environment is still lacking. Many investigations reported in the literature have focused on the interaction between metals in urban storm water and different soil systems on specific locations [*4-6*]. This is often complicated, because the metal may originate from a number of diffuse sources, which results in complex travelling paths for the water and a large volume to sample ratio.

The aim of this study is to isolate and study the interaction between zinc runoff water from roofing materials through a controlled laboratory setup using realistic metal concentrations. According to the knowledge of the authors, no investigation has earlier combined metal-containing runoff water with soil column studies to evaluate the zinc retention capacity of soil, changes in chemical speciation and bioavailability during passage through the soil, and the ecotoxicity of the runoff water after soil passage (the so-called percolate). The work forms part of a large European industrial collaboration project with the aim to increase the knowledge on runoff rates, bioavailability and ecotoxicity of metal dispersion from zinc-based construction materials used in the society. The approach combines three interdisciplinary sciences, corrosion science, soil science and environmental toxicology for investigating the interaction between dispersed metals and the environment.

The focus of this paper is to illustrate the structure of the project by describing the most important steps of runoff water after leaving the roof and passing through the soil. By taking zinc as the example, the following key issues will be described in logistic order (see Figure 1).

Before soil passage

- Atmospheric corrosion and zinc runoff
- Concentration and bioavailability of zinc runoff

During soil passage

- Important physico-chemical processes during interaction between metal runoff and soil
- The design of an experimental set-up for percolation studies
- Experimental results on zinc retention

After soil passage

- Experimental results on percolate water and its bioavailability
- Massbalance of the column experiment

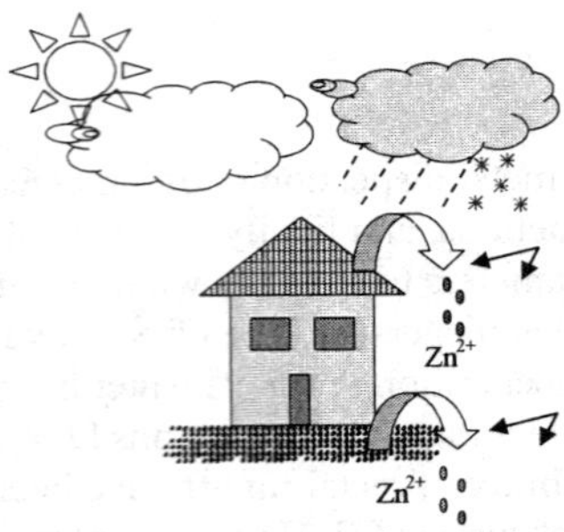

Figure 1- *Study of changes of metal concentration, chemical speciation, bioavailability and ecotoxicity of zinc runoff in contact with a soil system.*

Before Soil Passage

Atmospheric Corrosion and Zinc Runoff

Zinc forms relatively porous and voluminous patina layers, primarily composed of basic zinc carbonate $Zn_5(CO_3)_2(OH)_6$ and basic zinc sulphates $Zn_4SO_4(OH)_6 \times nH_2O$ (n=1-5), during urban atmospheric exposures [*7*]. One of the predominating atmospheric pollutants that governs the corrosion attack of zinc is sulphur dioxide (SO_2). A strong correlation between the zinc corrosion rate and the SO_2 concentration has earlier been demonstrated through a number of dose-response functions [*8*]. As a result of the substantial reduction in SO_2 concentration seen during the last decades in urban environments of Europe and North America, a significant decrease in the zinc corrosion rate has been observed. During a rain event, soluble and poorly adhesive phases of the corrosion patina can dissolve and be released from the surface by rain water. Not only the corrosion rate but also the zinc runoff rate has been substantially reduced as a result of lower pollutant levels. A correlation between SO_2 and the runoff rate has recently been presented in the literature [*8*]. The equation can easily be used to transform runoff rate data from present study (with an SO_2-concentration of 3-4 $\mu g/m^3$) to other more polluted urban test sites. Annual zinc runoff rates in the range of 2.5 to 3.5 g/m^2 have been determined for zinc sheet of varying age exposed in Stockholm. The runoff rate of zinc has been between 40 and 65% (for SO_2-concentrations higher than 50 $\mu g/m^3$) of the corrosion rate, at least during the first five years of exposure [*9*].

Zinc in urban storm water and runoff water originates also from natural deposition by dry and wet deposition. Annual natural deposition rates of zinc in Stockholm are typically 0.02 g Zn/m^2 [*1, 10*]. During entry into the environment, the runoff water from a roof will eventually be mixed with urban storm water from, *e.g.*, roads and sidewalks. This results in dilution of the metal concentration in the runoff water, and in concomitant complexation, sorption, ion exchange and other processes through which the metal can interact with the ambient environment along its way to the recipient.

Concentration and Bioavailability of Zinc Runoff

Information of metal concentration is one of several crucial parameters to consider when evaluating ecotoxicity of metal dispersion. The reason is that a concentrated dose of metal during a short exposure period may be much more harmful for living organisms and plants than the same dose extended over a long period. As can be seen in Figure 2 (left part), the variation in zinc concentration between different samplings within the present study varies from 1 to 14 mgZn/L. The data are based on runoff water from zinc sheet of varying age and based on all sampling periods (including several rain events) during a two-year exposure period in Stockholm [*11*]. Figure 2 (right part) exhibits the distribution of the zinc concentrations and shows that the average zinc concentration in the runoff water is approximately 5 mg/L.

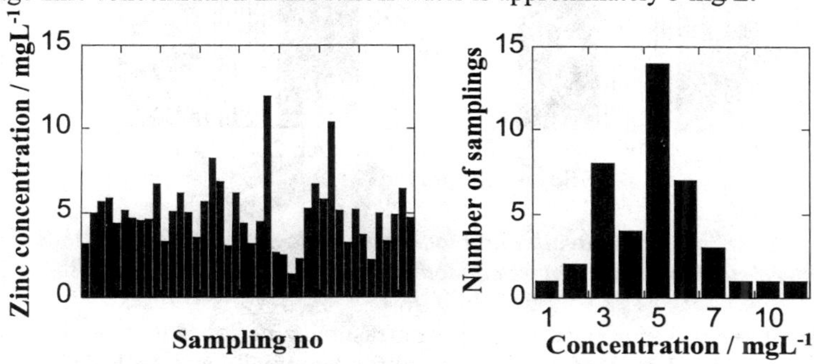

Figure 2 - *Total zinc concentration in all samplings of runoff water during a two-year exposure period in Stockholm (left), and distribution (n=41) of zinc concentration (right) in the same samplings.*

The concentrations mentioned so far concern total concentrations of zinc in runoff water, irrespective of chemical speciation. However, when discussing ecotoxicity it is important to assess the bioavailable fraction of the total zinc concentration. For metals, the most bioavailable form is often the free ion, which is hydrated in water solutions ($Zn(H_2O)_6^{2+}$ in the case of zinc, see Ref. 3). In the present study the bioavailability of zinc was evaluated using a metal-specific assay (Biomet®) with the bacteria *Alcaligenes eutrophus*. This is a genetically modified bacteria strain that produces light in the presence of bioavailable zinc. The procedure of the biosensor test is described in more detail elsewhere [*3*]. The Biomet® test has been verified through ecotoxicity tests of zinc in runoff water by using the internationally recognised standard 72-hour algal growth inhibition test with *Raphidocelis subcapitata* [*3*].

Biosensor test results are presented in Figure 3 from runoff water collected from new zinc sheet and galvanised sheet panels during two sampling periods. The results show that most of the zinc in the investigated runoff waters is bioavailable. The zinc sheet and the galvanized steel sheet show very similar behaviour, although a difference in both total zinc concentration and bioavailable zinc concentration is seen between the two sampling periods. The difference in total zinc concentrations between the two sampling periods is attributed to normal variability caused by different precipitation volumes, dry periods between rain events etc. The difference in the bioavailable zinc fraction is probably due to organic material found on the panels during the August sampling (pollen). This causes organic complexes to form with zinc ions, which results in less bioavailable forms of zinc than the free hydrated zinc ion.

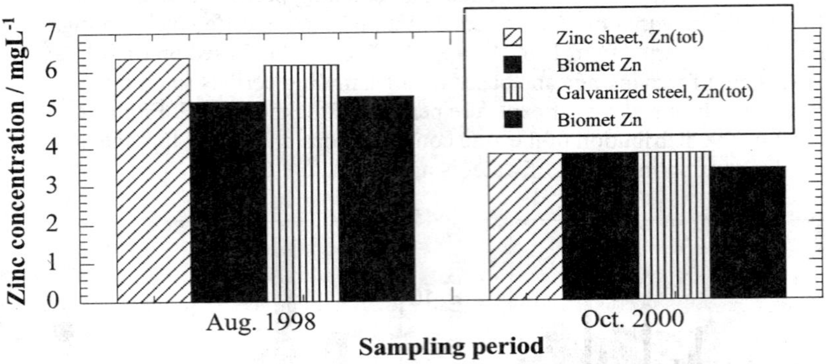

Figure 3- *Total (dashed) and bioavailable (black) zinc concentration for zinc sheet and galvanized steel in runoff waters collected during two sampling periods.*

At this point it seems appropriate to compare measured zinc concentrations in runoff water with selected concentrations reported for ecotoxicity effects. For water a defined deficiency boundary has been set to be less than 1 µg Zn/L, and an ecotoxicity boundary at zinc concentrations higher than 1000 µg Zn/L [*10*]. Ecotoxicity tests with the algae *Raphidocelis subcapitata* show EC_{50} values (*i.e.*, the concentration of zinc where the growth rate of the algae is reduced with 50%) from 30 to 60 µg/L Zn^{2+} [*3*]. For the crustaceans *Daphnia Magna*, the acute LC_{50} value (*i.e.*, lethal concentration causing 50% mortality) have been found ranging from 5 to 2300 µg Zn/L, depending on various test conditions, abiotic (nonbiological) factors and the circumstance that different life stages of the species have been used. For the rainbow trout *Oncorhynchus Mykiss* the acute LC_{50} values vary from 90 to 7200 µg Zn/L, due to the same factors as mentioned above [*12*]. These large variations in results stress the importance of choosing a test that is sensitive, relevant and with species that are not influenced by too many uncontrolled parameters. This requirement is achieved with the algae *Raphidocelis subcapitata*.

During Soil Passage

Important Physico-Chemical Processes during Interaction between Metal Runoff and Soil

Soil, as used in this context, consists of a thin layer of loose earth materials composed of weathered minerals (oxides and layer silicates) and decaying organic matter (humus), with all solid phases and water in close contact. Due to its complex composition, the soil can interact with metals through a variety of physico-chemical processes, as briefly discussed below [*13*]. Irrespective of process, the resulting ecotoxicity of the retained metal is often reduced in comparison to the free, hydrated, metal ion.

The solid phase (clay or humus) carries in most cases an excess of negative charge. This leads to a surface potential that determines the distribution of cations, including metal ions and anions in the liquid phase. Two opposing forces, electrical attraction and diffusion back into the bulk solution, act on cations, which have to counterbalance the negative surface charge. In addition to the electrostatic attraction, ions might bond to the solid phase by chemisorption processes. The latter type of bonding is much stronger.

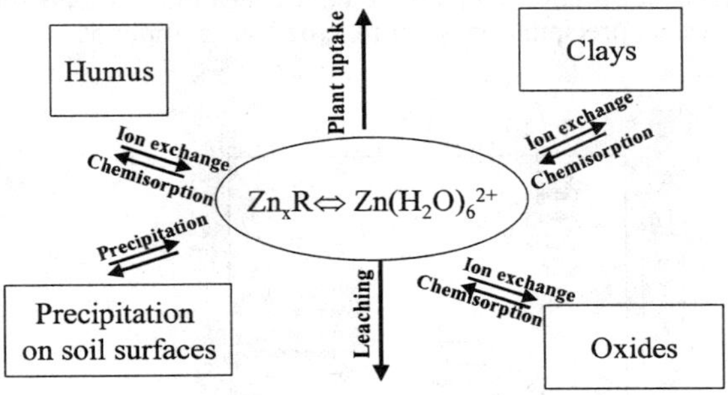

Figure 4- *Interactive processes determining retention and bioavailability of zinc.*

Figure 4 is a principal figure that shows the main processes for metal retention in the soil and complexation in solution (Zn_xR), as well as showing that these processes determine the bioavailability of zinc. Metal retention processes in soil can be divided into the following main groups:

- chemisorption,
- cation exchange, and
- precipitation of secondary minerals.

Electrostatic sorption in soil is controlled by either fixed charge constituents (*e.g.*, clay minerals such as hydrous silicates that develop negative charge as a result of lattice substitutions) or variable charge constituents (*e.g.*, the surface charge of humus and iron-, manganese-, aluminium- or titanium oxides that varies with pH of soil water). At a soil pH lower than 7, clay minerals adsorb far smaller quantities than other main sorbents of the soil, *e.g.* humus.

Cation exchange is governed by the number of ions exchangable (cation exchange capacity, $CEC_{relative}$) to zinc ions. Silicates, oxides and humus all contribute to the CEC. In the top soil layer, humus is the main contributor to CEC, whereas in deeper soil layers, mineral is the main contributor to CEC together with humus. The overall exchange process is diffusion controlled, and equilibrium is often attained within an hour.

Cations bound to the surface of minerals can diffuse into the interior of the solid phase. The relative diffusion rate depends, among others, on the ionic diameter of the cations. In the interior, these cations can neutralize negative charges and remain fixed in appropriate positions. The processes are described as irreversible and cause metals to become immobilized.

Other processes, such as microbial activity and CO_2 –dissolution, may influence soil pH to some extent, and thereby indirectly influence the metal retention capacity. To summarize, Figure 5 displays different zinc retention mechanisms as a function of reaction time and solubility of the adsorbate. The figure shows that surface precipitation (of, *e.g.*, hardly soluble sulfides, carbonates, phosphates or oxides) only occur at very high metal concentrations, in fact, far higher than those usually found in runoff water. For this reason, precipitation is not regarded as an important retention process within this context.

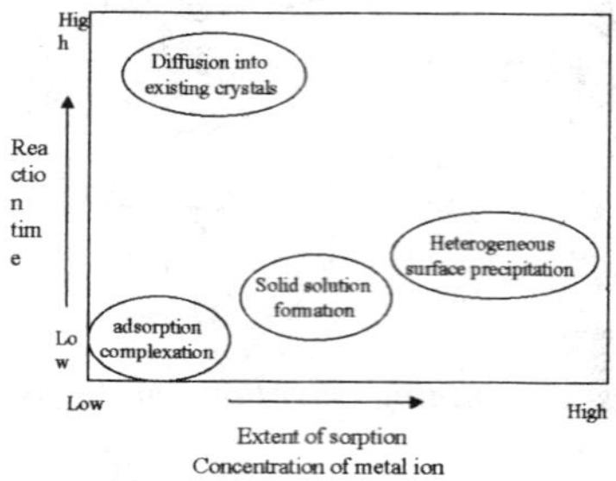

Figure 5- *Possible fates of ionic adsorbates in soil as a function of reaction time and metal concentration.*

Having discussed conditions of the metal runoff and possible processes that govern the interaction between metal runoff and soil, we next discuss a possible design of an experimental setup for soil percolation studies.

The Design of an Experimental Setup for Percolation Studies

Runoff Water- In order to obtain a well-defined runoff water for the percolation study, artificial rain was used with a composition that resembles precipitation in the central and southern part of Sweden, see Table 1 [*14-15*]. The use of artificial rain for simulating runoff water has been verified elsewhere [*2*].

Table 1 - *Composition of artificial rain used in the percolation study.*

SO_4-S (mg/L)	Cl^- (mg/L)	NO_3-N (mg/L)	NH_4-N (mg/L)	Na^+ (mg/L)	K^+ (mg/L)	Mg^{2+} (mg/L)	Ca^{2+} (mg/L)	pH
1.17	0.36	0.56	0.56	0.23	0.12	0.12	0.20	4.3

The pH of the artificial rain was adjusted to 6.2 with NaOH, in agreement with runoff water collected from zinc roofs [*1*]. Earlier results from the present field exposure program and the biosensor tests [*1*] suggest that between 90 and 99% of zinc in the runoff is present in a bioavailable form at this pH. Based on these results, the zinc content in the artificial rain was added as $Zn(NO_3)_2 \bullet 6\, H_2O$ (quality purum) with a zinc concentration of 5 mg/L. As discussed before, see Figure 2, this concentration is representative of field data obtained within the present exposure program.

The rain intensity, finally, was arbitrarily set at 2 mm/h, representative of a medium or light rain intensity in Stockholm, Sweden [*2*].

Soil- The soil material used in this study was sampled one year prior to the experiment from a depth of 0-20 cm in Kalmthout, Belgium. This soil represents one out of five evaluated soils within the European collaboration project. Despite different soil characteristics, all soils exhibit a similar tendency towards zinc retention during the simulated three and a half years. Soil samples were sieved (< 2 mm) under ambient moisture conditions, dried at 20°C, homogenised and stored at room temperature prior to investigation. The pH was measured in a water extract at 8°C with a combined glass electrode (Radiometer GK2401C 291-3-065). Total content of organic carbon (C_{org}) was determined using a LECO CHN-932 analyser. The cation exchange capacity (CEC) was calculated from the Ca, Mg, Na and K content of the soil (determined by extraction with 0.1 M $BaCl_2$ and analysed with atomic absorption spectroscopy, Analyst 300 Perkin Elmer). The natural zinc concentration in the Kalmthout soil was determined as, the exchangeable concentration of zinc, defined as zinc extractable using the complexing agent EDTA (disodiumsalt), and the total concentration of zinc, obtained through extraction with 7 M HNO_3. All extractions were performed for 24 hours. The results of soil characteristics are found in Table 2.

Table 2 - *Characteristics of Kalmthout soil used for the percolation studies.*

Soil Data	Clay (%)	C_{org} (% dry weight)	pH $_{H2O\ 8°C}$	CEC (cmol/kg)	Ca $_{BaCl2}$ (mg/100g)	Mg_{BaCl2} (mg/kg)	Na $_{BaCl2}$ (mg/kg)	K_{BaCl2} (mg/kg)	Zn_{EDTA} (mg/kg)	$Zn_{7M\ HNO3}$ (mg/kg)
	1	4.1	4.93	2.6	38.2	44.2	4.1	132.7	18.6	30.2

Kalmthout soil is regarded as a low retention soil for zinc, characterized by a clay content of only 1% (compared to typically 10-20%), a medium fraction of organic material, 4.1% (typically 2-10%), low pH 4.9 (typically 6-7), and low amount of CEC 2.6 cmol/kg (typically 10). As the soil pH is below 6-7 the soil oxides will have a positive charge and thereby will not contribute to the CEC. Average concentrations of total zinc content in some soils from Western Europe are found in Table 3. Such data, however, can vary widely from site to site.

Table 3 - *Mean concentrations of total zinc content in unpolluted soils from Western Europe [10,16].*

Country	*Belgium*[16]	*Germany*[16]	*Sweden*[10]
mg Zn/kg	*57*	*83*	*55*

Soil Column Design and Analysis

Changes in bioavailability and ecotoxicity when zinc runoff enters the Kalmthout soil were evaluated with soil column experiments, Figure 6. A similar set-up has earlier been used for clay soils [*17-18*]. The diameter of each glass column is 32 mm and the height is 140 mm. In order to avoid contamination, all columns, preparation and storage vessels were immersed in 10% HNO_3 for at least one day, rinsed three to four times with ultra-pure water (<18 $M\Omega cm^{-1}$) and sealed before use. The soil had an initial water holding capacity of ~15% and was packed in each column by gravity. Each soil core was about 7 cm in height, corresponding to a mass of 55g soil. A sintered glass-filter disc (Duran P4; pore size diameter 9-15 μm), a 0.2 μm filter (Gelman Supor), and a glass-fibre filter (Gelman A/E 1.0 μm) were placed between the bottom cover of the column and the soil core.

Figure 6 - *Experimental setup showing the soil cores, collecting vessels and the pump providing artificial rain to the soil cores.*

The artificial runoff water entered each soil core through a glass filter ensuring a uniform wetting of the whole soil core surface. A peristaltic pump was used to keep a constant rain intensity of 2 mm/h (equivalent to a supply of 1.5 mL/h for this column size). The intensity by which the artificial runoff water enters the soil was not found to affect the retention of zinc in the Kalmthout soil. The reason for this is that adsorption of zinc on organic material is a very fast reaction. When the adsorption sites on organic material are fully occupied and adsorption on oxides plays a greater role for the retention of zinc, the intensity will probably affect the retention of zinc by the soil. Duplicate control tests were performed, which resulted in excellent agreement. The result suggests that no preferred routes for the runoff water were created during the percolation experiments.

The investigations were performed at +8°C with aerobic conditions (achieved by a permanent vacuum of −100 mbar at the outlet of the columns) in a dark room. Samples were collected once a day and pH and total zinc concentration, determined by ICP-AES (inductively coupled plasma-atomic emission spectroscopy, ARL ICP 3520B) were analysed for each sample. Prior to total zinc concentration measurements the samples were acidified with HNO_3 to reach a pH of approximately 2. Total organic content (TOC) and bioavailability were evaluated for some of the percolate water samples. TOC was determined with a TOC-5000A Shimadzu analyser. Inbetween analyses, the samples were stored at +8°C.

The column experiments were conducted for about one month. The total volume of added artificial runoff was 1200 mL corresponding to 1600 mm of rain and a total quantity of 6.1 mg zinc. This volume is equivalent to three and a half years of rain in Stockholm.

Experimental Results on Zinc Retention

Figure 7 displays the zinc retention capacity of Kalmthout soil, *i.e.*, the ability to retain zinc within the soil core. The results of this figure are presented as a function of rain amount percolating through the soil. Figure 7 is based on data of zinc concentration of the percolated water, which is further discussed in a subsequent section (Figure 10, right). The data includes a natural leaching of zinc from the soil.

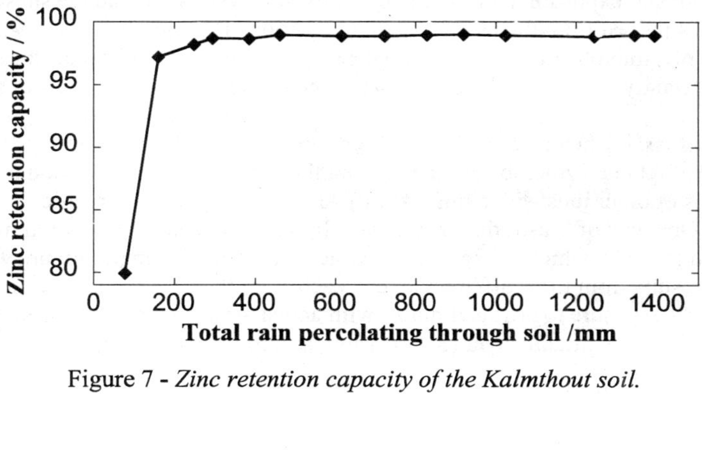

Figure 7 - *Zinc retention capacity of the Kalmthout soil.*

Apart from the initially low zinc retention capacity during the first 200 mm of rain, a capacity of 99% is observed throughout the remaining exposure of up to 1400 mm of continuous rain (corresponding to about three and a half years of natural rain in Stockholm).

The observed zinc retention is due to one or several of the previously discussed physico-chemical processes that operate during zinc interaction with the soil, probably a combination of ion exchange and chemisorption. As stated before, the formation of metal-organic complexes is a most plausible process at low zinc concentrations. A hint for possible causes of the initially low zinc retention is given in Figure 8, which displays the total organic content (TOC) in the percolate water as a function of rain amount. The behaviour of TOC follows the zinc concentration in the percolate water (see Figure 10, left), exhibiting a high value during the first 200 mm of rain, which drops to lower values with extended rain up to 1500 mm. This implies that complexation between zinc and mobile organic ligands are formed, thereby mobilizing zinc, which is transported through the soil to the percolate water.

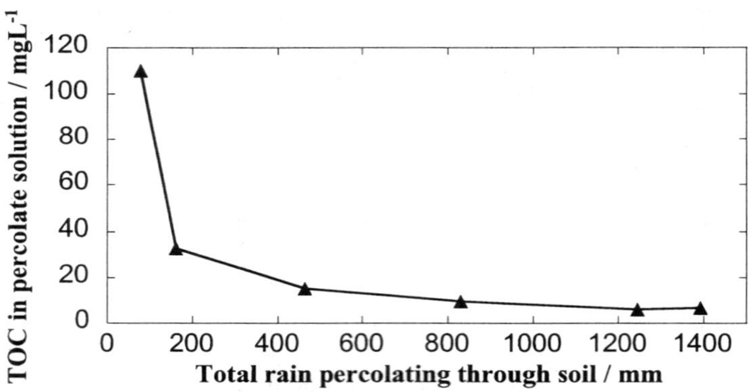

Figure 8 -*TOC concentration in percolate solutions.*

With prolonged exposure, the soil obviously exhibits very strong active sites for zinc uptake, as judged from the increasing zinc retention. Through changes in, *e.g.*, temperature, pH, microbial activity and dry deposition of organic substance, the soil system is constantly altering its TOC and, hence, new zinc-complexing abilities are formed.

An obvious issue when discussing zinc retention in soils is the possible mobilization of retained zinc, for instance during the increase in concentration of hydrogen ions or other ions with a high affinity towards inorganic or organic compounds. One way of illustrating this issue is by studying zinc mobilization from a soil with retained zinc, which is exposed to extracts of varying strength. Figure 9 summarizes results from three different layers, each 1 cm thick, of the column soil after completed exposure to artificial runoff with added zinc. The layers represent the topsoil layer (0-1 cm), middle layer (3-4 cm) and bottom layer (6-7 cm).

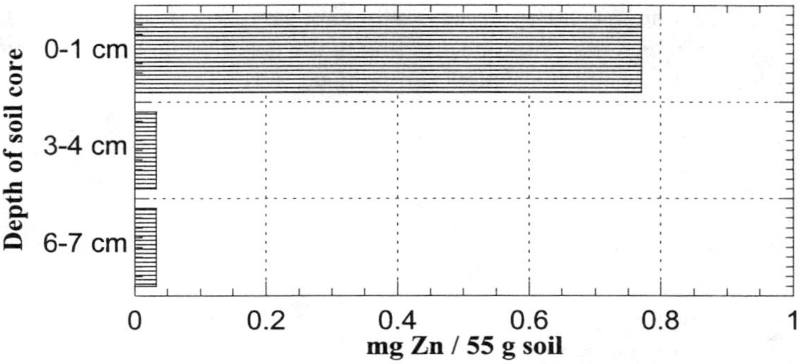

Figure 9 - *Zinc mobilized from column soil by extraction with artificial rain at pH 4.3.*

When each soil layer was extracted with artificial rain (5 g soil mixed with 40 ml artificial rain, after 24 hours shaked for 1 hour, centrifuged and analysed), the results reveal that 0.77 mg zinc per 55 g soil was extracted from the top layer and 0.033 mg from the other layers. As will be shown in a subsequent section (Massbalance of Kalmthout soil) 55 g soil has retained totally 5.9 mg zinc (7.6 mg subtracted with the natural zinc content of 1.7 mg). 0.77 mg zinc then corresponds to 13% of the totally retained zinc for the top layer and 0.5% for the middle and bottom layer. Leaching experiments have shown that 0.06 mg zinc per 55 g soil leaches naturally from Kalmthout soil. This is similar to the amount found at a depth of more than 3 cm in the soil core. Therefor the conclusion can be drawn that zinc from the runoff water has been retained in the first 3 cm of the soil core, at the most. If we assume that this top 3 cm layer has reached the breakthrough capacity for zinc retention, 24 g of soil (equals to 3 cm) would result in a breakthrough capacity of three and a half years. A worst-case scenario for breakthrough of a kilogram of Kalmthout soil would then be about 140 years.

To conclude: after passage of zinc-containing artificial runoff, equivalent to three and a half years of rain in Stockholm, more than 80% of the zinc in the runoff water is immobilized by various processes and retained in a top soil layer with a thickness of less than 3 cm.

After Soil Passage

Experimental Results on Percolate Water and its Bioavailability

As previously discussed, the zinc retention capacity of the soil proved to be between 80 and 99%, hence allowing only a small portion of introduced zinc in the rain or the natural zinc content of the soil to pass through the soil core during a continuous rain event. This is illustrated in Figure 10 (left) showing a significant decrease in zinc concentration (~80%) already in the first sampling (from 5000 μg/L to about 1000 μg/L) after which the zinc concentration in the percolating water becomes even lower (50-140 μg/L) during subsequent sampling periods.

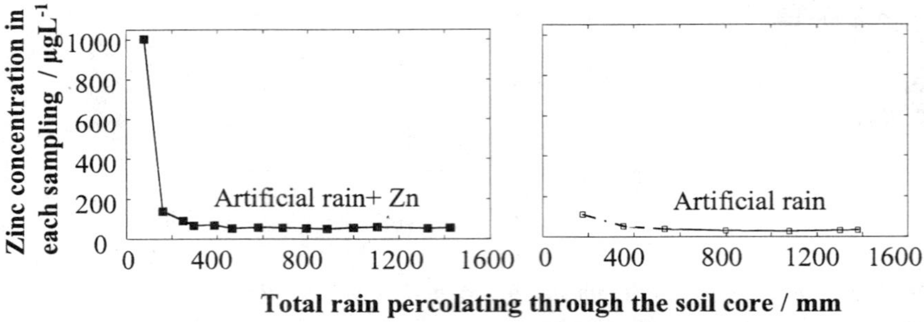

Figure 10 - *Zinc concentration in percolated water samples during a continuous rain event with (left) and without (right) zinc added to the artificial runoff.*

However, a continuous rain event without any zinc addition, showed a natural, and relatively constant leakage (30-100 μg/L) of zinc from the soil core itself, Figure 10 (left).

The bioavailability of zinc in the percolate water was evaluated using the metal-specific assay (Biomet®), previously described. The bioavailable fraction of zinc in the introduced runoff was approximately 100%. Since the detection limit of the test is approximately 200 μg Zn/L, tests could only be performed on the first sampling volume percolating the soil core (see Figure 10). The results for this sampling period are presented in Figure 11, showing 34% of the total zinc concentration (1000 μg/L) to be bioavailable. The difference between total and bioavailable zinc concentration as well as the relatively high zinc concentration during the first sampling volume (< 77 mm) is probably explained by mobile organic ligands from the soil acting as complexing agents for zinc. It is then assumed that the zinc-organic complexes are less bioavailable than the free hydrated zinc ion. Computer model calculations with WHAM (Windermere Humic Aqueous Model) will be performed in order to verify these results.

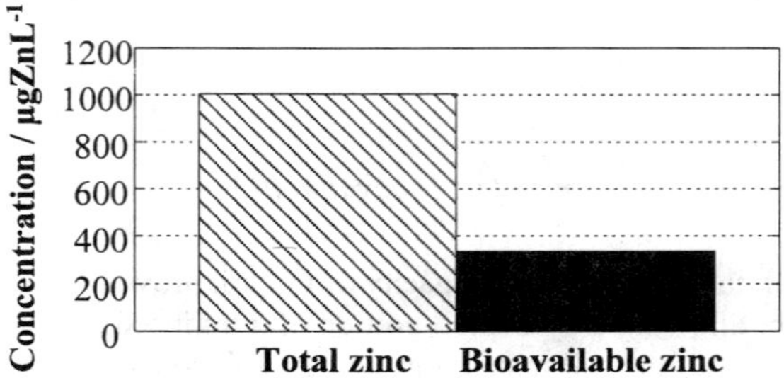

Figure 11 - *Comparison between the total and the bioavailable zinc concentration in percolate water after 77 mm of continuous rain.*

Massbalance of Kalmthout Soil from the Column Experiment

In order to verify that all zinc in the system is considered during the experiment a massbalance calculation was performed, the results are presented in Figure 12.

55 g of the Kalmthout soil has a natural zinc content of 1.7 mg (Figure 12, left, Table 2) of which 1.0 mg zinc is exchangeable.

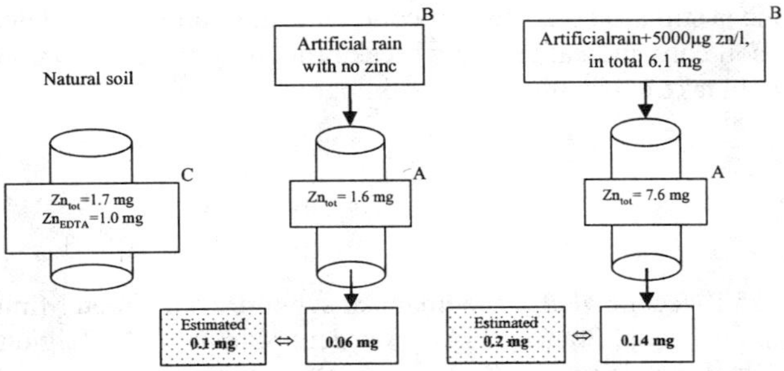

Figure 12 - *Massbalance of 55 g soil exposed to artificial rain (middle) and artificial rain with 5mg Zn/L (right).* Zn_{tot} *= extraction with 7 M* HNO_3. Zn_{EDTA}*= exchangeble fraction of Zn measured by EDTA extraction of the soil.*

Input data for the massbalance are the zinc content measured after extraction of the soil core after percolation (A) and introduced amount of zinc to the soil (B). This difference is then compared with the natural content of zinc in the soil (C). The discrepancy is an estimate of the amount of zinc that should have passed the soil core and be present in the percolate water. This value is then compared with the actual measurements of zinc in the percolate water. The results are in excellent agreement for both tests.

Conclusions

The experimental set up used in the study works very well for simulating interactions between runoff water and soil. The following main conclusions can be drawn,

- When artificial runoff water with a total concentration of 5000 µg/L zinc percolate through a soil core of 55g Kalmthout soil, the percolate water shows zinc concentrations of approximately 60 µg/L. This zinc concentration also includes naturally leached zinc from the soil.
- The zinc retention capacity of the Kalmthout soil, which is regarded as a low retention soil is 99 % and continued to be so even after three and a half years of constant exposure.

- Initially high total zinc concentrations in the percolate water are explained by mobile organic ligands in the soil functioning as complexing agents for zinc ions and thereby mobilizing zinc.
- The fraction of bioavailable zinc in percolate water is significantly lower compared to runoff water.
- 13% of zinc retained in the top 1-cm layer of the soil was mobilised when extractions with artificial rain were performed. In the centre and bottom layers of the core only 0.5% the zinc was mobilised.
- Retained zinc is still after 3.5 years of continuous rainfall not found deeper than 3 cm of the soil core, indicating that a breakthrough of the zinc retention capacity would take about 140 years per kilogram soil.

Acknowledgment

The authors gratefully acknowledge the financial support from Union Minière, Belgium; Sollac Usinor Group, France; SSAB, Sweden; Sidmar N.V., Belgium; Cockerill Sambre, Belgium; Norzink A.S., Norway; Outokumpu Zink O.Y., Finland; Zinc Info Norden AB, Sweden; Wirtschaftsvereinigung Metallw e.V., Germany and Grillo Werke AG, Germany. We are also grateful to Ass. Prof. Jon Petter Gustafsson at Dept. of Soil Science, KTH for discussions of results and ideas regarding the experimental set-up, to Folke Fredlund, Dept. of Chemistry, KTH for performing ICP-analysis, Kent Andersson and Gunilla Hallberg, Dept. of Soil Sciences, Swedish University of Agricultural Sciences Uppsala, Sweden. Dagobert Heijerick, Dept. of Environmental Toxicology and Aquatic Ecology, Univeristy of Gent, Belgium for performing Biomet tests and fruitful discussions on environmental toxicology.

References

[1] Odnevall Wallinder, I., Leygraf, C., Karlen, C., Heijerick, D., and Janssen, C.R., "Atmospheric corrosion of Zinc-Based Materials: Runoff rates, Chemical Speciation and Ecotoxicity Effects", *Corrosion Science,* Vol.43, No.5, 2001, pp.809.

[2] He, W., Odnevall Wallinder, I., and Leygraf, C., "A Laboratory Study of Copper and Zinc Runoff during First Flash and Steady-State Conditions", *Corrosion Science,* Vol.43, 2001 pp. 127-146.

[3] Karlén, C., Odnevall Wallinder, I., Heijerick, D., Leygraf, C., and Janssen, C.R., "Runoff Rates and Ecotoxicity of Zinc Induced by Atmospheric Corrosion" *The Science of the Total Environment",* in press.

[4] Barraud, S., Gautier, A., Bardin, J. P., and Riou, V., "The Impact of Intentional Stormwater Infiltration on Soil and Groundwater", *Water Science Technology,* Vol. 39, No. 2, 1999, pp.185-192.

[5] Christensen, J., Jensen, D., and Christensen, T., "Effect of Dissolved Organic Carbon on the Mobility of Cadmium, Nickel and Zinc in Leachate Polluted Groundwater", *Water Res.*, Vol. 30, No. 12, 1996, pp. 3037-3049.
[6] Mason, Y., Ammann, A., Ulrich, A., and Sigg, L., "Behaviour of Heavy Metals, Nutrients, and Major Components during Roof Runoff Infiltration", *Environmental Science and Technology,* Vol.33, 1999, pp. 1588-1597.
[7] Odnevall, I., and Leygraf, C., "Reaction Sequences in Atmospheric Corrosion of Zinc", *Outdoor Atmospheric Corrosion, ASTM STP 1239*, W. W. Kirk and H. H. Lawson, Eds., American Society for Testing and Materials, Philadelphia, PA, 1995, pp. 215-229.
[8] Odnevall Wallinder, I., Verbiest, P., He, W., and Leygraf, C., "The Influence of Patina Age and Pollutant Levels on the Runoff Rate of Zinc from Roofing Materials", *Corrosion Science*, Vol. 40, No. 11, 1998, pp. 1977-1982.
[9] Odnevall Wallinder, I., Verbiest, P., Janssen, C. R., and Leygraf, C., "Environmental Effects of Zinc Runoff from Roofing Materials of Different Age as a result of Atmospheric Corrosion", 14th International Corrosion Congress, 26 Sept.- 1 Oct. 1999, Cape Town, South Africa
[10] Landner, L., and Lindeström, L., "Zinc in Society and in the Environment" ISBN 91-630-6871-0, 1998, pp. 109.
[11] He, W., Odnevall Wallinder, I., and Leygraf, C., "Runoff Rates of Zinc- a Four-Year Field and Laboratory Study", *Outdoor Atmospheric Corrosion, ASTM STP 1421*, H. E. Townsend, Ed., American Society for Testing and Materials, West Conshohocken, PA, 2002.
[12] Van Tilborg, W. J. M., and Rozendaal, V., "Integrated Criteria Document Zinc", BMRO-VNO, Rozendaal, The Netherlands, 1999.
[13] Mcbride, M. B., "Environmental Chemistry of Soils", Oxford University Press, Inc., ISBN 0-19-507011-9, 1994.
[14] UN/ECE International co-operative Programme on Effects on Materials, including Historic and Cultural Monuments, Report no. 21, "Convention on Long-Range Transboundary Air Pollution ", Norwegian Institute for Air Research, Lillestrøm, Norway, 1997.
[15] Granat, L., Swedish Environmental Protection Agency Report 3942, *in Swedish* 1990.
[16] Frink, C. R., "A Perspective on Metals in Soils", *Journal of Soil Contamination,* Vol.5, No.4, 1996, pp. 329-359.
[17] Jonsson, E., "Retention of heavy metals in Soil" SLU Report ISSN 1102-1381, *in Swedish* 1999.
[18] Zysset, M., and Berggren, D., "Retention and Release of Dissolved Organic Matter in Podzol B Horizons: Column and Batch Studies", *European Journal of Soil Science,* in press.

Wenle He,[1] Inger Odnevall Wallinder,[1] and Christofer Leygraf[1]

Runoff Rates of Zinc - A Four-Year Field and Laboratory Study

---

Reference: He, W., Odnevall Wallinder, I., and Leygraf, C., "**Runoff Rates of Zinc - A Four-Year Field and Laboratory Study**," *Outdoor Atmospheric Corrosion, ASTM STP 1421*, H. E. Townsend, Ed., American Society for Testing and Materials International, West Conshohocken, PA, 2002.

Abstract: As part of a major European industrial collaboration project, zinc runoff rates from new and aged zinc panels have been investigated during exposure to natural rain (up to four years) in Stockholm, Sweden. The average runoff rate was 3.1 g/m^2,y for new zinc and 3.2 g/m^2,y for 40 year old zinc, respectively. New and old zinc panels were also exposed to artificial rain in which the previously established so-called first flush effect could be studied in more detail. The zinc concentration in the runoff was high during the initial first flush region (100-400 μM), and decreased to lower values during the subsequent steady-state region (50-150 μM). Measured zinc concentration in the runoff (both during first flush and steady-state) increased with rain acidity and with decreased rain intensity at a given rain volume. The zinc concentration (during first flush) was also observed to increase with length of the dry period preceding a rain event.

Keywords: roofing material, zinc, atmospheric corrosion, runoff, first flush, pH, intensity

Introduction

The present study forms part of an extensive investigation of runoff rates from zinc-containing materials, primarily intended for roofing and other building applications. It has been initiated by a majority of European producers of galvanized steel with the general aim to generate necessary data for future risk assessments of the environmental effects of zinc dispersed from diffuse sources.

[1]PhD student, Assistant Professor and Professor, respectively, Div. Corrosion Science, Royal Institute of Technology, Drottning Kristinas v. 51, SE-100 44 Stockholm, Sweden.

During a rain event, metal ions from any corrosion products will be dissolved into the runoff water, released from the surface, transported away until they either are captured by a reacting surface or reach a recipient within the eco-system. Several extensive field exposure studies have shown that the zinc corrosion rate has undergone a substantial reduction in major cities of Europe and North-America during the last decade. The main cause is a reduction of acidifying pollutants, including sulfur dioxide (SO_2), and to some extent also nitrogen dioxide (NO_2)[*1-2*]. Runoff rates of zinc, on the other hand, have by far not been investigated to the same extent as corrosion rates. Zinc runoff results from the present project, and based on up to two years of rain exposure, have been published in a series of publications [*2-4*] and have recently also been summarized and analyzed in a separate thesis [*5*]. Selected results have also been presented in these proceedings [*6*].

This paper is an extension of previous investigations and based on combined field and laboratory exposures of new zinc and zinc previously exposed during 40 years. The field exposure was conducted in an urban test site in Stockholm, Sweden, during up to four years in natural rain. The laboratory exposures were performed with artificial rain and permitted the influence to be studied of individual parameters, such as rain pH, rain intensity and runoff volume on zinc runoff rates.

Experimental

Field Exposure

The four-year field exposure was conducted from November 1996 to November 2000 in Stockholm, Sweden. This urban site is characterized by modest surrounding traffic intensity, an SO_2-level of 3-4 $\mu g/m^3$, a mean precipitation acidity of pH 4.7 and a total average annual precipitation amount of around 500 mm/y. The exposure of panels was performed on a rack with an inclination of 45°, facing south, and located on top of a building within the campus area of the Royal Institute of Technology (KTH).

A new zinc (99.97% purity) sheet and an old zinc sheet were cut into panels, sized 25*12 cm^2. The old zinc sheet had been exposed during 40 years in the city of Pontoise, France. The new zinc panels were abraded with 800P SiC paper and degreased in acetone and isopropyl alcohol before being exposed. Both panels, sealed by tape on the backsides, were mounted by Plexiglas fixtures onto the exposure rack. The Plexiglas fixtures were equipped with inclined gutters into which the runoff water from each panel was collected and transported to containers (polyethylene). The containers were changed about 20 times a year.

Laboratory Exposure

Precipitation was simulated by artificial rain (composition, see Table 1) in a rain chamber. Previous studies have proven that the actual artificial rain fairly well simulates natural rain in central and southern parts of Sweden [*7-8*]. The rain chamber includes compressed filtrated pure air and a peristaltic pump, by means of which artificial rain was generated in the following intensity ranges: <1 mm/h (drizzle), 3-8 mm/h (light rain), and approximately 20 mm/h (moderate rain). The acidity of the rain was adjusted from pH 4.3 to pH 4.8 and to pH 3.8 using additions of sodium hydroxide (NaOH) and perchloric acid ($HClO_4$), respectively. In addition, the effect of varying length of dry periods in ambient laboratory air (one day, one week and one month) in-between rain events was investigated at a fixed pH (4.3) and intensity (~5 mm/h). Zinc panels of varying age (0.5 and 40 years) intended for the laboratory studies were sized 6*6 cm^2. The 0.5-year old panel is from here on denoted new. Duplicate panels were used for each parameter investigated. The fixture (45° inclination) and runoff collection was performed in a similar fashion as in the field exposures. All laboratory tests are compiled in Figure 1.

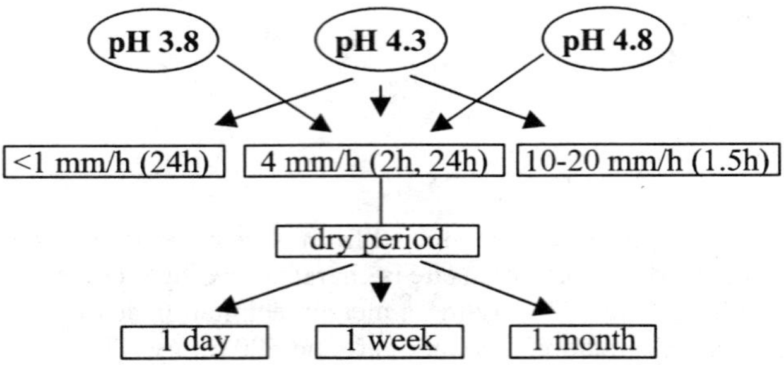

Figure 1 – *Laboratory investigations using artificial rain.*

Table 1 - *Composition of artificial rain.*

SO_4^{2-}-S mg/L	Cl^- mg/L	NO_3^--N mg/L	NH_4^+-N mg/L	Na^+ mg/L	K^+ mg/L	Mg^{2+} mg/L	Ca^{2+} mg/L	pH
1.17	0.36	0.56	0.56	0.23	0.12	0.12	0.20	4.3

Analysis of Runoff Water

The total zinc content in the runoff water was determined by means of atomic absorption spectrometry (AAS, Instrumental Laboratory model IL 551). Before being analyzed, the runoff water was acidified to pH < 3 using HNO_3. The detection limit for total Zn content is 40 ppb (μg/L). The yearly runoff rate and the runoff rate during each sampling period were evaluated and expressed as g/m^2, y.
A pH meter (PHM 92, Radiometer) was used for pH measurements of rain and runoff solutions.

Results and Discussion

Field Exposure

Annual runoff rates of zinc from one new and one 40-year old zinc panel obtained during the investigated four-year period are displayed in Figure 2 together with corresponding annual precipitation quantities (calculated from the sampling volume). The difference in runoff rate during individual years of exposure is only minor between the two panels. Similarities in runoff rates between new and aged zinc panels have previously been attributed to the bulky and porous nature of zinc corrosion products and the sensitivity to proton-induced dissolution of the phases in the corrosion products [*9*]. From this follows that the dissolution rate is mainly governed by the access of protons onto the dissolving corrosion product layer. These highly porous layers have a large effective surface area and, hence, a large ability to, both absorb significant amounts of water, and deposited acidifying atmospheric constituents. The new and the 40 year old zinc surface investigated in this study were covered by porous and bulky corrosion products primarily composed of basic zinc carbonate ($Zn_5(CO_3)_2(OH)_6$) with local areas of basic zinc sulfate ($Zn_4SO_4(OH)_6 \cdot 4H_2O$). More details on the effect of corrosion product composition on the runoff rate are given in Refs [*3,9-10*].

As a result of varying environmental conditions caused by, *e.g.*, differences in rainfall quantities and pH, the yearly runoff rate varied between 2.5 and 3.5 g/m^2 for both panels throughout the four-year exposure period. The figure shows no direct correlation between annual runoff rate and annual precipitation quantity. The runoff water pH was considerably higher (6.8±0.2) compared to the impinging rainwater pH (4.1-5.8), indicating a substantial buffering capacity of the zinc corrosion products.

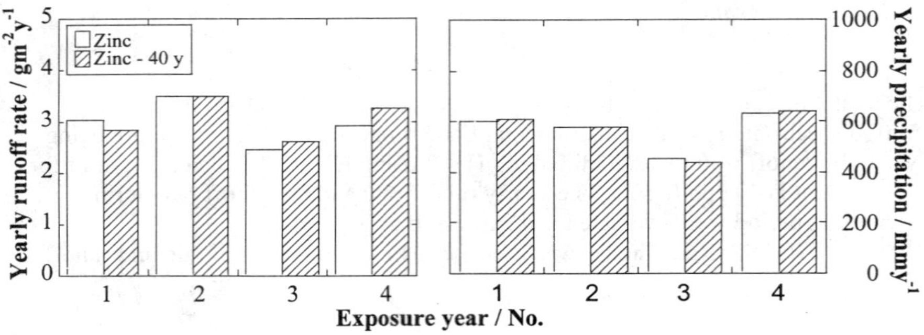

Figure 2 – *Yearly runoff rates of zinc from one new and one 40-year old panel and corresponding precipitation quantities during the 4-year urban field exposure.*

In agreement with previous findings on zinc [*3-4,9-11*] and copper [*9,12-13*], the long-term results during the whole four-year urban exposure clearly show that the accumulated zinc runoff quantity from both panels increased linearly with time (R^2=0.99; Figure 3, left).

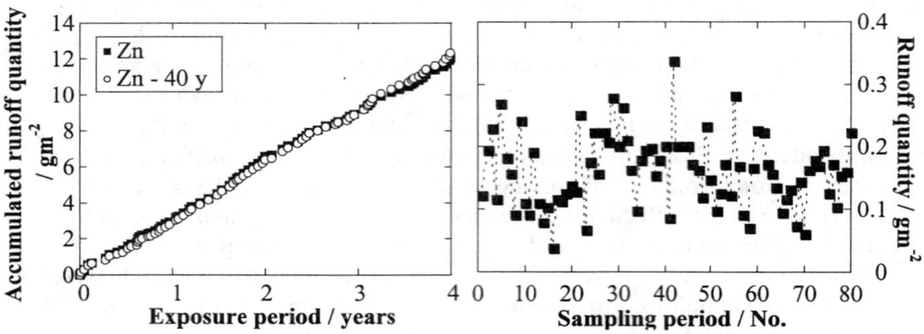

Figure 3 – *Accumulated zinc runoff quantity for the new and 40-year old zinc panel (left figure) and the runoff quantity of zinc during each sampling period from new zinc (right figure) during the 4-year urban exposure.*

The slope of the curves corresponds to an average runoff rate during the total exposure period of 3.1 g/m^2,y and 3.2 g/m^2,y for new and 40 year old zinc, respectively. Previous investigations have shown SO_2 to be a major factor determining the corrosion rate, and hence indirectly also the runoff rate [*2-3*]. As a result of low concentrations of SO_2 in Stockholm (~3 µg/m^3), the measured runoff rates are comparable to runoff rates obtained in rural atmospheres, and considerably lower compared to urban sites with higher SO_2 concentrations. This is in accordance with recent runoff rate data of zinc from zinc panels (inclination 30° from horizon, facing south or west) exposed at an urban (27 µg SO_2/m^3) and a rural site (< 1 µg SO_2/m^3) in the United States, showing runoff rates of 6.4 g/m^2,y and 3.5 g/m^2,y, respectively [*11*]. The runoff rate obtained at the rural test site agrees well with the results obtained in the present study and also with a previously obtained relationship between the runoff rate and the SO_2 concentration [*2-3,10*].

Even though the long-term trend (over years) is a relatively constant runoff rate (Figure 3, left), short-term studies (over days or weeks) showed considerable variation between different sampling periods, as illustrated for new zinc in Figure 3, right. Each sampling period is of course unique and characterized by, *e.g.*, the actual rainfall quantity and frequency as well as the prevailing environmental conditions in-between rain events. However, even at a given collected rain water volume, the zinc runoff quantity showed considerable variations (Figure 4). The figure illustrates the situation for sampling periods characterized by a collected rain volume (45° inclination) in the interval 20±1 L/m^2. Some of the sampling periods were preceded by long and dry periods and a subsequent single high-intensity rain event, whereas other periods included many low-intensity rain events or a few medium-intensity rain events. From Figure 4 is obvious that other parameters, besides rainfall volume, governs the runoff quantity. The effect of precipitation parameters, such as pH, intensity and volume and length of dry periods in-between rain events, was therefore investigated in separate laboratory investigations.

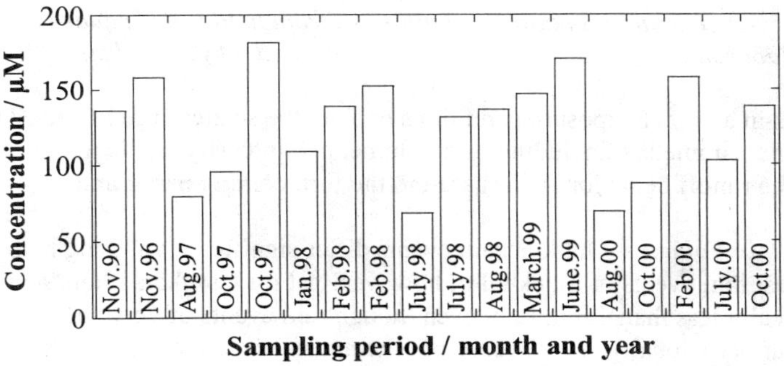

Figure 4 - *Variation in zinc runoff concentration of different sampling periods from a new zinc panel with a collected rain volume of 20±1 L/m^2. Sampling month and year is given in each stack.*

Laboratory Exposure

Resemblance between Outdoor Conditions and the Laboratory Simulation – Previous investigations of copper have shown that artificial rain used in the laboratory, and characterized by realistic rain volumes, intensities and composition, successfully can be used to simulate runoff rates in natural outdoor exposure conditions [*5,7*]. Because outdoor conditions only provide information on the combined effect of a number of influencing factors, the use of artificial rain permits the study of single precipitation parameters and their influence on the runoff rate of a metal. The resemblance in runoff rate data between field and laboratory results is displayed in Figure 5. Here, the runoff rate from each sampling period during the natural outdoor exposure of new and 40 year old zinc (filled squares) has been plotted together with runoff rates obtained during selected artificial rain exposures with varying intensity, amount, pH and dry periods (circles).

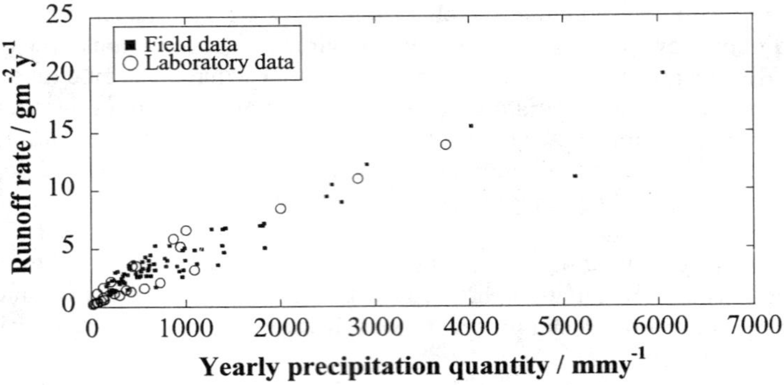

Figure 5 - *A comparison between runoff rates from field (filled squares) and laboratory (circles) investigations of new and 40 year old zinc.*

Variations in ambient exposure conditions explain the scatter in field data. The effect of single parameters, including rain volume, pH, intensity and length of dry periods, on the runoff behavior is discussed in the forthcoming paragraphs.

Effect of Rain Volume – Outdoor rain events differ significantly in length and intensity. Most rain events in Stockholm, however, last for less than 1 hour and produce typically less than 4 mm rain, even though rain events of very short or longer duration occur occasionally [*14*]. Because of the numerous rain events of short duration and low volume, the effect of the first water portion flushing the surface after a dry period has a marked influence on the metal runoff rate. Figure 6 illustrates the effect of rainfall volume with data obtained at fixed pH (4.3) and intensity (5 mm/h) of zinc panels with a corrosion product layer primarily composed of $Zn_5(CO_3)_2(OH)_6$. It should be stressed that all laboratory investigations of runoff kinetics, based on single rain events, showed analogous results. Figure 6 exhibits a high zinc concentration in the runoff from rain events with only a small volume (1.1 L/m^2). With increasing rain volume the zinc concentration in the runoff decreases until it reaches a steady-state region with a more or less constant zinc concentration. The so-called first flush effect has previously been described in the literature during runoff from metal roofs [*7*] and during urban stormwater runoff from buildings [*15-17*].

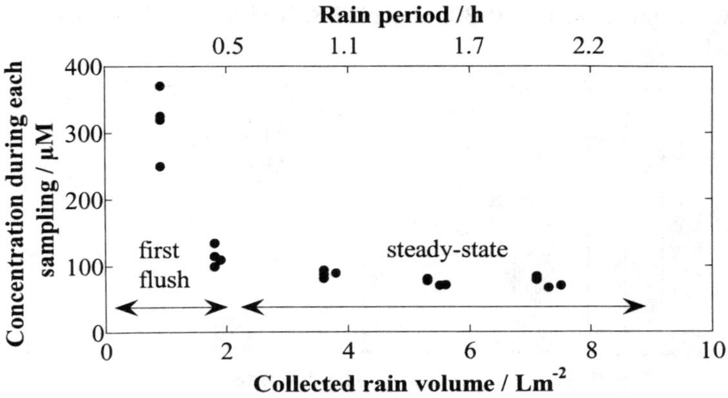

Figure 6 – *Zinc runoff concentration for each sampling as a function of rain volume during single rain episodes (pH 4.3 and intensity 5 mm/h) from 40-year old zinc panels.*

The zinc concentration in the steady-state region was approximately 75±5 µM, a concentration that remained the same also up to 18 hours (~70 L/m^2) of continuous rain (not displayed in the figure). The first flush effect has previously been explained by the capacity of corrosion products to absorb and retain water and also to retain gaseous and particulate pollutants deposited during dry periods between rain events. During the initial stage of a subsequent rain event, the concentration of dissolution-stimulating species is very high which causes an accelerated runoff rate [*5,7,18*].

The effect of a few precipitation parameters on the magnitude of the first flush effect is presented in the following paragraphs. When comparing effects of different precipitation parameters on the runoff rate, it is necessary to use data from the steady-state region of a rain event rather than from its first flush region.

Effect of Rainfall pH – Any metal surface exposed to atmospheric conditions experiences wetting and drying cycles during which the pH of the water layer changes considerably, and hence also the extent of corrosion attack. Similarly, any metal surface experiences rainwater of varying pH. The chemical composition and the acidity of precipitation usually exhibit a large variation between different events. A substantial variation in composition also occurs within a single rain event with a decrease in concentration of ionic species, such as H^+, SO_4^{2-} and NO_3^-, during the initial portion of a rainfall as a result of scavenging [*19*]. In the following, the effect of rain pH will be illustrated during single rain events of constant composition (see Table 1) and intensity (5 mm/h). Figure 7 shows that the dissolution of corrosion products is markedly influenced by rain pH, both during the first flush and during the subsequent rainfall portion corresponding to the steady-state region. The figure shows that an increased proton activity lowers the stability of zinc corrosion products, hence increasing the released zinc concentration. The average zinc concentration during steady-state was approximately 130±30μM at pH 3.8, 70±15μM at pH 4.3 and 50±5μM at pH 4.8.

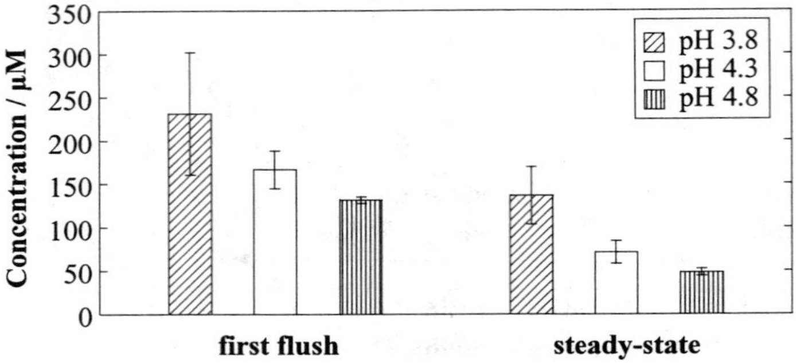

Figure 7 – *Zinc runoff concentration from new zinc during first flush (1.1 L/m²) and steady-state regions of single rain events of varying pH at constant intensity (5 mm/h).*

A correlation could be discerned between the concentration of dissolved zinc and hydrogen ions in the steady-state region, as displayed in Figure 8. In this figure all dissolved zinc was assumed to be present as Zn^{2+}-ions. This assumption was justified by previous investigations by the authors showing that 95-99% of the total zinc release in the runoff (pH<6) is present as hydrated Zn^{2+}-ions immediately after leaving a roof surface [*20*]. The calculated regression between Zn^{2+}-ions and H^+-ions, expressed as

$$[Zn^{2+}] = 0.55[H^+] + 49\ \mu M$$

shows a correlation coefficient (R^2) of 0.89. This regression is in excellent agreement with previous data based on short-term exposures of zinc in Washington DC ($[Zn^{2+}] = 0.53[H^+] + 49\ \mu M$) [*21*] and laboratory investigations using deionized water and acidic species ($[Zn^{2+}] = 0.48[H^+] + 59\ \mu M$) [*22*]. A ratio of 0.5 between the concentration of Zn^{2+} and H^+ implies that two hydrogen ions are required to dissolve one zinc ion into solution. The calculated regression curve intercepts at 49 μM, a concentration that agrees very well with the solubility of zinc corrosion products in deionized water (48 μM) in equilibrium with CO_2 at a pH of 5.6, as described by Cramer et al. [*9*]. Both regressions are based on rain conditions. In a wider perspective these results are also consistent with the strong correlation found between the runoff rate and the SO_2-concentration [*2-3*].

The higher zinc concentration observed during first flush release can also be explained by solubility arguments. During the dry period neutral zinc salts with high solubility constants, *e.g.* $ZnSO_4$ or $Zn(NO_3)_2$, are frequently formed through dry deposition of, *e.g.*, sulfur- or nitrogen containing species [*11,18*]. These are easily dissolved during the first flush release, whereby less soluble zinc salts are formed, including zinc hydroxysulfates and zinc hydroxycarbonates. Their solubility properties will govern the dissolution rate during steady-state release.

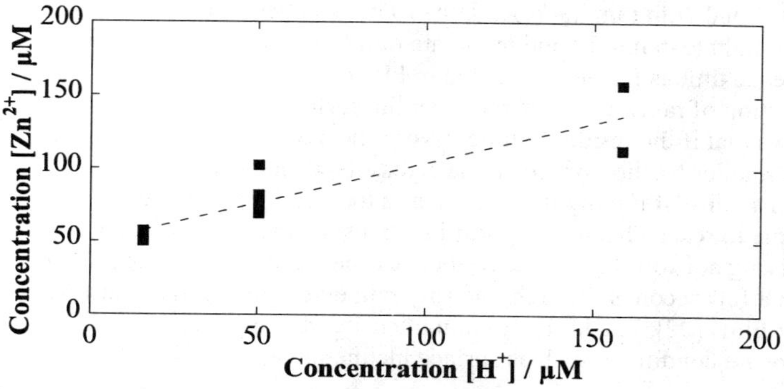

Figure 8 – *Correlation between zinc ion concentration and hydrogen ion concentration during steady-state release from new zinc.*

Effect of Rainfall Intensity – A sufficient rain volume is required during a rain event to deliver dissolution-stimulating species to the zinc surface and dissolved species to be flushed from the surface. The time-period during which the rainwater is in contact with the surface, the residence time, is one major factor that determines the rate at which dissolved species of corrosion products are dissolved and released from the surface. The residence time is primarily governed by the rain intensity. At a given rain volume, a low-intensity rain event results in an increased surface residence time period during which a larger portion of the corrosion products can be dissolved. This is illustrated in Figure 9 from a single rain event of pH 4.3 impinging on a 40-year old zinc panel. The figure shows a significant decrease in zinc concentration with increased rain intensity, both during first flush (2.2 L/m^2) and during steady-state conditions.

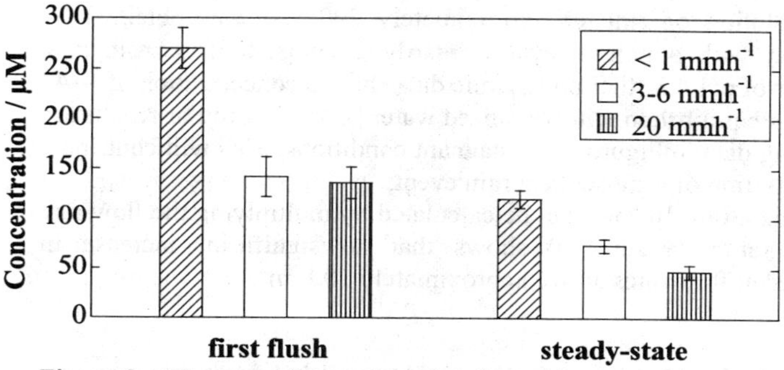

Figure 9 – *Zinc runoff concentration from 40-years old zinc sheet during first flush (2.2 L/m^2) and steady-state release from single rain events (pH 4.3) of constant rain volume and varying intensity.*

The average zinc concentration during steady-state was approximately 120±9µM at an intensity of less than 1 mm/h, 73±7µM at an intensity of between 3 and 6 mm/h, and 47±7µM at an intensity of approximately 20 mm/h. The time period to reach the steady-state condition varied with rain intensity from 10 hours, over 1 hour to 0.5 hour for drizzle (< 1 mm/h), light (3-6 mm/h) and moderate rain (20 mm/h), respectively.

The effect of residence time is further illustrated in Figure 10 showing the zinc concentration as a function of rainwater flow rate over the surface of a 40-year zinc panel. The figure shows that if the residence time is very short at a given rain volume, i.e., a fast flow rate, a smaller fraction of the patina is dissolved and *vice versa*. At a given composition and acidity of impinging rain water, a too short residence time implies insufficient time to reach chemical equilibrium between the zinc corrosion product and the dissolving aqueous layer. The residence time has been estimated in the literature to vary from a few seconds for high intensity rain events to several thousand seconds for thin water films [*22*]. In addition to rain intensity, the residence time depends on other exposure conditions such as surface inclination and panel temperature.

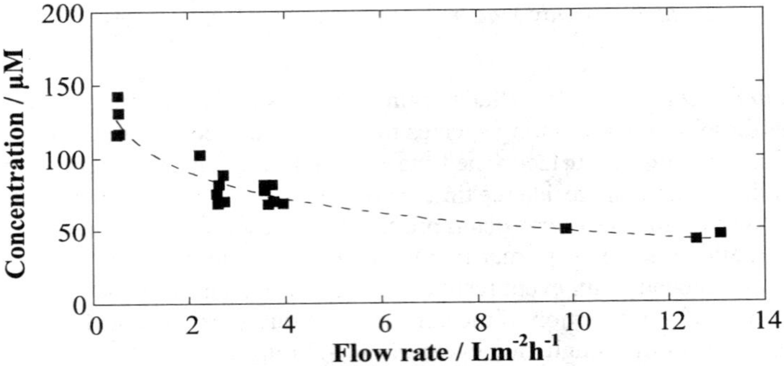

Figure 10 – *The concentration of released zinc ions in the steady-state region as a function of flow rate of impinging rainwater (pH 4.3) on aged zinc panels.*

A steady-state dissolution of zinc of approximately 200 µM was obtained by immersing a zinc sheet with a patina layer primarily composed of a basic zinc carbonate into rainwater of pH 4.3 (thermodynamic data show a concentration of ~140 µM for zinc carbonate in equilibrium with deionized water [*23-24*]). Similar results are obtained by extrapolating data in Figure 10 to stagnant conditions. This will contribute to the first flush concentration of a subsequent rain event.

The runoff rate during a rain episode can be calculated by multiplying the flow rate with the zinc concentration. Figure 10 shows that no significant increase in concentration is obtained at flow rates above approximately 10 L/m^2.

Effect of Dry Periods in-between Rainfall Events – Atmospheric corrosion is an on-going process that occurs even during relatively dry conditions outdoors in periods in-between rain events as a result of e.g., hygroscopic corrosion products and dry deposited species keeping water trapped within the corrosion patina. The result is an increasing residence time during which phases of the patina can be dissolved. This entrapment is facilitated by the porous and bulky nature of the corrosion patina formed on zinc and by cracks and defects in the patina formed during long periods of dry conditions. The effect of length of dry periods in-between rain events of constant intensity and pH (3-6 mm/h, pH4.3) on the zinc release is illustrated in Figure 11 for 40-year old zinc panels.

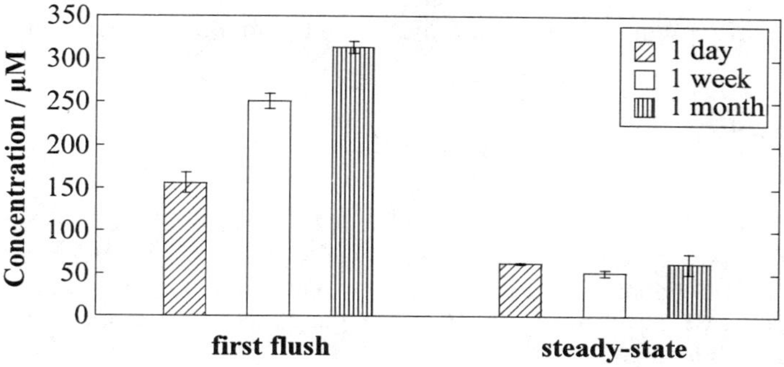

Figure 11 – *Zinc runoff concentration from 40- year old zinc panels during first flush (1.1 L/m^2) and steady-state of single rain events of constant pH and intensity (pH 4.3, 5 mm/h) and preceded by dry periods of varying length.*

The zinc concentration in the first flush region (1.1 L/m^2) is approximately twice as high after a 1-month dry period compared to a one-day dry period in-between two rain events. It is obvious that an increasing length of the dry period in-between two rain events increases the zinc concentration of the first flush but not in the steady-state region.

Conclusions

The present study is based on new and 40 year old zinc panels exposed to natural rain in an urban test site located in Stockholm, Sweden, for up to four years and to artificial rain obtained in the laboratory. The following conclusions were drawn.

- Good agreement was obtained between zinc runoff rates from both new and old zinc during natural rain and artificial rain exposure.
- The zinc concentration in runoff was high during the initial part of the rain volume with observed concentrations in the range 100-400 µM, depending on present exposure conditions. It decreased with extended rain volume until it reached steady-state values in the range 50-150 µM, again depending on exposure conditions. This so-called first flush effect has been attributed to deposited gaseous and particulate pollutants during periods in-between rain events causing an increased concentration of dissolution-stimulating species during the initial stage of the subsequent rain event.
- The zinc concentration in runoff increased with rain acidity (both during first flush and steady-state conditions), with decreased rain intensity (during first flush and

steady-state), and increased length of dry period preceding a rain event (during first flush).

- The annual runoff rate from new and old zinc obtained during different years varied between 2.5 and 3.5 g/m^2, with an average runoff rate throughout the total 4 year period of 3.1 g/m^2,y for new zinc and 3.2 g/m^2,y for old zinc, respectively.

Acknowledgments

The authors gratefully acknowledge the financial support from Union Minière, Belgium; Sollac Usinor Group, France; SSAB, Sweden; Sidmar N.V., Belgium; Cockerill Sambre, Belgium; Norzink A.S., Norway; Outokumpu Zink O.Y., Finland; Zinc Info Norden AB, Sweden; Wirtschaftsvereinigung Metallw e.V., Germany and Grillo Werke AG, Germany.

References

[1] Landner, L., and Lindeström, *Zinc in Society and in the Environment*, Swedish Environmental Research Group, Borås, ISBN 91-630-6871-0, 1998.

[2] Odnevall, I., Verbiest, P., He, W. and Leygraf, C., "The Influence of Patina Age and Pollutant Levels on the Runoff Rate of Zinc from Roofing Materials," *Corrosion Science*, Vol. 40, No. 11, 1998, pp. 1977-1982.

[3] Odnevall Wallinder, I., Verbiest, P., Janssen, C.R., and Leygraf, C., "Environmental Effects of Zinc Runoff from Roofing Materials of Different Age as a Result of Atmospheric Corrosion," *14th International Corrosion Congress*, Cape Town, 1999.

[4] Odnevall Wallinder I., Verbiest, P., He, W., Leygraf, C., "Effects of Exposure Direction and Inclination on the Runoff Rates of Zinc and Copper Roofs," *Corrosion Science*, Vol. 42, 2000, pp. 1471-1487.

[5] He, W., *Licentiate Thesis*, Royal Institute of Technology, Stockholm, Sweden, ISBN 91-7170-527-9, 2000.

[6] Odnevall Wallinder, I., and Leygraf, C., "Environmental Effects of Metals Induced by Atmospheric Corrosion," *ASTM STP 1421*, A.B. Smith and C.D. Jones, Eds., American Society for Testing and Materials, West Conshohocken, PA, 2002.

[7] He, W., Odnevall Wallinder I., Leygraf, C., "A Laboratory Study of Copper and Zinc Runoff during First Flush and Steady State Conditions," *Corrosion Science*, Vol. 43, 2001, pp. 127-146.

[8] Granat, L., "Luft-och Nederbördskemiska Stationära Nätet inom PMK", *Naturvårdsverket Rapport 3942*, 1990. (in Swedish)

[9] Cramer, S. D., McDonald, L. G., Spence, J. W., "Effects of Acidic Deposition on the Corrosion of Zinc and Copper," *12 th International Corrosion Congress*, Houston, 1993.

[10] He, W., Odnevall Wallinder I., Leygraf, C., "A Comparison between Corrosion Rates and Runoff Rates from New and Aged Copper and Zinc as Roofing Material," *Water, Air and Soil Pollution*, in press.

[11] Cramer, S. D., Matthes, S. A., Holcomb, G. R., Covino, B. S. Jr., and Bullard, S. J., "Precipitation Runoff and Atmospheric Corrosion," *Proceedings, Corrosion,* Orlando Florida USA, March, 2000, paper No. 00452.

[12] Odnevall Wallinder, I. And Leygraf, C., "A Study of Copper Runoff in an Urban Atmosphere," *Corrosion Science*, Vol. 39, No. 12, 1997, pp. 2039-2052.

[13] Odnevall Wallinder, I. and Leygraf, C., "Corrosion Rates and Runoff Rates of Copper – A Comparison,"*Proceedings, 14th International Corrosion Conference*, Cape Town, South Africa, 26 Sep.-1 Oct. 1999.

[14] Harald Grip och Allan Rodhe, *Vattnets Väg från Regn till Bäck*, Lagerblands Tryckeri, Karlshamn 1985. (in Swedish)

[15] Yaziz, M. I., Gunting, H., Sapari, N. and Ghazali, A.W., "Variations in Rainwater Quality from Roof Catchments," *Water Research*, Vol. 23, No. 6, 1989, pp. 761-765.

[16] Good, J. C., "Roof Runoff as a Diffuse Source of Metals and Aquatic Toxicity in Storm Water," *Water Science and Technology*, Vol. 28, No. 3-5, 1993, pp. 317-321.

[17] Quek, U. and Förster, J., "Trace Metals in Roof Runoff," *Water, Air and Soil Pollution*, Vol. 68, 1993, pp. 373-389.

[18] Zheng, Y., Wang, Z., and Zhang, F., "Influence of Acid Deposition on Atmospheric Corrosion of Zinc," *Journal of Environmental Science* (China), Vol. 3, No. 2, 1991, pp. 53-60.

[19] Graedel, T.E., "Corrosion-Related Aspects of the Chemistry and Frequency of Occurrence of Precipitation," *Journal of the Electrochemical Society*, Vol. 133, No. 12, 1986, pp. 2476-2482.

[20] Karlén, C., Odnevall Wallinder, I., Heijerick, D., Leygraf, C. and Janssen, C. R., "Runoff Rates and Ecotoxicity of Zinc Induced by Atmospheric Corrosion," *The Science of the Total Environment*, in press.

[21] Cramer, S. D., Carter, J. P., Linstrom, P. J., and Flinn D. R., "Environmental Effects in the Atmospheric Corrosion of Zinc," *ASTM STP 965,* S.W. Dean and T. S. Lee, Eds., American Society for Testing and Materials, Philadelphia, 1988, pp. 229-247.

[22] Stiles, D. C. and Edney, E.O., "Dissolution of Zinc into Thin Aqueous Films as a Function of Residence Time, Acidic Species, and pH," *Corrosion* Vol. 45, No. 11, 1989, pp. 896-901.

[23] Spence, J. W. and Haynie, F. H., "Derivation of a Damage Function for Galvanized Steel Structures: Corrosion Kinetics and Thermodynamic Considerations," *Corrosion Testing and Evaluation: Silver Anniversary Volume, ASTM STP 1000*, R. Baboian and S.W. Dean, Eds., American Society for Testing and Materials, Philadelphia, 1990, pp. 208-224.

[24] Spence, J. W., Haynie, F. W., Lipfert, F. W., Cramer, S. D., and McDonald, L. G., "Atmospheric Corrosion Model for Galvanized Steel Structures," *Corrosion* Vol. 48, No. 12, 1992, pp. 1009-1019.

Inger Odnevall Wallinder,[1] Tapio Korpinen,[2] Rolf Sundberg,[3] and Christofer Leygraf[1]

Atmospheric Corrosion of Naturally and Pre-Patinated Copper Roofs in Singapore and Stockholm – Runoff Rates and Corrosion Product Formation

---

Reference: Odnevall Wallinder, I., Korpinen, T., Sundberg, R., and Leygraf, C., **"Atmospheric Corrosion of Naturally and Pre-Patinated Copper Roofs in Singapore and Stockholm – Runoff Rates and Corrosion Product Formation,"** *Outdoor Atmospheric Corrosion, ASTM STP 1421*, H. E. Townsend, Ed., American Society for Testing and Materials International, West Conshohocken, PA, 2002.

Abstract: Runoff rates of copper have been determined during a one-year field exposure in the city of Singapore, Republic of Singapore, and a four-year exposure in the city of Stockholm, Sweden, from brown and green pre-patinated copper sheet and naturally patinated copper sheet being fresh and 130 years old. Except for being geographically distant, the urban sites are characterized by large differences in climatic and environmental conditions. Measured annual runoff rates from fresh and brown pre-patinated are 1.1-1.6 g/m^2 and 5.5-5.7 g/m^2, in Stockholm and Singapore, respectively. Naturally aged copper sheet being 130 years old and green pre-patinated copper sheet show slightly higher (1.6-2.3 g/m^2), but comparable runoff rates in Stockholm. In Singapore, runoff rates from green pre-patinated copper sheet are 8.4-8.8 g/m^2. Differences in runoff rates between Singapore and Stockholm are primarily a result of more abundant rainfall, being five to eight times larger in Singapore. Comparable runoff rates between fresh and brown-patinated copper sheet and between green naturally patinated and green pre-patinated copper sheet at each site are related to similarities in patina morphology and composition.

Keywords: Copper, pre-patination, atmospheric corrosion, patina, runoff rate, corrosion products, cupric ion

---

[1]Assistant Professor and Professor, respectively, Department of Materials Science and Engineering, Division of Corrosion Science, Royal Institute of Technology (KTH), Dr. Kristinas v. 51, SE-100 44 Stockholm, Sweden.

[2] Project Manager, Outokumpu Copper Partner AB, SE-721 09 Västerås, Sweden.

[3] Consultant, Scandinavian Copper Development Association, SE-721 09 Västerås, Sweden.

Introduction

A large number of materials, including, e.g., tiles, pre-painted steel, copper, stainless steel and various zinc-based products, such as pre-painted galvanized steel and coated zinc-aluminum alloys, are often used as roofing materials on buildings. The selection of a material depends on a number of factors such as tradition and history, environmental conditions, weight, life-length, durability, cost and maintenance. From an architectural point of view, colouring and forming properties also have great importance for the selection of a suitable material.

The architectural use of copper on buildings dates back to the thirteenth century in Scandinavia and many medieval churches in Europe used copper roofs. Today, copper is still a common material for roofs, wall-claddings, down-pipes and gutters, primarily in Scandinavia, Germany, Italy, Switzerland and Austria. In other European countries and worldwide, the use of copper for these applications is less common.

During atmospheric exposure of a newly laid copper roof [*1-5*], a patina is developed on the copper surface changing its appearance from the metallic lustre to a brownish finish within a short period of time (days to weeks, depending on exposure conditions). A greenish patina layer will eventually cover this layer within months or years of exposure as a result of the interaction with air pollutants such as sulfur dioxide (SO_2) and chloride (Cl^-). The main constituents of the patina formed in urban atmospheres are the copper(I)oxide, cuprite (Cu_2O), which forms initially, usually combined with a mixture of basic copper sulfates, e.g., posjnakite ($Cu_4SO_4(OH)_6xH_2O$) and brochantite ($Cu_4SO_4(OH)_6$) appearing in the green patina. In heavily SO_2-polluted urban environments, other basic copper sulfates, such as antlerite ($Cu_3SO_4(OH)_4$) and $Cu_{2.5}(OH)_3SO_4x2H_2O$ have been detected in the patina. In chloride-rich environments, e.g., marine atmospheres, the cuprite layer is covered by a green-bluish layer of a basic copper chloride, atacamite ($Cu_2Cl(OH)_3$). Depending on the air pollutant level, the time-period to develop the green patina layer is typically many years or decades. Pre-patinated copper sheets have therefore been developed to be used on new roofs and to repair old roofs on ancient buildings to avoid an initial period of a new-looking appearance [*6-8*]. A few examples of re-laid and newly installed pre-patinated copper roofs are on the gothic cathedral of Linköping in Sweden, the Helsinki Cathedral in Finland and the Warsaw library in Poland. Few studies exist in the literature on patina formation and effects of mechanical wear from sliding snow and frost on pre-patinated copper sheet [*8*].

Dispersion of metals from buildings and other outdoor constructions has been subject for discussion in environmental and legislative work in Europe and the US during the last decade. As a result, a number of metal regulations and restrictions have been implemented on, e.g., copper. A number of studies have therefore been implemented to fill the gap of knowledge and to provide accurate and recent data on metal dispersion rates. Easily soluble and poorly adhesive phases of the patina and corrosion products being dissolved during atmospheric exposure can, during a rain-event, either be removed from the surface in the runoff water, or re-precipitate to the adhering corrosion product layer. Any rain event results in a high copper concentration in the first rainwater volume impinging the surface, the first flush, after which the concentration decreases and becomes relatively constant. Previous investigations have shown the magnitude of the first flush, and hence the total released copper quantity, to depend on the corrosion layer morphology and the prevailing environmental conditions such as air pollutant levels, dry and wet deposition, length of dry periods in-between rain events, rain amount, pH and intensity [*9-10*]. The runoff rate from roofs of different directions and inclination has shown to be largely influenced by the prevailing wind conditions determining the rainwater volume impinging each surface and deposition of corrosion stimulating atmospheric constituents [*11*].

The aim of this paper is to supply accurate runoff rate data of copper from naturally and pre-patinated copper sheet exposed in two significantly different environmental conditions, a tropical and maritime site in the south-Asian region (Singapore) and a site in the northern part of the tempered climatic zone having a mixture of the maritime Atlantic conditions and the continental climate of the inner parts of east Europe (Stockholm). The results will be interpreted and discussed in view of the first flush phenomena, the fraction of cupric ions in the runoff water and formation of corrosion products in the patina.

Experimental

Materials and Exposure Conditions

The following panels were freely exposed during one year in Singapore and during four and a half and two years in Stockholm, see Table I.

Table 1-*Material and field exposure duration.*

Material	**Abbreviation**	**Singapore** *Exposure period*	**Stockholm** *Exposure period*
Fresh copper sheet	*Cu*	April-98 - May-99	April-96 - Oct.00
Nordic Brown	*NB*	April-98 - May-99	Nov-98 - Oct.-00
Nordic Green	*NG*	April-98 - May-99	April-96 - Oct.00
Copper sheet - 130 years	*old*	-	April-96 - Oct.00

Copper sheet (99.98% Cu, 0.02% P) and pre-patinated copper sheets (Nordic Brown and Nordic Green) were supplied by Outokumpu Copper Partner AB, Sweden. The brown-patinated copper sheet is chemically oxidized using sodium chlorite at 70 °C during 2-24 hours to obtain an adherent copper oxide layer with a thickness of typically 1 μm. The composition of the patina is predominantly copper(I)oxide (cuprite-Cu_2O) and copper(II)oxide (tenorite-CuO). The same chemical oxidation is also applied on the green-patinated sheet followed by a subsequent application of a mixture of basic copper salts to obtain the greenish layer formed on top of the oxidized brown surface. The main crystalline constituents of the initial green patina are a basic copper nitrate (gerhardite–$Cu_2(OH)_3NO_3$) and a basic copper sulfate (posnjakite – $Cu_4SO_4(OH)_6xH_2O$). The total thickness is typically 18 μm. The 130 year old copper panel, exposed in Stockholm, originates from the church of Riddarholmen in Stockholm, Sweden. The patina is predominantly composed of an inner layer of cuprite and an outer green layer of a basic copper sulfate (brochantite-$Cu_4SO_4(OH)_6$).

No pre-treatments of the panels were made before exposure except for the fresh copper sheet and the brown pre-patinated sheet, being degreased in acetone and iso-propylic alcohol.

Single-sided copper sheet panels (300 cm^2) were attached with Teflon screws onto Plexiglas fixtures exposed at an angle of 45 degrees from the horizon on test racks facing south. Duplicate panels of each material were exposed in Singapore and single panels in Stockholm. The backside and the edges of each panel were sealed with tape, to prevent from corrosion. In parallel, Plexiglas fixtures were used as inert surfaces to collect rainwater to determine pH and background deposition of copper. Impinging rainwater was collected and transported via Plexiglas gutters and down-pipes of silicon rubber into acid-cleaned flasks. Samplings were typically made every second week, depending on weather conditions. All flasks collected in Singapore were acidified on site and transported to Sweden for analyses.

The test sites represent very different environmental conditions. As a result of the geographical location of Singapore, the climate is characterized by small changes in temperature, high relative humidity and abundant rainfall, typically every day. In comparison, the urban test site of Stockholm, Sweden, shows a large span in temperature and relative humidity. The frequency and amount of rainfall are also much lower in Stockholm compared to Singapore. A comparison between the environmental conditions at the test sites is exemplified in Table 2 for the period April 1998 to May 1999. Environmental data (T (temperature), RH (relative humidity), SO_2, NO_2, PM10 (aerosol particles < 10 μm)) are compiled from data supplied by the Stockholm Environment and Health Protection Administration and the Singapore Productivity and Standards Board, respectively. Details on pH and rainfall quantities are given in Results and Discussion.

Table 2 - *Environmental conditions during a one-year exposure in Singapore and Stockholm – a comparison.*

Test site	T / °C		RH / %			SO_2 / μgm^{-3}	NO_2 / μgm^{-3}	PM10 / μgm^{-3}
April-98 to May-99	min	max	min	max	mean			
Singapore	23	26	60	100	84	22	36	34
Stockholm	-17	31	26	99	77	3	44	36

The test sites are located on top of a seven-storey building within the campus area of Stockholm and on top of a two-storey school building in Singapore (Figure 1).

Figure 1-*Test sites in Stockholm (left) and Singapore (right).*

Analyses

The pH of the collected runoff water (Stockholm) and rainwater (Singapore and Stockholm) was measured after each sampling period. The total copper concentration was measured with flame atomic absorption spectroscopy (AAS) after acidifying the runoff water with HNO_3 to a pH less than 3. The copper concentration (typically between 1 and 10 mg/L) was determined as the average value of 15 measurements. The detection limit of the current set-up is 0.02 mg/L.

An ion-selective electrode (ORION Model 9629 ionplus™ Series) was used to determine the cupric ion content in non-acidified runoff water at the Stockholm site. The results are given as the average concentration from five independent measurements.

The composition of the corrosion product layer formed in Singapore was investigated by means of a combination of scanning electron microscopy with energy dispersive spectroscopy, SEM/EDS (morphology and semi-quantitative elemental composition), X-ray photoelectron spectroscopy, XPS (elemental and chemical state information), glow discharge optical emission spectroscopy, GDOES (elemental depth profiling), and X-ray diffraction, XRD (phase identification of crystalline phases). Analyses were made using the following instruments: SEM/EDS (JEOL 820 microscope equipped with a Tracor Series II micro-analytical EDS-system); XPS (Kratos Axis HS Photoelectron spectrometer equipped with an Al Kα-anode (1486.6 eV). Wide scans and detailed scans were made on Cu 2p, C 1s, O 1s, S 2p, N 1s.); GDOES (GDS 750A emission spectrometer), XRD (Siemens D500 X-ray diffractometer. Diffractograms were recorded on panel surfaces from $2\theta=6°$ to 70° using an interval of 0.050° and a counting time of 1s/step).

Results and Discussion

In the following, runoff rates will be displayed as the total copper quantity being released from surfaces positioned 45 degrees from horizon, facing south. From a corrosion point of view, this position is considered to reflect the most severe condition [*12*], which emphasizes the importance of using a weighed runoff rate for real buildings of varying geometry. In addition, each sampling of runoff water reflects an average measure of the copper content from a few or several rain events, depending on the rainfall quantity, and hence constitutes an averaged effect of the first flush. Note that the data are collected when the rainwater leaves the metal surface, without considering the effect of changes in chemical speciation or dilution when entering the environment. Because of similarities in patina composition and runoff rates, the following results are shown first for fresh and brown-patinated copper sheet, and subsequently for aged and green pre-patinated copper sheet.

Atmospheric Corrosion and Runoff Rates of Fresh Copper Sheet and Brown Pre-Patinated Copper Sheet

Figure 2 displays the average annual runoff rate of copper being released from the corrosion layer of naturally patinated fresh copper sheet (Cu) and brown pre-patinated copper sheet (NB) exposed in Singapore and Stockholm. The error bars reflect the highest and lowest yearly runoff rate between duplicate panels during the one-year exposure in Singapore and variations in yearly runoff rates from single panels during the four and a half-year and two-year exposures in Stockholm for fresh copper sheet and brown pre-patinated sheet, respectively.

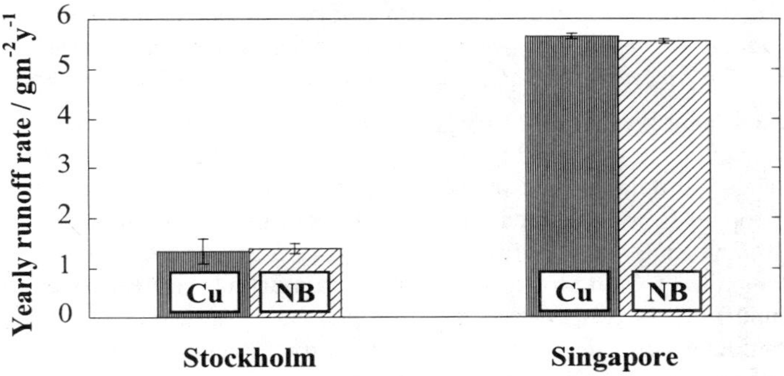

Figure 2 - *Yearly runoff rates of copper from fresh copper sheet (Cu) and brown pre-patinated copper sheet (NB) exposed during one year in Singapore and during four and a half and two years of exposure in Stockholm, respectively.*

Even though the figure shows substantial differences in yearly runoff rates between Stockholm and Singapore, very small differences can be seen between the fresh copper sheet and the brown pre-patinated copper sheet at each individual test site. The runoff rate varies between 1.1 and 1.6 g/m^2,y and between 1.3 and 1.5 g/m^2,y in Stockholm for the fresh copper sheet and the brown pre-patinated copper sheet, respectively. Corresponding runoff rates in Singapore vary between 5.6 and 5.7 g/m^2,y for the fresh copper sheet and between 5.5 and 5.6 g/m^2,y for the brown pre-patinated copper sheet. Variations in runoff rates at each site are primarily connected to different rainfall quantities during each year of exposure. Higher runoff rates in Singapore compared to Stockholm can partly be explained by significantly higher precipitation quantities as a result of high-intensity daily rain events, and partly by notable differences in environmental conditions, such as higher humidity and concentration of corrosion stimulating atmospheric constituents (*c.f.* Table 2). The total yearly precipitation quantity is five to eight times higher in Singapore compared to Stockholm (Figure 3). In Stockholm, the rain frequency is much lower, typically 100 days per year with a rainfall quantity of more than 1 mm.

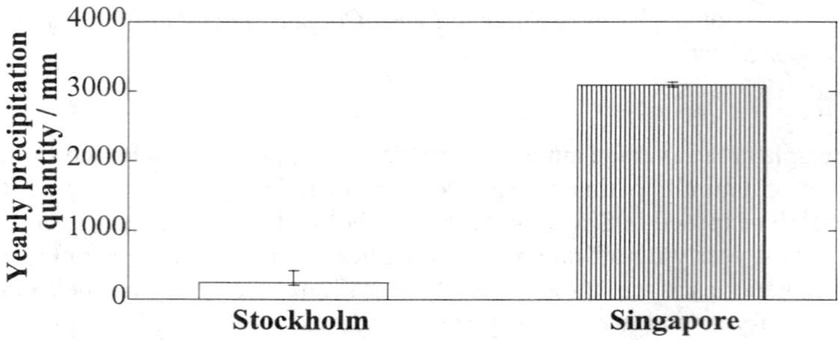

Figure 3 – *Total yearly precipitation quantities in Stockholm and Singapore. The error bars reflect the highest and the lowest rain quantity, impinging duplicate panels during the one-year exposure in Singapore and between different years during the four and a half-year exposure in Stockholm.*

Higher air pollutant levels result in higher concentration of corrosive species in precipitation as well as in increased deposition of corrosive stimulants into the surface water layer. These stimulants interact with the copper surface and thereby enhance the dissolution of copper from the patina layer. An effect of this is seen in the rain acidity showing a pH between 3.9 and 4.9 (median value 4.4) in Singapore, and between 4.2 and 6.4 (median value 5.0) in Stockholm. A lower rain pH has earlier proven to increase the runoff rate of copper from naturally patinated copper panels [*9-10*]. Higher pollutant levels also result in larger deposition rate of corrosive species onto the copper surface, which enhances the formation rate of easily soluble phases of the patina. During a subsequent rain event, these easily soluble corrosion products can easily be flushed from the surface. In agreement with previous results, the released copper quantity in the collected runoff water from different sampling periods varies for all panels at both test sites as a result of varying rain quantity and variations in environmental conditions. These variations are illustrated in Figure 4 for naturally patinated fresh copper sheet and the brown pre-patinated sheet of each sampling period during the one-year exposure in Singapore. The figure also shows a striking resemblance in released copper quantities between the two copper sheet panels. Similar results were obtained at the Stockholm test site.

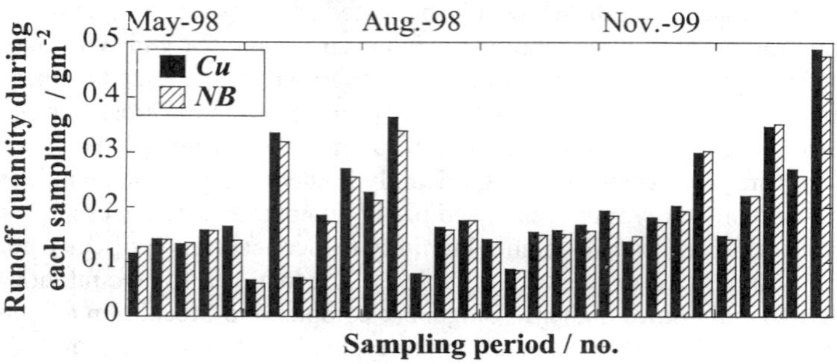

Figure 4 - *Copper runoff quantity from fresh and brown pre-patinated copper sheet in each sampling period during the one-year exposure in Singapore.*

Previous investigations have shown the released metal quantity from a copper surface to vary significantly between different sampling periods, even at comparable precipitation quantities. This has been shown to be the result of a combination of environmental conditions prior to a rain event and the characteristics of the patina layer, hence determining the magnitude of the first flush [*9*]. This effect cannot directly be seen in Figure 4 since each bar compiles data from several rain events during a sampling period. However, an indirect effect can be discerned when plotting the total copper concentration as a function of collected rain quantity for all sampling periods from fresh and brown-patinated copper at both test sites (Figure 5).

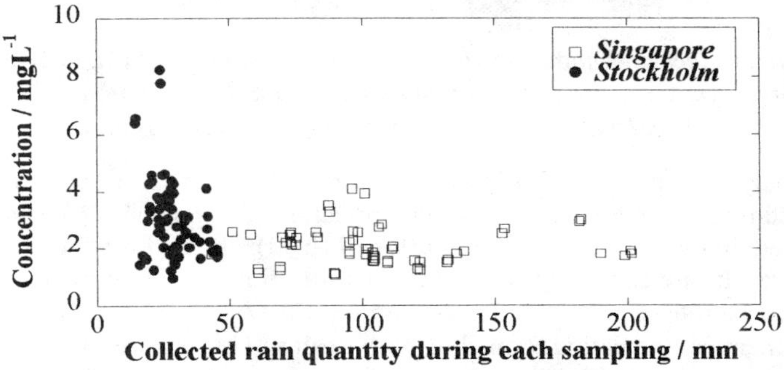

Figure 5 - *Copper concentration as a function of collected rain quantity during each sampling period from fresh and brown pre-patinated copper during the one-year exposure in Singapore and the two-year exposure in Stockholm.*

The data show that sampling periods with a low collected rain quantity result in a high copper concentration in the runoff water (Stockholm) compared to sampling periods with a high collected rain quantity (Singapore). A similar dependence is seen during a rain event, with high copper concentrations in the first flush volume, followed by a more constant concentration with increasing rain volume.

An explanation for the close resemblance in runoff rates between fresh and brown-patinated copper sheet is evident when investigating the formation and composition of the patina. The development of the patina on the fresh copper sheet exposed in Singapore followed the same pattern as has previously been described for copper sheet exposed in Stockholm and other urban atmospheres [*1*]. As a result of a much more humid climate and higher levels of SO_2 in Singapore ($\approx$20 $\mu g/m^3$) compared to Stockholm ($\approx$3 $\mu g/m^3$), *c.f.* Table I, a more rapid formation of a cuprous oxide layer and the subsequent formation of basic copper sulfates was seen on the fresh copper sheet exposed in Singapore compared to Stockholm.

Already within a few days, the whole surface was completely covered with a brownish patina layer. The surface of the brown pre-patinated copper sheet changed its appearance from shiny brown to a frosted surface finish. After one year of exposure in Singapore, the surfaces of the naturally patinated fresh copper sheet and the brown-patinated sheet still showed a similar appearance, being brownish without any visible green layer, although frequent crystallites with an elemental composition of Cu, O and S had formed locally within a layer predominantly composed of Cu and O (Figure 6).

Figure 6 - *Secondary electron image showing crystallites composed of Cu, O and S on top of a layer composed of Cu and O on fresh and brown-patinated copper sheet freely for one year in Singapore. (1cm=10μm)*

Compositional analyses with XPS of the outermost surface layer of the patina, on both naturally patinated fresh copper sheet and brown pre-patinated copper sheet, showed copper to be present in two oxidation states, Cu(I) and Cu(II), and sulfur to be present as sulfate. These results are consistent with phase identifications made by means of XRD, showing brochantite ($Cu_4SO_4(OH)_6$) and cuprite (Cu_2O) to be the main crystalline components of the patina formed on both fresh and brown pre-patinated copper sheet exposed in Singapore. The presence of tenorite (CuO), identified as a constituent of unexposed brown-patinated copper, was not with XRD seen after one year of exposure. The close resemblance in patina formation and composition was also evident from depth profiling studies, by means of GDOES, showing a few micrometer thick patina layer, primarily composed of copper and oxygen with sulfur in the outermost surface layer (Figure 7).

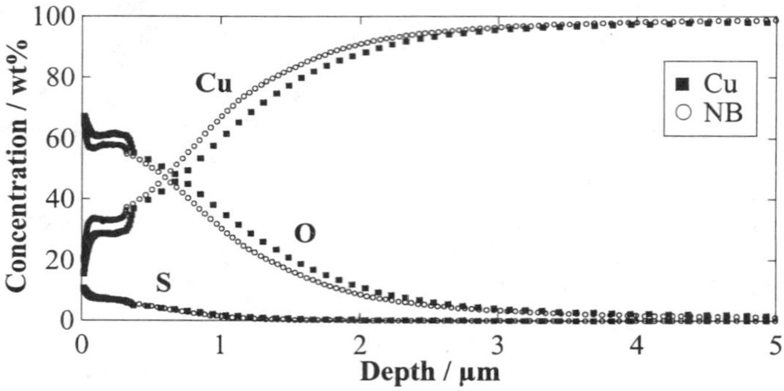

Figure 7 – *GDOES depth profile of patina formed on fresh (squares) and brown pre-patinated (circles) copper sheet exposed one year in Singapore.*

The visual changes in appearance of the patina formed on fresh and brown-patinated copper sheet observed in Singapore was also seen at the Stockholm test site. However, the time to develop a completely surface covering brownish corrosion layer on fresh copper was longer, typically one month. Both panels showed cuprite to be the dominating constituent of the patina also after one year of unsheltered exposure.

Separate platelets composed of Cu, O, and S were only observed on a few locations. These platelets could be associated with the presence of posnjakite ($Cu_4SO_4(OH)_6\ H_2O$) in very low quantities, as indicated by XRD. This slow formation rate of a basic copper sulfate on unsheltered exposed surfaces is consistent with previous results obtained in Stockholm, showing weak XRD-signals of crystalline posnjakite, a precursor for brochantite ($Cu_4SO_4(OH)_6$) formation, after one and two years of exposure [*11*]. Similar results have also been obtained in a separate field exposure of unsheltered copper in Stockholm, showing posnjakite to be present after two years of exposure and brochantite to be the dominating crystalline phase of the patina after four and eight years of exposure [*2*].

Atmospheric Corrosion and Runoff Rates of Naturally Aged Copper Sheet (130 Years) and Green Pre-Patinated Copper Sheet

Annual average runoff rates of copper released from the patina layer of green pre-patinated copper sheet (NG) exposed in Stockholm and Singapore and 130 year old naturally patinated copper (old) exposed in Stockholm are displayed in Figure 8. Yearly runoff rates from the green pre-patinated sheet varied between 1.8 and 2.3 g/m^2,y in Stockholm and between 8.4 and 8.8 g/m^2,y in Singapore. For comparison, yearly runoff rates between 1.6 and 2.0 g/m^2,y were determined in Stockholm from the 130 year old copper sheet with naturally formed green patina.

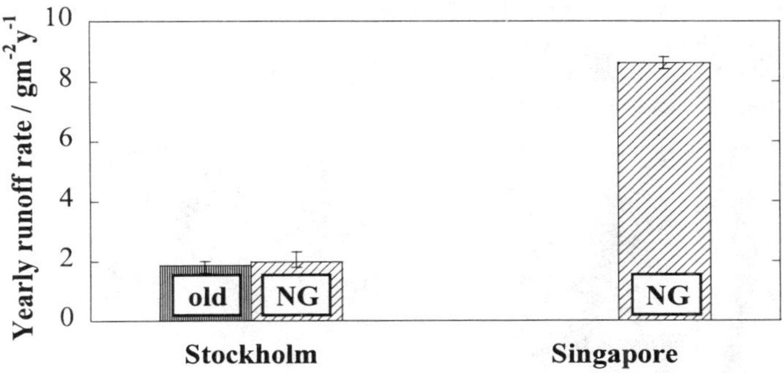

Figure 8 - *Yearly runoff rates of copper from 130 years old copper sheet (old) and green pre-patinated copper sheet (NG) freely exposed during one year in Singapore and during four and a half years of exposure in Stockholm, respectively.*

In agreement with the results from fresh and brown pre-patinated copper sheet, the runoff rates are significantly higher for green pre-patinated copper sheet exposed in Singapore compared to Stockholm, primarily a result of higher rainfall quantities and higher pollutant levels. Runoff rates of copper from the naturally patinated 130 year-old copper sheet are comparable with the results from the green pre-patinated copper panels exposed in Stockholm. This can be explained by similar characteristics of the patina layer. Both panels show a spatially heterogeneous and porous corrosion layer, which facilitates for corrosive species and water to be absorbed within the layer

The surface morphology of unexposed green pre-patinated copper sheet is shown in Figure 9 (left). Area analyses and depth profiles of the patina using EDS and GDOES showed mainly Cu, O, N and S. The presence of oxidized copper (oxidation state I and II), S present as sulfate and N present as nitrate was confirmed with XPS. A phase characterization using XRD showed cuprite (Cu_2O), gerhardite ($Cu_2NO_3(OH)_3$) and posnjakite ($Cu_4SO_4(OH)_6xH_2O$) to be the main initial crystalline components in the green patina prior to exposure. However, the presence of amorphous phases of the patina cannot be excluded.

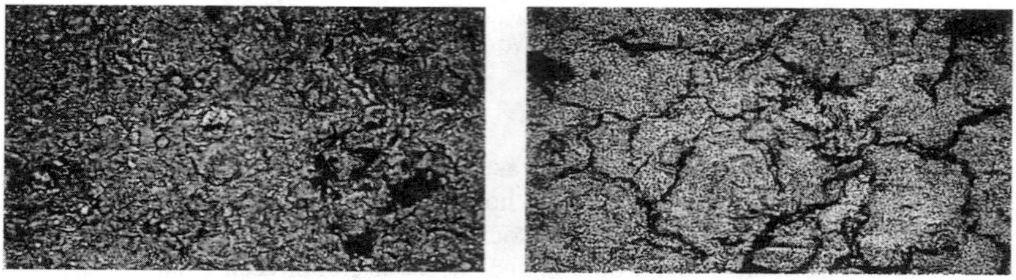

Figure 9 - *Surface morphology of green pre-patinated copper sheet before (left) and after (right) a one-year exposure in Singapore (1cm equals 30 μm).*

Within one year of exposure in Singapore, the surface of the green pre-patinated copper sheet had changed from light green on top of a pre-oxidized brownish surface into a surface with dark green stains (the green layer appears greyish in the figure) on top of a brownish surface layer (Figure 10).

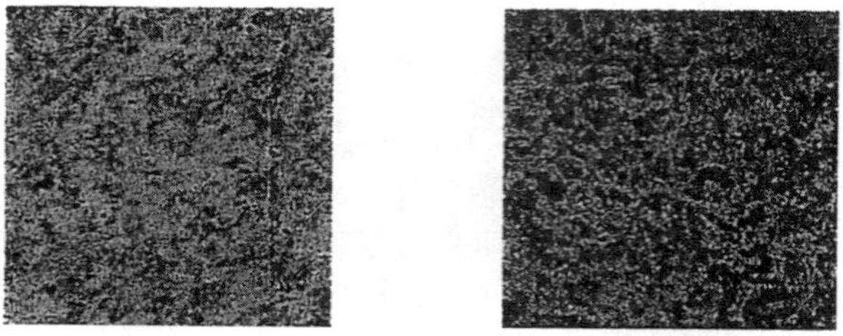

Figure 10 - *Visual appearance of pre-patinated copper sheet before (left) and after (right) a one-year exposure in Singapore.*

A similar behavior has previously been observed as a result of thinning of the patina layer due to mechanical processes in connection with snow and frost [*8*].

The surface morphology of the green pre-patinated panel exposed during one year in Singapore is also shown in Figure 9 (right). Both EDS analyses and GDOES depth profiles showed Cu, S, and O to be the main elements of the relatively porous corrosion layer. These elements were identified by XPS as Cu being in oxidation state I and II and S as sulfate. Except for an inner layer of cuprite, the only crystalline phase identified in the green patina layer was brochantite ($Cu_4SO_4(OH)_6$), i.e., similar phases as being identified on fresh and brown pre-patinated copper sheet exposed in

Singapore. The crystalline basic copper nitrate, being an initial component of the artificial patina, was not detected in the corrosion product layer after one year of exposure. The depth profile showed a relatively thick patina (15-20 μm) with S primarily present in an outer 5 μm thick layer.

Green pre-patinated copper sheets exposed in Stockholm showed similar tonal variations and corrosion product identifications. However, the formation of a basic copper sulfate covering the surfaces was much slower and only posnjakite, the precursor to brochantite, was identified. The transformation of the artificial patina into a natural patina was investigated by means of XPS. The results are given in Figure 11, showing a sudden change in surface composition, expressed as the sulfate/nitrate ratio, as a function of time.

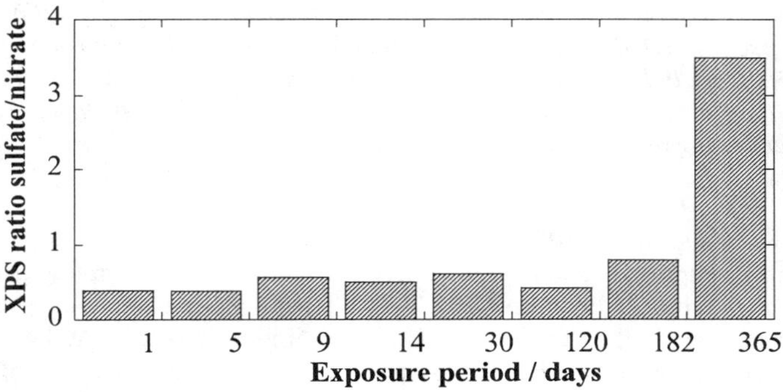

Figure 11 - *Changes in surface composition expressed as the sulfate/nitrate atomic ratio on green pre-patinated copper sheet during one-year of exposure in Stockholm.*

The results show the surface composition to change from nitrate-containing phases during the first three months of exposure to sulfate-containing corrosion products with prolonged exposure time. The figure shows a relatively slow transformation in surface composition as a result of low concentration levels of SO_2. However, with increasing SO_2-concentration, the transformation is anticipated to occur much faster.

Chemical Speciation of Copper Runoff Water from Naturally and Pre-Patinated Copper Sheet.

All data presented so far reflect the total amount of copper released from a surface without considering the chemical speciation, a prerequisite when investigating the potential environmental effect of copper runoff from roofing material. The free hydrated cupric ion ($Cu(H_2O)_6^{2+}$) is regarded to be the most toxic form of copper [*13*]. The fraction of cupric ions in the runoff water, immediately after leaving the copper surface, is compiled in Figure 12 for naturally patinated (Cu, old) and pre-patinated (NB, NG) copper sheet exposed in Stockholm.

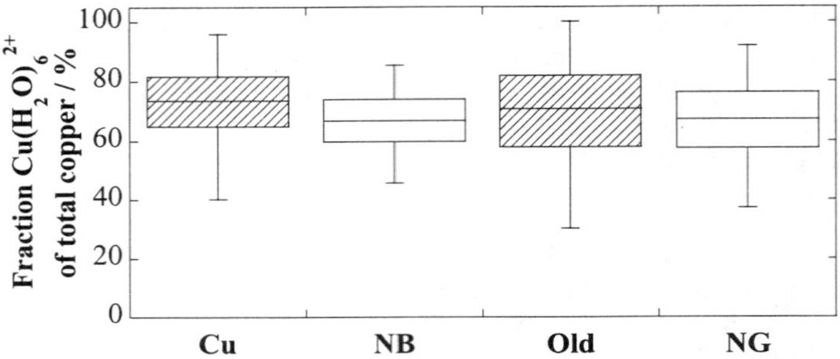

Figure 12 - *Fraction of the free cupric ion in runoff water from naturally patinated and pre-patinated copper sheet exposed in Stockholm. Each box comprises 50% of all data from 60 different sampling periods (Cu, old, NG) and 30 different sampling periods (NB) with the median value displayed as a line. Remaining data are located within the range of the error bars, showing the highest and the lowest value for each panel.*

The figure shows a significant difference between the total amount of copper and the fraction of the free cupric ion and a large variety between different sampling periods. The fraction varies typically between 40 and 99%, depending on content of reactive species, such as pollen, in the runoff water. However, it is anticipated that the chemical speciation will change dramatically as a result of the high complexing ability of the cupric ion with inorganic and organic matter upon entry into the environment.

Conclusions

Runoff rates of copper have been determined from naturally patinated copper sheet (fresh and 130-year old) and pre-patinated copper sheet (Nordic Brown and Nordic Green) exposed during one year in Singapore, Republic of Singapore and during four years in Stockholm, Sweden. Formation and composition of corrosion products in the patina has been determined using a multi-analytical approach.

The following main conclusions have been drawn

- Yearly runoff rates from fresh copper sheet are 1.1-1.6 g/m^2 and 5.6-5.7 g/m^2 in Stockholm and Singapore, respectively.
- Yearly runoff rates of copper from brown pre-patinated copper sheet are comparable to fresh copper sheet, being 1.3-1.5 g/m^2 and 5.5-5.7 g/m^2 in Stockholm and Singapore, respectively.
- Cuprite (Cu_2O) and brochantite ($Cu_4SO_4(OH)_6$) are the main constituents of the brownish patina formed on fresh and brown pre-patinated copper sheet exposed in Singapore.

- Cuprite and single platelets of posnjakite ($Cu_4SO_4(OH)_6xH_2O$) are the main constituents of the brownish patina formed on fresh and brown pre-patinated copper sheet exposed in Stockholm.
- Yearly runoff rates from green pre-patinated copper sheet are 1.8-2.3 g/m^2 and 8.4-8.8 g/m^2 in Stockholm and Singapore, respectively.
- Yearly runoff rates of copper from a 130-year old copper sheet are comparable to green pre-patinated copper sheet exposed in Stockholm, being 1.6-2.0 g/m^2.
- Cuprite and brochantite are the main constituents of the patina formed on green pre-patinated copper sheet exposed in Singapore.
- In Stockholm, the transformation of a nitrate-rich corrosion product layer on green pre-patinated copper sheet is much slower compared to Singapore, as a result of low concentration levels of SO_2.
- Higher runoff rates in Singapore compared to Stockholm are the result of five to eight times higher rainfall quantities and approximately seven times higher SO_2 air concentration level.
- The fraction of the cupric ion varies typically between 40 and 99% for both naturally patinated and pre-patinated copper sheet exposed in Stockholm.

Acknowledgments

Dr. Ingegerd Annergren at Gintic, Singapore, is highly appreciated for performing the field exposure in Singapore, Lena Wickman, Outokumpu Copper Partner, Västerås, for arranging the field exposure in Singapore, Pekka Taskinen and Tuija Suortti at Outokumpu research OY, Finland for performing the XRD analyses and Sofia Hänström at the Swedish Institute for Metals Research for performing GDOES-analyses. Finally, we are grateful to M.Sc. Wenle He for help with metal analyses and maintenance of the test exposure site in Stockholm.

References

[1] C. Leygraf and T.E. Graedel, *Atmospheric Corrosion*, John Wiley & Sons, Inc., New York, 2000, ISBN 0-471-37219-6.

[2] Krätschmer, A. and Stöckle, B., "Results from XRD Analysis of Copper Corrosion Products," UN/ECE Report No. 32, Bayerisches Landesamt fur Denkmalpflege, München, Germany, 1998.

[3] Odnevall, I. and Leygraf, C., "Atmospheric Corrosion of Copper in a Rural Atmosphere," *Journal of Electrochemical Society*, Vol. 142, 1995, p. 3682.

[4] Fitzgerald, K.P., Nairn, J., and Atrens, A., "The Chemistry of Copper Patination," *Corrosion Science*, Vol. 40, No. 12, 1998, pp. 2029-2050.

[5] Graedel, T.E., "Copper Patinas Formed in the Atmosphere – A Qualitative Assessment of Mechanisms," *Corrosion Science*, Vol. 27, 1987, p. 721.

[6] Vernon, W.H.J., "The Open-Air Corrosion of Copper. Part III.- Artificial Production of Green Patina," *Journal of Institute of Metals*, 1932, p. 153.

[7] Fitzgerald, L.D., "Protection of Copper Metals from Atmospheric Corrosion," *Atmospheric factors affecting the corrosion of engineering metals, ASTM STP 646*, S. K. Coburn, Ed., American Society for Testing and Materials, 1978, pp. 152-159.

[8] Mattsson, E., and Holm, R., "Properties and Applications of Pre-Patinated Copper Sheet," *Sheet Metal Industries*, 1967, pp.1-5.

[9] He, W., Odnevall Wallinder, I. and Leygraf, C., "A Laboratory Study of Copper and Zinc Runoff During First Flush and Steady State Conditions," *Corrosion Science*, Vol. 43, No. 1, 2000, p.127.

[10] Cramer, S. D., McDonald, L.G. and Spence, J.W., "Effect of Acidic Deposition on the Corrosion of Zinc and Copper," *Proceedings 12th International Corrosion Congress*, Houston, USA 1993, p. 722.

[11] Odnevall Wallinder, I. and Leygraf, C., "Seasonal Variations in Corrosion Rate and Runoff rate of Copper Roofs in an Urban and a Rural Atmospheric Environment," *Corrosion Science*, Vol. 43, No. 12, 2001, pp. 2379-2396.

[12] Hechler, J.J., Boulanger, J., Noel, D. and Pinon, C., "Corrosion Rates, Wetness, and Pollutants on Exterior of Building," *Journal of Materials in Civil Engineering*, Vol. 5, No. 1, 1993, p. 53.

[13] Allen, H.E., and Hansen, D.J., "The Importance of Trace Metal Ion Speciation to Water Quality Criteria," *Water and Environmental Research*, Vol. 68, 1995, p. 42.

S. D. Cramer,[1] S. A. Matthes,[1] B. S. Covino, Jr.,[1] S. J. Bullard,[1] and G. R. Holcomb[1]

Environmental Factors Affecting the Atmospheric Corrosion of Copper

Reference: Cramer, S. D., Matthes, S. A., Covino, Jr., B. S., Bullard, S. J., and Holcomb, G. R., "**Environmental Factors Affecting the Atmospheric Corrosion of Copper,**" *Outdoor Atmospheric Corrosion, ASTM STP 1421*, H. E. Townsend, Ed., American Society for Testing and Materials International, West Conshohocken, PA, 2002.

Abstract: Precipitation runoff experiments were conducted on copper panels at sites polluted with sulfur dioxide (Washington, DC), an unpolluted site (Albany, OR), and an unpolluted marine site (Newport, OR). Copper patinas had complex surface chemistry representing a sequence of transitional mineral phases terminating in brochantite at the polluted sites, atacamite at the unpolluted marine site, and copper oxides at the unpolluted site. Precipitation runoff rates on large panels were 3.3 (0.37), 1.7 (0.19), and 1.7 (0.19) g Cu/m^2y (μm Cu/y) for the Washington, DC, Albany, OR, and Newport, OR, sites, respectively. Higher rates were observed for small panels and appeared due to the reduced impact of residence time effects. Zinc corrosion products on the panel at Newport depressed the copper runoff until the zinc corrosion products were dissipated. Copper runoff at the Washington, DC, site was partitioned into contributions from dry deposition of acidic gases, and wet deposition of strong acids and weak acid (carbonic acid) using geochemical modeling software. The software also predicted precipitation of brochantite during winter when sulfur dioxide levels were highest. An empirical atmospheric corrosion model with early parabolic corrosion kinetics for copper and long-term linear kinetics, and including synergy between properties of the copper patina and the environment, was described. Applied to the patina on a 100-year old roof, the model suggests that only 15 % of the total mass loss is retained in the corrosion film after 100 years of exposure.

Keywords: copper, runoff, atmospheric corrosion, modeling, corrosion model, wet deposition, dry deposition, sulfur dioxide, cuprite, brochantite, atacamite

Precipitation runoff of metal ions from a boldly exposed metal surface is increasingly recognized as an important component of the atmospheric corrosion of metals for its effect on materials damage [*1-3*] and impact on the environment [*4-11*]. While the boldly exposed copper patina may contain fewer compounds and be thinner than the sheltered patina, corrosion rates in bold exposures often are higher [*12*]. Public agencies are examining the role of point sources of copper ions such as roofs, flashing, and gutters on drinking water and storm water [*13*].

[1] Chemical Engineer, Chemist, Research Chemist, Research Chemist, and Materials Engineer, respectively, Albany Research Center, US DOE, 1450 Queen Ave. SW, Albany OR 97321.

A broad base of knowledge exists on the effects copper metallurgy and the environment have on the formation of copper patinas [*12,14-16*]. Copper patina growth has been described as parabolic during the early stages of formation [*16*]. Copper ions interact with the environment to form a sequence of increasingly insoluble mineral corrosion products concluding in terminal phases characteristic of the environment [*5-6,12,14-16*]. These often are brochantite for sulfur dioxide polluted environments and atacamite for unpolluted marine environments. Cuprite is a major constituent of the corrosion film that is overlaid by this mineral sequence [*5-6,12,14-16*].

Precipitation runoff studies for boldly exposed copper surfaces suggest that the copper removal rate is linear with time, and characteristic of the environment to which the copper surface is exposed [*2,4-6,17,19*]. Evidence suggests that runoff removal rates are not strongly affected by rain intensity except at very low and high rain intensities [*5*]. The most important factor in determining the copper runoff rate is precipitation volume, with precipitation pH, the presence of acidic gases, the time between precipitation events, and the orientation of the copper surface other important factors [*2,4-6,17,19*].

This paper presents precipitation runoff rates for copper in bold exposures in polluted, unpolluted, and unpolluted marine environments. In doing so, it examines environmental factors that affect the runoff rate and the impact of runoff on corrosion kinetics.

Experimental Methods

Exposure Sites and Material

Mass loss and precipitation runoff measurements were made using panels of electrolytic tough pitch copper (UNS-C11000) 1.59 mm (1/16-in.) thick, a material that is used for

Table 1 B *Atmospheric corrosion test site and annual average meteorological and air quality parameters.*[1]

Site	Type	Runoff panels	Precip. rate,[2] L/m^2 y	Precip. pH	Temp., E C	Wind speed, m/s	SO_2 $\mu g/m^3$
Steubenville OH	polluted	no	853	4.06	11.7	1.3	56.5
Washington DC	polluted	yes	958	4.21	14.3	3.8	27.2
Chester NJ	polluted	no	1306	4.23	9.0	2.3	17.3
Research Triangle Park NC	polluted	no	1003	4.40	14.9	1.4	6.3
Newcomb NY	polluted	no	1056	4.38	5.7	0.7	4.4
Albany OR	unpolluted	yes	1084	5.78	11.2	3.1	0.0
Newport OR	unpolluted marine	yes	1822	6.13	10.5	3.8	0.0

[1] ND = Not determined. [2] cm/y = $0.1 L/m^2y$

architectural applications. The mass-loss panels were 10x15 cm (4x6 in.) in size, masked on one side with electroplating tape to confine corrosion to the exposed side. The exposed area was 0.01587 m^2. Precipitation runoff panels were 30x61cm (12x24 in.) in size, also masked on one side with electroplating tape. The exposed area was 0.1839 m^2. Prior to exposure the panels were chemically cleaned in a 33 vol % solution of hydrochloric acid at room temperature for 1-3 minutes, rinsed with deionized water, rinsed with methanol, and dried. The mass loss panels were weighed to 0.1 mg.

The exposure sites are described in Table 1. Mass loss panels were exposed at all sites. All panels were exposed skyward and inclined at an angle of 30 degrees with the horizon (as measured along the long axis). All panels faced due south except for those in Oregon; they faced west into the prevailing winds. After exposure, the mass loss panels were stripped of corrosion product in a 33 % hydrochloric acid solution for 1-3 minutes, rinsed with deionized water, rinsed with methanol, dried, and weighed to 0.1 mg. The stripping solution containing the corrosion products was chemically analyzed to determine the mass of copper in the corrosion film. Corrections were made to both mass loss and corrosion film measurements for copper metal removed by the stripping solution.

Panels were exposed at several polluted sites in the eastern United States with significant levels of sulfur dioxide, an unpolluted marine site on the Oregon coast, and at an unpolluted site in the Willamette Valley of western Oregon. Precipitation runoff measurements by direct analysis of the runoff from the large copper panels were made at Washington, DC (polluted site), Albany, OR (unpolluted site), and Newport, OR (unpolluted marine site). The Washington, DC, site was on a second floor rooftop; the Newport site was within 100 m of the ocean at the Yaquina Bay Bridge (Figure 1); the Albany site was in an agricultural area 83 km from the ocean. Precipitation runoff losses were also estimated from mass loss measurements at Washington, DC, and Newcomb, NY. The Newcomb site was in a forest clearing within the Adirondack National Park.

Figure 1 - *Newport (OR) marine test site.*

Corrosion Film Chemistry

All analyses of copper patina chemistry were made *in situ* on skyward side exposures of the panels. Samples were cut from both mass loss and runoff panels. Minerals in the patina were identified by X-ray diffraction (XRD) using a Philips APD 3720 diffractometer. Patina surface chemistry was analyzed by X-ray photoelectron spectroscopy (XPS) using a Surface Science Laboratory SSX-100 ESCA spectrometer, argon ion sputtering at a rate of 1.25 and 1.5 Δ/s, and a 600 μm analysis spot size; a flood gun was used to neutralize charging of the marine sample. SEM images and X-ray maps of polished patina cross-sections were made using a LEO Stereoscan S440 SEM and Oxford eXLII energy dispersive spectrometer.

Analytical scanning electron microscopy (ASEM) analysis of the polished patina cross-sections to produce composition depth profiles were made using the LEO Stereoscan and an Oxford 3PC wavelength dispersive spectrometer with four crystals.

Precipitation Runoff Measurements

All precipitation washing a runoff panel surface was collected in clean 40 L polypropylene containers at monthly intervals, except during periods of high rainfall when as many as three collections could be made in a month. Incident precipitation was collected concurrently in an Aerochem Metrics wet/dry collector following National Acid Deposition Program (NADP) protocols [*20*]. Collection of precipitation runoff and incident precipitation was done so as to minimize sample evaporation losses and contamination. Precipitation runoff and incident precipitation samples were filtered to remove insoluble particulates that had washed into the sample. The samples were analyzed for the NADP list of acid rain ions (H, Ca, Mg, K, Na, NH_4, NO_3, SO_4, and Cl) and selected metal ions (Cu, Zn, and Fe) by ion chromatography, inductively-coupled plasma (ICP) emission spectroscopy, and pH meter. Standards were periodically run, including a low ionic strength pH standard, to assure the accuracy of the analyses. No analysis was made for carbonate and bicarbonate ions. Instead, they were estimated from sample compositions using geochemical modeling software developed by the U.S. Geological Survey [*21*].

Geochemical Modeling

Geochemical modeling software [*21*] was used to examine dissolution and precipitation processes on the runoff panel at the Washington, DC, site where there were contributions to the runoff from wet and dry deposition of acidic species. Geochemical modelling was not used for the Albany, OR, and Newport, OR, sites where all runoff could be attributed to weak acid (carbonic acid). The geochemical modeling software was based on equilibrium chemistry and mass and charge balance constraints to interpret the chemical events occurring on the runoff panel surface and contributing to the precipitation runoff chemistry.

Each runoff collection consisting of the incident precipitation and the precipitation runoff sample was analyzed separately, Table 2. Speciation of solution constituents was computed assuming equilibrium with atmospheric carbon dioxide. Inverse modeling was used to identify mineral dissolution and precipitation processes that would explain the runoff chemistry and pH. The modeling software required a list of mineral phases likely to be found on the runoff panel to begin the calculations. Using this list, the software found a series of models consisting of combinations of mineral chemistry that satisfied the observed chemistry. These models were not unique, but were differentiated by practical considerations of corrosion film and runoff chemistry, and by the internal consistency of model trends.

Model results, such as shown in Table 2 for Sample 10, were treated in the following way. Neutral copper salts that dissolved were assumed to have formed on the panel surface during periods of dry deposition by reaction with acidic gases. Basic copper minerals (primarily cuprite) that dissolved were assumed to have been neutralized by strong acids and weak acid (carbonic acid) present in wet deposition. The weak acid contribution was determined by subtracting the contribution of the strong acids,

Table 2 B *Geochemical inverse model results for Sample 10 (March), Washington, DC.*

Model input:				
! solution composition:	H, Ca, Mg, K, Na, NH_4, NO_3, SO_4, Cl, Cu for incident precipitation and precipitation runoff			
! possible phases removed from panel	Cu_2O, $CuSO_4$, $Cu_4SO_4(OH)_6$, $CaSO_4$, $MgSO_4$, $Ca(NO_3)_2$, $Mg(NO_3)_2$, NaCl, KCl, $MgCl_2$, $CaCl_2$, CO_2, NH_3			
Model output:				
! speciation	H^+	Cl	Mg^{+2}	SO_4^{-2}
	OH^-		$MgSO_4$	$CuSO_4$
		Cu^+	$MgHCO_3^-$	$CaSO_4$
	CO_2	Cu^{+2}		$MgSO_4$
	HCO_3^-	$Cu(OH)_2$	NH_4^+	$NH_4SO_4^-$
	$CaHCO_3^+$	$CuSO_4$	$NH_4SO_4^-$	$NaSO_4^-$
	$MgHCO_3^+$	$CuOH^+$	NH_3	HSO_4^-
	CO_3^{-2}			KSO_4^-
	$NaHCO_3$	K^+	NO_3^-	$CaHSO_4^-$
		KSO_4^-		
	Ca^{+2}		Na^+	
	$CaSO_4$		$NaSO_4^-$	
	$CaHCO_3^+$		$NaHCO_3$	
	species	moles removed from panel, *m*	redox	
! mole transfers				
	Cu_2O	4.70×10^{-4}	oxidation	
	$CuSO_4$	4.10×10^{-4}		
	$Cu_4SO_4(OH)_6$	-3.09×10^{-4}		
	$CaSO_4$	none		
	$MgSO_4$	6.00×10^{-6}		
	$Ca(NO_3)_2$	8.88×10^{-5}		
	$Mg(NO_3)_2$	none		
	NaCl	2.72×10^{-5}		
	KCl	5.86×10^{-6}		
	$CaCl_2$	1.24×10^{-5}		
	$MgCl_2$	none		
	CO_2	1.18×10^{-4}		
	NH_3	-1.11×10^{-4}	reduction	

computed as the change in hydrogen ion concentration between the incident precipitation pH and pH 5.6. Precipitation of copper minerals, specifically brochantite, was shown in the model calculations as a negative dissolution. Brochantite precipitation was always accompanied by dissolution of neutral copper salts and cuprite, with a portion of the copper ions released then consumed by the precipitation process.

Results

Mass Loss Measurements

Mass loss curves and corrosion film (expressed as mass of copper in the corrosion film) curves as a function of exposure time were constructed for a series of single-sided copper exposures begun at Washington, DC, and Newcomb, NY. Initial exposures were staggered over a four year period beginning in March, June, September and December. Exposure times ranged from three months to three years. The one year and three-year curves were nonlinear, reflecting the increasingly protective patina typical of copper. These data were used to compute precipitation runoff losses by the difference between mass loss and the mass of copper in the patina, as an additional measurement of runoff loss.

Patina Chemistry

Patina chemistry results are shown in Table 3. XRD showed cuprite (Cu_2O) as the principle crystalline constituent of the patina after three year exposures at polluted, unpolluted, and unpolluted marine sites. Brochantite ($CuSO_4{\bullet}3Cu(OH)_2$)and, in a few cases, prosnjakite ($CuSO_4{\bullet}3Cu(OH)_2{\bullet}H_2O$) were also present at the polluted sites. Tenorite (CuO) and a trace of chalconatronite ($Na_2Cu(CO_3)_2{\bullet}3H_2O$) were present at the unpolluted site. Nantokite (CuCl), atacamite ($CuCl_2{\bullet}3Cu(OH)_2$), and a trace of chalconatronite were present at the unpolluted marine site.

XPS showed the extreme outer surface of the patina contained high amounts of elemental carbon and organic compounds. The surface was sputtered to a depth of about 1000 Δ to remove this carbon and enter a region of relatively stable elemental composition. However, at this depth, the surface composition was complex and had a composition quite different from that shown by XRD. The elemental composition (in atomic pct) of the patina at 1000 Δ is given in Table 4. Sulfur in the outer patina surface increased with increasing sulfur dioxide, being highest for patina from Steubenville, OH, and lowest for Research Triangle Park, NC.

The complex nature of the outer surface is reflected XPS peak data, Table 4. At the polluted site, the 931.9 eV Cu peak suggests copper is present as CuS. The 162.1 eV S peak for sulfide supports this, but there is a167.6 eV sulfite/sulfate peak as well. The 530.2 eV O peak represents metal oxides. The 284.4 eV C peak is for elemental carbon, but the additional peak at 188.3 eV suggests carboxyls such as formic and acetic acid. By comparison, the patina from the unpolluted site is less complicated with elemental carbon (284.2 eV C peak), metal oxide (529.9 eV O peak), and hydroxide/carbonate (531.5 eV O peak). The 931.9 eV Cu was not reconciled with the patina and environmental chemistry. The patina at the marine site was also less complicated. The 933.2 eV Cu peak suggests CuO and is supported by the 531.2 eV O peak for metal oxides. The 935.0 eV Cu peak suggests $CuCl_2$ and is supported by the 199.4 eV Cl peak for $CuCl_2$. The 285.2 eV C peak is for elemental carbon.

Table 3 – *Patina chemistry by XRD.*

Site type	XRD analysis [1]
polluted	Cu_2O p $CuSO_4{\bullet}3Cu(OH)_2$ p-s $CuSO_4{\bullet}3Cu(OH)_2{\bullet}H_2O$ s
unpolluted	Cu_2O p CuO s $Na_2Cu(CO_3)_2{\bullet}3H_2O$ tr
unpolluted marine	Cu_2O p CuCl s $CuCl_2{\bullet}3Cu(OH)_2$ s $Na_2Cu(CO_3)_2{\bullet}3H_2O$ tr

[1] p = primary; s = secondary; tr = trace.

Table 4 – *Patina chemistry by XPS.*

Site type	XPS Ar sputter time (thickness)	XPS analysis, atomic pct (major peak binding energies, eV)				
		Cu	S	C	O	Cl
polluted	835 s	19-29	7.2-12.4	6.2-9.8	52-66	0.6-1.0
	(1040 Δ)	(931.9)	(162.1)	(284.4)	(530.2)	(198.9)
			(167.6)	(288.3)		
unpolluted	835 s	47		3.2	50	
	(1250 Δ)	(931.9)		(284.2)	(529.9)	
			0.4		(531.5)	0.2
unpolluted	747 s	32		11.5	46	6.2
marine	(1120 Δ)	(933.2)	0.0	(285.2)	(531.2)	(199.4)
		(935.0)			(532.9)	

A back-scattered electron SEM photomicrograph of a polished patina cross-section from the Newport OR marine site is shown in Figure 2(a) along with X-ray maps for copper (b), oxygen (c), and chlorine (d). Molar volume ratios for cuprite (1.64) and atacamite (4.00) [*22*] suggest by inspection the corrosion product overlying the rough copper metal interface, Figure 2(a), does not represent all of the copper generated by corrosion directly beneath it. The X-ray maps showed Cu and O were the only constituents of the interior corrosion product. In contrast, the exterior corrosion product was modified by the environment (weathering) and contained a high concentration of chloride, Figure 2(d). The weathered layer comprised the outer 1/3 to 1/2 of the corrosion product. Dust particles, particularly those containing Ca, were occasionally present in the weathered portion of the corrosion product. Zinc at low levels was also present in the weathered portion, but not the interior portion. The zinc was inadvertently deposited on the copper panel by a nearby zinc thermal-spray project. SEM photomicrographs and X-ray maps for the unpolluted site showed corrosion penetration similar to the marine site. The corrosion product interior contained only Cu and O. The layer of corrosion product modified by weathering was much thinner and contained a trace of chloride, but no other constituents.

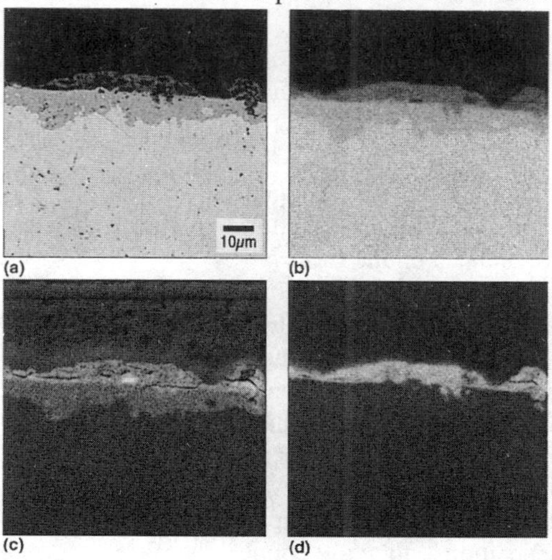

Figure 2 - *X-ray maps of 2.5 years marine corrosion film cross-section.*

Corrosion film depth profiles are shown in Figure 3 for typical patina from polluted and unpolluted sites. Copper metal is on the left side of this figure. Near the copper metal-patina interface is a region of constant composition at around 65-70 atomic % Cu. Were the oxygen profiles shown, this region matches up with oxygen values at 30-35 atomic % O and is characteristic of cuprite. The chemistry becomes more complex near the patina surface where hydrated and basic copper minerals typically form.

The patina chemistry results are in agreement with other investigations [*14-16*].

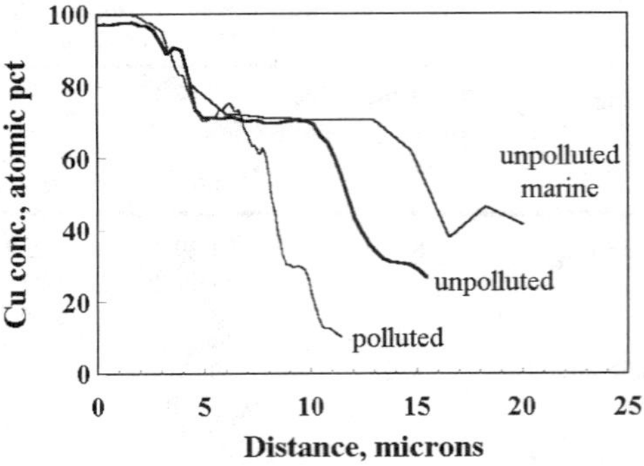

Figure 3 - *Typical ASEM line scans of corrosion film* cross-section.

Runoff Measurements

Wet and dry deposition to a panel surface consists of a series of discrete and unique events. Monthly collection of incident precipitation and precipitation runoff corresponds to a time and flow average of these events. Cumulative results of the precipitation runoff collection for Washington, DC, are given in Figure 4. Most of the collection curves can be represented by a straight line. This suggests a constancy of the chemical environment in which corrosion panels are immersed over a span of several years as least despite the unique nature of the individual events. Some of the curves are not linear, notably the chloride and sodium curves early in the exposure, and calcium later. This may be related to increasing surface roughness of the patina with exposure and how particles are retained on the surface.

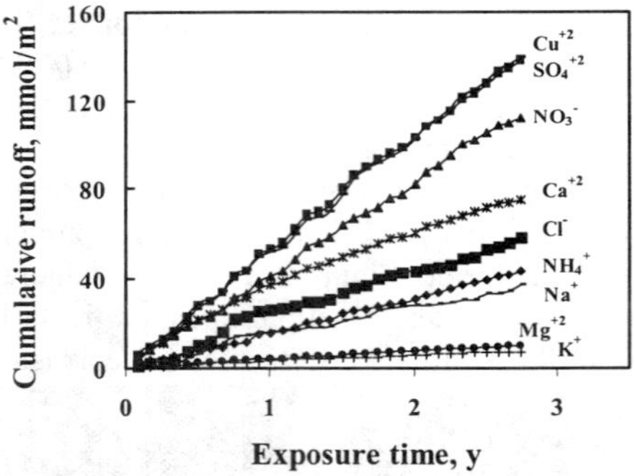

Figure 4 - *Precipitation runoff results for 33 month experiment at Washington, DC.*

Dry deposition was computed as the difference between the flux of chemical species to the runoff panel surface in a series of discrete rain events and the flux of species leaving the surface in precipitation runoff over the same time period. To determine how dry deposition to an inert surface compared with dry deposition to a copper surface, precipitation runoff was also collected from a 304 stainless steel panel. The

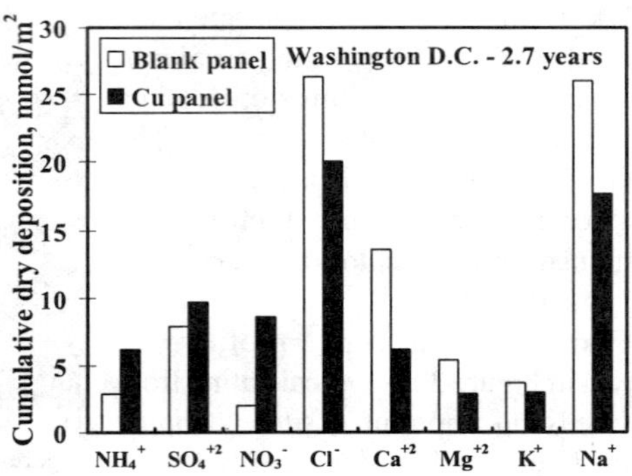

Figure 5 - *Dry deposition on copper and inert (blank) panel at Washington, DC.*

cumulative dry deposition to the two panels for the 33 month exposure period is shown in Figure 5. Dry deposition of ions associated with the acidic gases sulfur dioxide and nitric acid was much higher to the more reactive copper surface. The lower apparent Ca and Mg deposition to copper suggests that particulates, typically fly ash which should be deposited to both surfaces at the same rate, may be retained by the developing surface roughness of the copper panel. Other ions are present similarly on both panels.

Discussion

Copper Precipitation Runoff

Cumulative copper removal from the patina in precipitation runoff is shown for the polluted, unpolluted, and marine sites as a function of exposure time, Figure 6, and by cumulative precipitation volume, Figure 7. Seasonal variations are evident, particularly for the Albany and Newport sites where precipitation rates can be very low during the summer. Never the less, the runoff curves are essentially straight lines in time and in precipitation volume. The slope of the curves versus time, determined by a least squares fit of the data and having R^2 values of 0.94 and higher, is the copper removal rate. It was 3.3 (0.37), 1.7 (0.19), and 1.7 (0.19) g Cu/m^{2}y (µm Cu/y) for the polluted, unpolluted, and marine sites, respectively, Table 5. The slope of the curves versus precipitation volume, determined by a least squares fit of the data and having R^2 values of 0.98 and higher, is the flow-average concentration of copper in the precipitation runoff. It was 4320, 1240 and 670 µg/L for the polluted, unpolluted, and marine sites, respectively.

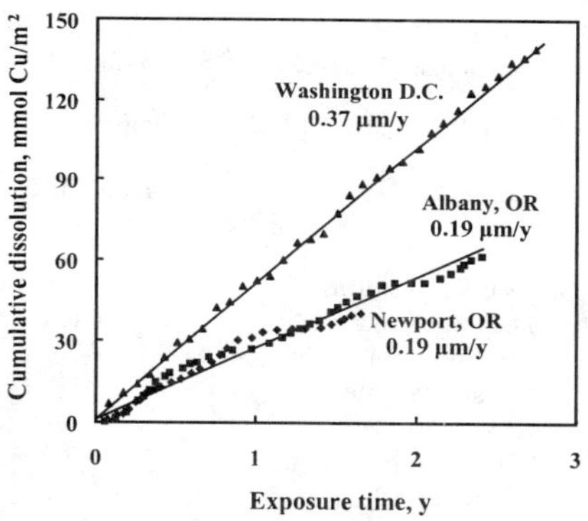

Figure 6 - *Copper runoff results at marine, unpolluted and polluted sites as a function of exposure time.*

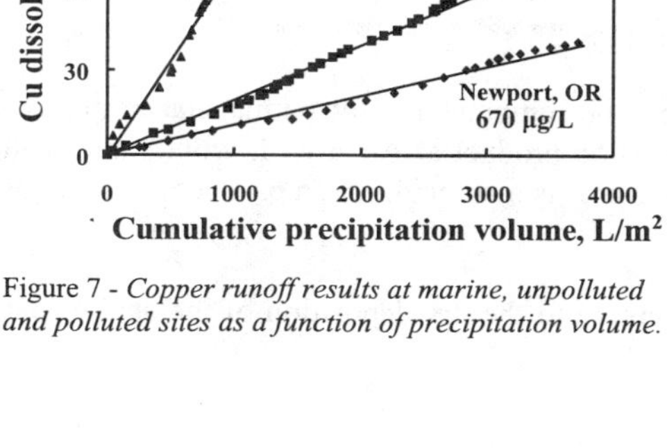

Figure 7 - *Copper runoff results at marine, unpolluted and polluted sites as a function of precipitation volume.*

While they may look different, Figures 6 and 7 are essentially the same. The x-axes of the two figures differ simply by the annual average precipitation rate for the site. They highlight the importance of precipitation volume on the rate soluble copper minerals are removed from the patina. For example, runoff removal rates were the same for Albany and

Newport. However, the runoff

Table 5 B *Copper concentration in runoff and runoff removal rate.*[1]

Site	Type	Cu conc. in runoff, µg/L	Runoff removal rate		Ref
			g Cu/m^{2}y	µm Cu/y	
Washington DC					
! runoff	polluted	4320 [2]	3.3	0.37	
! mass loss (1985-89)	A	5430	5.2	0.58	
Newcomb NY					
! mass loss (1985-89)	polluted	3530	3.7	0.42	
Albany OR	unpolluted	1240 [2]	1.7	0.19	
Newport OR	unpolluted marine	670 [2]	1.7	0.19	
Storrs CT (16 storms)					
! new roof, avg.	polluted	3630 [2]	5.2	0.58	[8-9]
first flush	A	–7000			[8-9]
steady state	A	–2000			[8-9]
! 70 yr old roof, avg.	polluted	1460 [2]	2.0	0.22	[8-9]
first flush	A	–3800			[8-9]
steady state	A	–1000			[8-9]
Stockholm Sweden					
! unaged Cu	polluted		1.3	0.15	[5]
! 100 yr old roof	A		2.0	0.22	[5]
Laboratory					
! unaged Cu	polluted		1.2	0.13	[5]
	unpolluted		0.7	0.078	[5]
	marine		1.7	0.19	[5]
! preaged/unaged Cu					
first flush	polluted	–9530			[5]
steady state	A	–2540			[5]

[1] All results from precipitation runoff chemistry except the two cases based on mass loss measurements.

[2] Flow-averaged concentration.

concentration was lower for Newport where one component of the patina, atacamite, has a substantially lower solubility than the corrosion product at Albany. It would appear the similarity of runoff removal rates at the two sites is a coincidence; a product of the higher precipitation rate at Newport.

Modeling Copper Runoff

The copper runoff panel was more receptive to the dry deposition of the acidic gases sulfur dioxide and nitric acid than the inert panel at the Washington, DC, site, Figure 5. Thus one component of the copper precipitation runoff is the reaction of acidic gases with the patina (and with basic particulates deposited on the patina) to form neutral salts. These could then be washed from the patina in during precipitation events. Acidic gases should readily penetrate pores in the patina, particularly as the patina weathers and porosity increases [5], increasing the concentration of neutral salts formed between precipitation events on more weathered patina surfaces.

A second component of the runoff is the reaction of strong acids in the precipitation with the patina. Hydrogen ion is highly mobile and the reaction with basic constituents at the patina surface should be rapid. Strong acids, computed as the hydrogen ion concentration in excess of precipitation in equilibrium with atmospheric carbon dioxide, appeared to be fully consumed as the pH of the incident precipitation at Washington, DC, was always below pH 5.6 while the pH of the runoff was always above pH 5.6.

Carbonic acid (weak acid) in precipitation runoff in equilibrium with atmospheric carbon dioxide is the third contributor to copper runoff. It supplies hydrogen ions above pH 5.6 for the continued dissolution of basic patina constituents. The weak acid component should be sensitive to precipitation residence time on the panel, and be affected by mixing within the water layer draining from the surface and by rain intensity. While incident precipitation would be in equilibrium with atmospheric carbon dioxide, reactions with basic patina constituents above pH 5.6 require replenishment of the carbonic acid; hence, carbon dioxide transport across the air-water interface. This also involves diffusion of carbon dioxide into a thickening water layer and then mixing of the layer to bring the carbonic acid in contact with patina constituents.

Basic particulates known to settle on the corrosion film, such as fly ash at the polluted Washington, DC, site or carbonate shell fragments in dust at the coastal Newport site, compete with basic patina constituent for hydrogen ion delivered in wet and dry deposition. These particles have the potential for lessening the impact of acid deposition. For purposes of calculation, the geochemical model does not distinguish between neutral calcium and magnesium salts deposited on the patina surface or deposited calcium and magnesium oxides that later successfully competed for acid species delivered by wet and dry deposition. Thus, the flux of acidic species to the patina may be substantially higher than that represented by the runoff of the copper ions shown in Figures 6 and 7.

Computer runs of the geochemical modeling software showed inputs and outputs similar to, but with substantially more detail, than shown in Table 2 for Sample 10 from Washington, DC. The speciation shown includes both ionic constituents and undissociated hydrated constituents in the runoff at concentrations greater than $10^{-9}m$. The heading "mole transfers" represents the mineral dissolutions and precipitations that occur on the patina surface as a consequence of the incident precipitation and precipitation runoff chemistry. In this particular run, the copper ions in the runoff are the net of the dissolving copper sulfate and cuprite and precipitating brochantite. Note also that oxidations and reductions may be required to satisfy the observed solution chemistry. Also, carbon dioxide has been absorbed by the runoff and ammonia has been lost. It should be noted again that the computer runs are not unique models of the system chemistry, but approximations based on the corrosion film and precipitation chemistry.

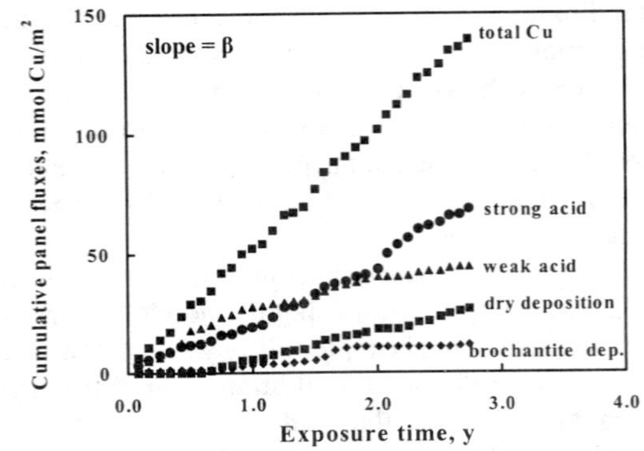

Figure 8 - *Dry deposition, strong acid, weak acid, and brochantite deposition effects on runoff as determined by geochemical modeling for Washington, DC.*

Calculations such as those represented by Table 2 can be used to partition the copper runoff into contributions from dry deposition, strong acid, and weak acid. This partitioning is shown in Figure 8 for Washington, DC. It shows the strong acid contribution that, while exhibiting some seasonal effect, is essentially linear with time. The weak acid contribution is higher in the early stages of patina development when the patina presumably contains minimal brochantite. The dry deposition contribution only increases following an induction period associated with early patina development. While the individual contributions may vary with time, the net copper runoff is seen to be linear with time. The slope of the net curve for copper is the precipitation runoff rate, β, and given here in units of g Cu/m^{2}y and μm/y. Cumulative runoff, βt, where t is the exposure time, can then be partitioned into its component parts and is represented by

$$\beta t = [dry\ deposition] + [strong\ acid] + [weak\ acid] \qquad (1)$$

with contributions from dry deposition, strong acid and weak acid. Only the weak acid contribution was observed at the unpolluted Albany and unpolluted marine Newport sites.

Brochantite precipitation in Washington, DC, is also shown in Figure 8 and was predicted by inverse modeling. Brochantite precipitation occurred during the winter months when monthly average sulfur dioxide levels were high at the test site, regularly exceeding 55 μg/m^3 (20 ppb) [*23*]. No brochantite precipitation was predicted for the spring, summer, and fall months when sulfur dioxide levels were lower, often well below 27 μg/m^3 (10 ppb) during late summer or fall [*23*], and monthly rainfall tended to be higher. These effects are indicated by a rising cumulative brochantite curve for the winter months and a flat curve corresponding to the spring, summer and fall months.

Regarding Equation 1, the dry deposition component of copper runoff should flush from the patina surface early in a precipitation event, tailing off as the event continued. This would contribute to the first flush effect observed in laboratory and field studies [*5,8-9*]. Also contributing would be the low rain pH observed at the beginning of precipitation events. The long-term steady state runoff concentration for single precipitation events is observed in both laboratory and field studies [*5,8-9*] to be substantially lower than the first flush concentration. With the dry deposition component eliminated in the first flush, Equation 1 then shows continued copper removal would be confined to the contributions from strong and weak acid as rain continues to fall.

Copper runoff concentrations are tabulated in Table 5 to illustrate the similarity of measurements form several studies. The Storrs, CT, study [*8-9*] shows a first flush Cu concentration for a new roof, averaged over 16 storms, of about 7000 μg/L and a steady-state concentration of about 2000 μg/L. The inter-storm variability was high for the first flush effect. Since the dry deposition contribution in Equation 1 depends in part on the elapsed time between storms, the inter-storm variability could be explained by dry deposition of acidic gases. The first flush effect was shown to increase with elapsed time between precipitation events by He et al. [*5*]. The results for a 70-year old copper roof in the Storrs, CT, study [*8-9*] were similar to those for the new roof, but the first flush and steady-state Cu concentrations were lower. This is in contrast to results that indicate higher flow-averaged Cu concentrations for 40 and 100 year old copper roofs [*6*]. Pre-aged and unaged copper panels showed a first flush concentration of about 9530 μg/L and a steady state concentration of 2540 μg/L [*5*].

The geochemical modeling software was used to compute mineral solubility as a function of pH for water equilibrated with atmospheric carbon dioxide, Figure 9. The range of flow-averaged and steady state copper concentrations (but not first flush concentrations) observed in precipitation runoff, Table 5, is the shaded area. Minerals above the shaded area are soluble in the runoff. Solubility curves crossing though the shaded area indicate regions of limited solubility. Minerals with curves below the shaded area would tend to be insoluble in the runoff. These include the minerals thought to be terminal phases in the sequence of increasingly insoluble minerals that form the patina on weathering, brochantite at the polluted sites, atacamite at the marine site [*12,14-16*]. The presence of significant concentrations of copper ions in the runoff, including weak acid contributions, suggests the mineral chemistry at the patina surface is complex and that minerals early in the solubility sequence are present on the patina surface. Furthermore, a mineral dissolution-reprecipitation process in concentrated solutions produced on drying appears to significantly add to the availability of soluble minerals. These soluble minerals are likely amorphous, represent transitional phases in the sequence of minerals in the patina surface, and are the major source of copper ions removed in precipitation runoff. The terminal mineral phases probably do not contribute significantly to the runoff. The presence of the quite soluble chalconatronite at the Albany and Newport sites would appear to be a temporary occurrence brought on by the precipitation event.

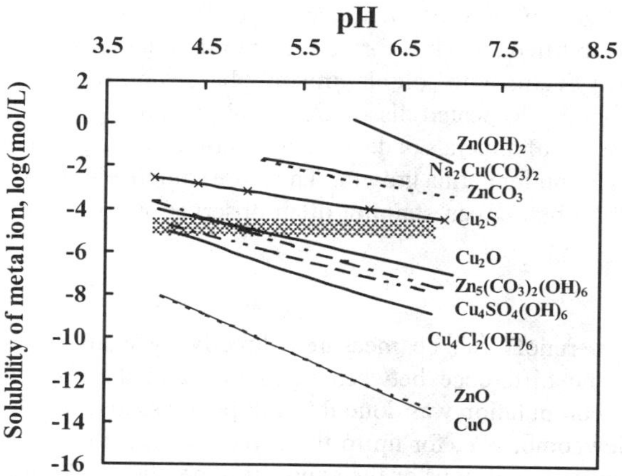

Figure 9 - *Solubility of some minerals in copper corrosion film in water in equilibrium with atmospheric carbon dioxide.*

Competing Ion Effects

The original copper runoff data for the Newport site is shown in Figure 10 as a function of time. The south approach to the Yaquina Bay Bridge, adjacent to the test site, was being thermal sprayed with zinc when the runoff experiment began. The original thermal-spray work began about 150 yards away from the test site and gradually approached the site. The final work concluded within 30 yards. Figure 10 shows significant concentrations of zinc were measured in the precipitation runoff from the copper panel as the project work approached the test site. Thermal spraying ended at the vertical dashed line. Several more months were required for zinc to wash out of the patina. In fact it never did, but only washed out of the patina surface. X-ray maps showed significant zinc in the weather patina (but not the cuprite). ASEM analyses of polished cross-sections showed as much as 8-10 atomic % zinc in the patina. XPS showed no zinc at the patina surface.

About a year later, at the two year mark, the contractors equipment was moved, stirring up zinc dust that was again detected in the copper runoff.

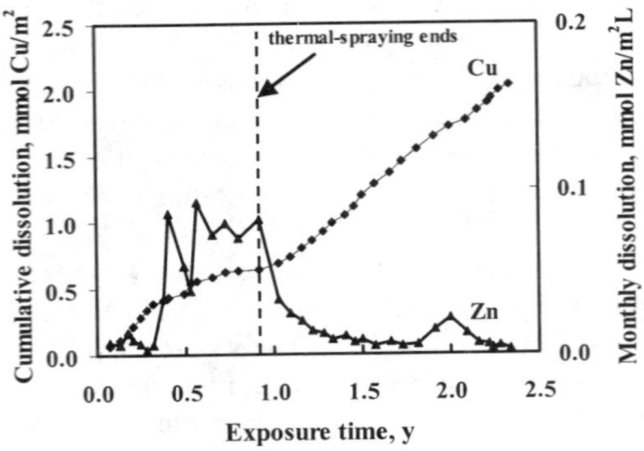

Figure 10 - *Copper runoff results showing effect of zinc overspray.*

Copper runoff losses from the patina were significantly depressed while zinc was present on the patina surface. This was observed from about 0.3 to 1.0 years and again at 2.0 years, Figure 10. Because zinc dust has a large surface area to volume ratio, the individual dust particles quickly oxidized to amorphous zincite (ZnO) and zinc hydroxide. Zinc hydroxide is more soluble than copper minerals, Figure 9. It successfully competed for hydrogen ions available from carbonic acid, raising the solution pH and depressing the copper dissolution. The slopes of the copper runoff curves were the same before and after copper dissolution was depressed. Because of this, the Newport runoff curve in Figure 6 and 7 was adjusted by removing affected region and the slope of the resulting curves was that for copper in the absence of competing alkaline minerals.

The ability of the patina to shield zinc within the weathered layer from the external environment suggests the layer is not modified once it is formed and nor does it dissolve in the runoff. This fits with a patina that begins with soluble mineral phases on its surface produced by outward migration of copper. Repeated dissolutions and precipitations sort through these phases producing a sequence of minerals of decreasing solubility, ending in the terminal phase. This phase is the foundation for patina growth. The more soluble phases on the surface contribute to runoff and the blue-green staining often observed when runoff streaks a structure surface.

Residence Time Effects

As noted earlier, precipitation runoff can be measured directly by analysis of collected solutions or indirectly as the difference between mass loss and the copper accumulated in the patina. The latter computation was done for multiple mass loss panels exposed at Washington, DC, and Newcomb, NY, for up to three years over a four year period. Regardless of the initial exposure conditions or the exposure time, the results for each site fell on a straight line with an R^2 of 0.95 for both Washington, DC, and Newcomb,

NY, Table 5. The fitted lines for the small mass loss panels and the experimental data and fitted curve for the larger runoff panel are shown in Figure 11. The smaller panel mass loss panel gave cumulative copper runoff values 58 % higher than the larger runoff panel. This was unexpected and all data and calculations were reviewed with no evidence of significant error.

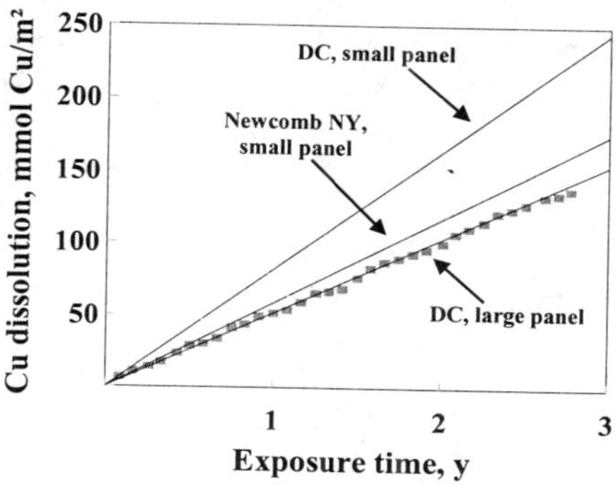

Figure 11 - *Copper result as a function of panel size.*

A plausible explanation is residence time and solubility effects on the weak acid contribution to the runoff. The argument is that the dry deposition and strong acid contributions are unaffected by panel size. The dry deposition contribution occurs during periods of dryness; the strong acid reaction is fast and should not be affected by mass transfer considerations. On the other hand, as runoff moves down the patina from a higher to lower position, the weak acid contribution will be reduced by the rising concentration of copper ion, depressing mineral dissolution. It will also be reduced by the thickening water layer draining the surface and only incrementally refreshed by direct precipitation. In the latter case, replenishment of consumed carbonic acid requires transfer of carbon dioxide across the air-water interface and then sufficient mixing to bring the fresh carbonic acid in contact with the patina. Higher on the surface the water layer will be thinner, carbon dioxide saturated direct precipitation will be less diluted by mixing, and equilibrium with atmospheric carbon dioxide more likely. Lower on the surface there may be a significant gradient in carbonic acid from the patina outwards and dissolution of patina constituents controlled by this gradient.

The smaller 15 cm (6 in.) long mass loss panels correspond to the up 25 % of the larger 61 cm (24 in.) runoff panel. The flow off the lower edge of the upper 25 % of the runoff panel (represented by the mass loss panels) contains 5430 μg Cu/L based on the mass loss runoff rate and the annual average precipitation rate for Washington DC, Table 5. Flow off the lower edge of the panel is four times the flow off the upper 25 % and contains 4320 μg Cu/L. The partitioned runoff contributions off the larger panel at the end of the experiment were 23 % dry deposition, 45 % strong acid and 32 % weak acid, Figure 8. The dry deposition and strong acid contributions (on an area basis) were the same on the upper 25 % of the panel as on the whole panel. The difference in the small panel and large panel runoff curves then requires the contributions to runoff from the upper 25 % (small panel) to be 12 % dry deposition, 31 % strong acid, with the weak acid contribution increased to 57 %. In other words, as the runoff leaves the upper 25 % of the panel, the contribution of weak acid to the total runoff decreases from 57 % to 32 % at the lower edge of the large panel. This means that about 70 % of the total weak acid contribution to the runoff from the large panel occurred on the upper 25 % of the panel where carbon dioxide saturated precipitation had a greater effect on removing patina constituents. It would suggest that less thinning of the patina occurs lower on a structure.

Empirical Model for Copper Atmospheric Corrosion

Copper runoff contributes to copper corrosion by continuing the corrosion process even after the patina is well developed. Spence and Haynie [*24*] have suggested an empirical model that can connect the effects of copper runoff with those of the corrosion process. The model was developed to describe the atmospheric corrosion involving the competing process of corrosion film formation and corrosion film dissolution and removal in precipitation runoff. It involves two parameters, one which might be considered a diffusivity for copper in the patina, α, and one which has been described here in some detail, the copper runoff rate, β. The basic equation is

$$dM/dt = \alpha/T = dT/dt + \beta \qquad (2)$$

where M is the total corrosion mass loss, T is the total mass of copper in the patina, and t is time. Substitution of α/T for the corrosion rate allows the patina to grow parabolically early in the corrosion process. The integrated form of the model is

$$M = \alpha/\beta\,[1 - \exp\{-\beta M/\alpha\}] + \beta t \qquad (3)$$

While mass loss cannot be solved for explicitly, the equation is relatively easy to use and fit to experimental data with spreadsheet software. The equation has the following properties, parabolic corrosion kinetics for short times and long-term linear kinetics reflecting a fully developed patina, i.e., the steady state patina. It can easily be put into dimensionless form using the following derived quantities: (1) the steady-state patina thickness $T_s = \alpha/\beta$; and (2) a characteristic time $t_c = \alpha/\beta^2 = T_s/\beta$. t_c is roughly the time when corrosion transitions from parabolic to linear kinetics. The dimensionless equation model is

$$M^* = [1 - \exp\{-M^*\}] + t^* \qquad (4)$$

It is a single curve and shown in Figure 12. The text in this figures shows how each of the derived quantities varies with changes in the parameters α and β. The model expresses the synergism of environmental factors and patina characteristics on the corrosion process. Combinations of the two parameters characterize the state of the corrosion process. When the dimensionless time is well below one, the kinetics are parabolic and the corrosion process is young with respect to the environment, regardless of the actual exposure time. When the dimensionless time is well above one, the kinetics are linear, the corrosion process is mature, and the patina is well developed, regardless of the actual exposure time.

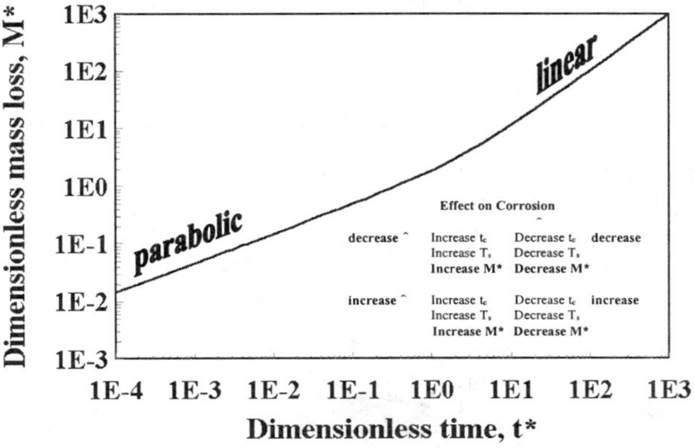

Figure 12 - *Empirical model for copper atmospheric corrosion showing effects of environmental and corrosion film parameters.*

This paper is not about applications of the empirical model except to illustrate the effect runoff has on total mass loss and the release of copper into the environment. The model was fit to three sets of data, two from the present study, i.e., three year mass loss and patina data from Washington, DC, and Newcomb, NY. The third set was information on a 100-year old roof in Stockholm[5] plus several assumptions. These were that after 100 years the steady-state patina thickness had been achieved and could be computed from available thickness and composition information. β was then varied until the model gave a film thickness at 100 years that was about 98 % the value computed from the composition and thickness information. The results from these efforts are presented in Table 6. The surprise was Washington, DC, where the characteristic time was only 2.8 years. Other than that the times and thickness are not unreasonable. The diffusivity is on the outside edge of values mentioned regarding parabolic atmospheric corrosion kinetics [*16*].

Table 6 B *Atmospheric corrosion model parameters.*

Site	Panel age, y	α		β	T_s	t_c
		μm²/y	cm²/s	μm Cu/y	μm Cu	y
Washington DC	3	0.95	0.29×10^{-15}	0.58	1.6	2.8
Newcomb NY	3	3.9	1.2×10^{-15}	0.42	9.3	22.1
Stockholm Sweden	100	13.9	4.4×10^{-15}	0.9	15.5	17.2

The 100-year old Stockholm rooftop provides a good opportunity to consider the relative effects of patina growth and runoff on total corrosion losses. This is done with the clear understanding that environmental factors can change substantially over 100 years. The mass loss, runoff, and patina (corrosion film) in Figure 13 were computed for the 100 year period using the parameter values in Table 6. The curves show the patina contains more than 50 % of the copper lost by corrosion during the first 15 years. Beyond this time runoff becomes the dominant destination for copper ions. After 100 years, according to the model, only 15 %

of the copper lost by corrosion was still contained in the patina. The balance, 85 %, was lost to the environment in precipitation runoff.

Conclusions

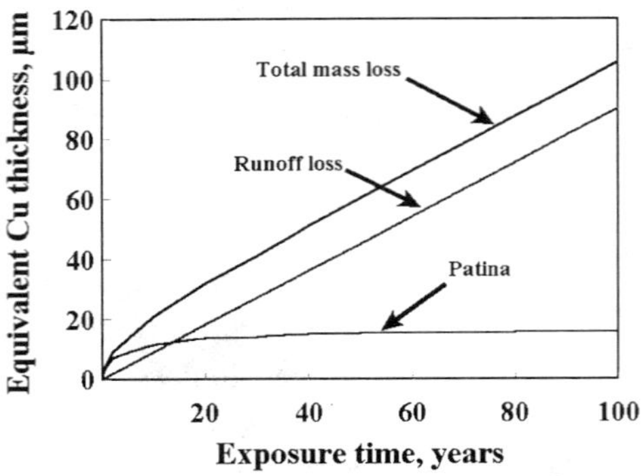

Figure 13 - *Comparison of corrosion film growth and runoff for 100-year old roof in Stockholm, Sweden.*

Cumulative copper precipitation runoff at an unpolluted marine site, two polluted sites, and an unpolluted site were linear with time and over a roughly three-year period indicating a constant rate of removal of copper from the patina. The runoff removal rate was characteristic of the site environment and of the patina composition. Precipitation runoff removal rates on large panels were 3.3 (0.37), 1.7 (0.19), and 1.7 (0.19) g Cu/m^{2}y (µm Cu/y) for the Washington, DC, Albany, OR, and Newport, OR, sites, respectively. Dissolution and precipitation of patina constituents tends towards a patina of increasingly insoluble mineral phases, with brochantite the more common terminal phase at polluted sites, atacamite the marine site, cuprite and tenorite at the unpolluted site. The dissolution/precipitation process primarily involves constituents near the outer surface of the patina that are early in the sequence of minerals forming the terminal phases. These transitional minerals are the primary source of the copper ions removed in precipitation runoff, rather than the terminal phases.

In a simplified way, the transitional minerals are neutralized by acidic gases delivered to the surface in dry deposition, and strong acids and weak acid in wet deposition. These effects can be partitioned into individual contributions to precipitation runoff using geochemical modeling. Brochantite precipitation during winter when sulfur dioxide concentrations were highest at the Washington, DC, site can also be predicted by geochemical modeling. More alkaline minerals such as zinc corrosion products can suppress dissolution of the copper patina. Weak acid contributions to precipitation runoff are reduced as the water layer on a surface thickens and increases in copper concentration. This reduces the area-averaged runoff removal rate for surfaces with longer inclined axes and results in greater thinning of the patina towards the upper edge of the surface.

An empirical atmospheric corrosion model with early parabolic corrosion kinetics for copper and long-term linear kinetics, and including synergy between properties of the copper patina and the environment, may be a useful approximation of copper atmospheric corrosion kinetics. Based on limited data, it shows the well-known characteristic of high retention of copper in the corrosion film early in the corrosion process. But in an application to a 100-year old roof, it suggests that the patina may contain only 15 % of the copper consumed in corrosion reaction. The balance is lost to the environment in precipitation runoff.

References

[1] Cramer, S. D., Matthes, S. A., Holcomb, G. R., Covino, Jr., B. S. and Bullard, S. J.,

APrecipitation Runoff and Atmospheric Corrosion,@ Paper 00452, *Corrosion/2000*, NACE International, Houston TX, 2000, 15 pp.

[2] Cramer, S. D., McDonald, L. G., and Spence, J. W., *Proceedings of the 12th International Corrosion Congress*, NACE International, Houston TX, Vol. 2, 1993, pp. 722-733.

[3] Spence, J. W., Haynie, F. H., Lipfert, F. W., Cramer, S. D., and McDonald, L. G., AAtmospheric Corrosion Model for Galvanized Steel Structures,@ *Corrosion*, Vol. 48, No. 12, 1992, pp.1009-1019.

[4] Odnevall Wallinder, I., and Leygraf, C., AA Study of Copper Runoff in an Urban Atmosphere,@ *Corrosion Science*, Vol. 39, No. 12, 1997, pp. 2039-2052.

[5] He, W., Odnevall Wallinder, I., and Leygraf, C., AA Laboratory Study of Copper and Zinc Runoff During First Flush and Steady-State Conditions,@ *Corrosion Science*, Vol. 43, 2001, pp. 127-146.

[6] He, W., Odenvall Wallinder, I., and C. Leygraf, AA Comparison Between Corrosion Rates and Runoff Rates From New and Aged Copper and Zinc as Roofing Materials,@ Water, *Air and Soil Pollution*, in press.

[7] Gromaire-Mertz, M. C., Garnaud, S., Gonzales, A., and Chebbo, G., ACharacterization of Urban Runoff Pollution in Paris,@ *Wat. Sci. Tech.*, Vol. 39, No. 2,1999, pp. 1-8.

[8] Boulanger, B., and Nikolaidis, N. P., AMobility and Aquatic Toxicity of Copper in an Urban Watershed,@ *J. of American Water Resources Association*, in press.

[9] Michels, H. T., Boulanger, B., and Nikolaidis, N. P., "Copper Roof Sotrmwater Runoff – Corrosion and the Environment," Paper No. 02xxx, Corrosion/2002, NACE International, Houston TX, 2002.

[10] Barron, T. S., "Architectural Uses of Copper – An Evaluation of Stormwater Pollution Loads and Best Management Practices," Palo Alto Regional Water Quality Control Plant, Palo Alto CA, 2000.

[11] Odnevall Wallinder, I., Leygraf, C., Karlen, C., Heijerick, D., and Janssen, C. R., AAtmospheric Corrosion of Zinc-Based Materials: Runoff Rates, Chemical Speciation and Ecotoxicity,@ *Corrosion Science*, Vol. 43, 2001, pp. 809-816.

[12] Leygraf, C., and Graedel, T. E., *Atmospheric Corrosion*, Wiley-Interscience, New York, 2000, pp. 140-148, 269-280.

[13] U. S. Environmental Protection Agency Drinking Water Standards Program website, http://www.epa.gov.safewater/standard/leadfs.html

[14] Graedel, T. E., Nassau, K., and Franey, J. P., ACopper Patinas Formed in the Atmosphere B I. Introduction", *Corrosion Science*, Vol. 27, No. 7, 1987, pp. 639-657.

[15] Graedel, T. E., ACopper Patinas Formed in the Atmosphere B II. Qualitative Assessment of Mechanisms", *Corrosion Science*, Vol. 27, No. 7, 1987, pp. 721-740.

[16] Graedel, T. E., ACopper Patinas Formed in the Atmosphere B III. A Semi-Quantitative Assessment of Rates and Constraints in the Greater New York Metropolitan Area", *Corrosion Science*, Vol. 27, No. 7, 1987, pp. 741-769.

[17] Cramer, S. D., and McDonald, L. G., AAtmospheric Factors Affecting the Corrosion of Zinc, Galvanized Steel, and Copper,@ *Corrosion Testing and Evaluation*, ASTM STP 1000, R. Baboian and S. W. Dean, Eds., American Society for Testing and

Materials, Philadelphia, 1990, pp. 208-224.

[18] Flinn, D. R., Cramer, S. D., Carter, J. P., Hurwitz, D. M., and Linstrom, P. J., AEnvironmental Effecs on Metallic Corrosion Products Formed in Short-term Atmospheric Exposures,@ *Materials Degradation Caused by Acid Rain*, ACS Symposium Series 318, R. Baboian, Ed., American Chemical Society, 1986, pp. 119-151.

[19] Odnevall Wallinder, I., Verbiest, P., He, W., Leygraf, C., AEffects of Exposure Direction and Inclination on the Runoff Rates of Zinc and Copper Roofs,@ *Corrosion Science*, Vol. 42, 2000 pp. 1471-1487.

[20] National Acid Deposition Program, *NADP/NTN Annual Data Summary: Precipitation Chemistry in the United States. 1994*, Natural Resource Ecology Laboratory, Colorado State University, Fort Collins CO, 1996, 256 pp.

[21] Parkhurst, D. L., *User=s Guide to PHREEQC B A Computer Program for Speciation, Reaction-path, Advective-Transport, and Inverse Geochemical Calculations*, Water Resources Investigations Report 95-4227, U. S. Geological Survey, Lakewood CO, 1995, 143 pp.

[22] Kubaschewski, O., and Hopkins, B. E., *Oxidation of Metals and Alloys*, Butterworths, London, 1962 pp. 5-14.

[23] Cramer, S. D., Carter, J. P., Linstrom, P. J., Flinn, D. R., "Environmental Effects in the Atmospheric Corrosion of Zinc," *Degradation of Metals in the Atmosphere*, ASTM STP 965, S. W. Dean and T. S. Lee, Eds., American Society for Testing and Materials, Philadelphia, 1988, pp. 229-247.

[24] Spence, J. W., and Haynie, F. H., ADerivation of A Damage Function for Galvanized Steel Structures: Corrosion and Kinetics and Thermodynamic Considerations,@ *Corrosion Testing and Evaluation*, ASTM STP 1000, R. Baboian and S. W. Dean, Eds., American Society for Testing and Materials, Philadelphia, 1990, pp. 208-224.

S. A. Matthes,[1] S. D. Cramer,[1] B. S. Covino, Jr.,[1] S. J. Bullard,[1] and G. R. Holcomb[1]

Precipitation Runoff From Lead

Reference: Matthes, S. A., Cramer, S. D., Covino, Jr., B. S., Bullard, S. J., and Holcomb, G. R., "**Precipitation Runoff From Lead,**" *Outdoor Atmospheric Corrosion, ASTM STP 1421*, H. E. Townsend, Ed., American Society for Testing and Materials International, West Conshohocken, PA, 2002.

Abstract: Lead ions may be introduced into the environment by the flow of precipitation runoff from the surface of lead structures such as gutters, roofs, piping, siding, and sculpture. Precipitation runoff is water from rain, dew, or fog that drains from a surface and contains air or water-born deposited reactants and soluble ions from the metal surface. Analysis of precipitation runoff from sites in Newport (marine unpolluted) and Albany (rural unpolluted), Oregon, was used to characterize these sites. Typical lead concentrations found in the precipitation runoff were between 0.7 and 3.7 mg/L compared with the United States EPA lead drinking water standard of 0 mg/L (with an action level of 0.015 mg/L). Corrosion film studies indicate that lead in the runoff is primarily from the solubility of cerrusite (lead carbonate) and hydrocerrusite (lead hydroxy carbonate). After an initial induction period, the measured release rate of lead ions to the environment was a constant 0.010 millimoles Pb per liter of precipitation runoff flowing over one square meter of lead surface (2.1 mg Pb/L) at both Albany and Newport. Cumulative corrosion film dissolution rates were 14.3 and 19.6 mmol Pb/m^{2}y for Albany and Newport, respectively. This corresponds to steady state lead corrosion rates of 0.26 and 0.36 μm/y respectively. Ionic species dry deposited onto the lead surface were determined from precipitation runoff data, giving valuable information concerning the impact of environmental constituents and pollution on lead corrosion.

Keywords: lead, runoff, atmospheric corrosion, wet deposition, dry deposition, cerussite, hydrocerussite, environmental effects, pollution, marine, rural

Introduction

Lead is of particular concern to the public, because of the adverse nature of even very small amounts on human health. The official view that lead is an extreme health hazard is reflected in the fact that the EPA permits 0 mg/L Pb in drinking water with an action

---

1 Chemist, Chemical Engineer, Research Chemist, Research Chemist and Materials Engineer, respectively, Albany Research Center, U.S. Department of Energy, 1450 Queen Ave. SW, Albany, OR 97321.

Figure 1 - *Center for the Visual Arts in Toledo, Ohio.*

level of 0.015 mg/L [*1*]. In spite of the concern over lead as a health risk, lead is frequently used as an architectural material. Older structures used lead as a desirable material for gutters, flashing, siding, and pipes. Modern structures may use considerable quantities of lead for aesthetic purposes [*2*], such as the Center for Visual Arts in Toledo, Ohio where part of the structure is sheathed in lead clad copper sheet (Figure 1).

Concern over the effects of lead has resulted in numerous studies detailing the amount of lead present in urban runoff [*3-4*]. In a study on the urban runoff pollution in Paris, Gromaire-Mertz [*4*] reported that the average lead concentration values of samples collected from roof runoff, yard runoff, and street runoff all exceeded the Level 2 water quality limit for France of 0.050 mg/L Pb (Level 2 is the permitted maximum concentration allowed for irrigation and animal uses). Some of the urban runoff samples collected even exceeded the maximum allowable concentration for lead in water discharges for industries (0.5 mg/L Pb). Gromaire-Mertz goes on to state that

> *"...roof runoff, far from being unpolluted, is an even greater source of pollutants than streets. Up to very high heavy metals concentrations measured on roof runoff samples, their direct discharges could have toxic effects on natural waters and their local infiltration would rapidly lead to soil contamination. The use of metals like zinc, lead, or copper for roof covering and gutters appear prejudicial for runoff quality."* [*4*]

Lead runoff results from the solubility of the corrosion film that develops on the metal surface when undergoing atmospheric corrosion. If the corrosion film is soluble, then lead will be found in the runoff. If the corrosion film is insoluble, then it acts as a protective barrier to the metal surface reducing corrosion and limiting the lead available in the runoff. In the past it was commonly believed that lead was corrosion resistant. In fact, as recently as 1963 the *Encyclopedia of Engineering Materials and Processes* reported that lead is noted for corrosion resistance and rapidly forms a protective film highly resistant to corrosion [*5*].

However, DuRose noted that while lead is very resistant to corrosion in New York City, this is not true at marine sites [*6*]. Tranter used infrared spectroscopy to study the composition of the corrosion film on lead in an urban environment, and found that initially the film consisted of lead hydroxy carbonates and lead oxide (hydroxide), but over 18 months became almost entirely lead sulfate and lead sulfite, Table 1 [*7*].

Lead runoff is part of the total corrosion mass loss as a function of time, t, represented by the equation

$$M(t) = T(t) + R(t) \tag{1}$$

where $M(t)$ = cumulative corrosion mass loss, g Pb/m^2
$T(t)$ = corrosion film mass, g Pb/m^2
$R(t)$ = cumulative loss of corrosion products in precipitation runoff, g Pb/m^2

Table 1 - *Composition of Lead Corrosion Film (urban environment)* [7]

Compound	Percent composition (by weight) versus Exposure period			
	3 days	4 months	9 months	18 months
$Pb_x(OH)_y(CO_3)_z$	45	-	-	-
$PbSO_3$	7	45	60	35
$PbSO_4$	3	45	40	65
$PbCO_3$	tr	10	2	1
PbO	45	-	-	-

The soluble corrosion products that account for lead ions in the runoff are represented by R(t) above. After formation of the corrosion film begins, the film is subject to removal by dissolution in precipitation runoff (water from rain, dew, or fog). In wet deposition, basic compounds (oxides, hydroxides, carbonates) will be dissolved by reaction with acidic species present in the precipitation due to acid rain or dissolved CO_2 (carbonic acid), and corrosion film is removed from the surface to drain away as runoff. In dry deposition, particulates and acidic gases in the atmosphere react with the corrosion film during periods when there is little or no runoff. Neutral salts produced by these reactions accumulate on the metal surface to be removed during the next period of precipitation.

The time derivative of Equation 1 shows that the corrosion rate (*dM/dt*) is equal to the rate of protective film growth plus the rate of corrosion film loss from runoff

$$dM/dt = dT/dt + dR/dt \tag{2}$$

At steady state, net corrosion film growth ceases and the corrosion rate is equal to the precipitation runoff rate, i.e.

$$dM/dt = dR/dt = \beta \tag{3}$$

where β is the steady state precipitation runoff rate, and

$$dT/dt = 0 \tag{4}$$

Recent results [*8*] for runoff from zinc surfaces suggest that precipitation runoff losses vary linearly with time and are relatively insensitive to seasonal variations in precipitation chemistry, air chemistry and meteorology. The present paper reports the results of precipitation runoff from lead surfaces exposed at two sites having different local environments.

Figure 2 - *Runoff collection site at Newport, OR.*

Experimental Design

Exposure Sites and Material

Lead precipitation runoff experiments were conducted at an unpolluted marine site in Newport, OR, Figure 2, and an unpolluted rural site in Albany, OR. The concentration of atmospheric SO_2 at both sites was zero, and is reflected in the high rainfall pH values. Table 2 lists the environmental conditions of the collection sites during the experiment. The Newport site (at the Yaquina Bay Bridge) is within 100 m of the Pacific Ocean, and the Albany site is in the Willamette Valley, a rural area 83 km from the ocean. Flat panels of high purity lead sheet (> 99.9 pct Pb), measuring 0.305 × 0.610 m (1 × 2 ft), were mounted in polyethylene trays that collected all precipitation washing the exposed (skyward) side of the panels. The exposed area of the runoff panel was 0.184 m^2. The panels were degreased with methanol and dried prior to installation. The groundward side was masked with electroplating tape limiting runoff contributions to only the exposed (skyward) side. The trays (with panels) were inclined at 30° to the horizon and faced west into the prevailing winds for both Newport and Albany.

Table 2 - *Average annual environmental properties of the runoff collection sites.*

Exposure site	Site type	Precipitation rate cm/y	(L/m^2y)	Precipitation pH	Temperature °C
Albany OR	Rural	108	(1084)	5.78	11.2
Newport OR	Marine	182	(1822)	6.13	10.5

Precipitation Runoff Measurements

Precipitation runoff samples were collected at least monthly for 2.3 years (September 1996 to January 1999). The collection system minimized evaporation losses and contamination. Up to three collections per month were made during periods of heavy rainfall. Incident precipitation was also collected on the same schedule as the runoff samples using an Aerochem Metrics wet/dry collector following National Acid Deposition Program (NADP) protocols [9]. Incident precipitation and runoff were filtered to remove undissolved particulates and analyzed for the acid rain ions (H^+, Ca^{+2}, Mg^{+2}, K^+, Na^+, NH_4^+, NO_3^-, Cl^-, and SO_4^{-2}) and selected metal ions (Pb, Cu, Zn, Fe) by pH meter, by ion chromatography and by inductively coupled plasma (ICP) emission spectroscopy. Low ionic strength pH standards were used for pH meter calibration, and calibration verification standards were used to ensure accurate metals analysis. The results provide data on the total lead concentration, and do not consider any chemical speciation of the lead ions.

Corrosion Film Chemistry

Corrosion films from both the Albany and the Newport lead panels were analyzed for mineral species by X-ray diffraction of the corrosion film on the panel and using powders scraped from the panel (Philips APD 3720 diffractometer). Back-scattered electron SEM (LEO Stereoscan S440 with Oxford eXLII energy dispersive spectrometer) photomicrographs were taken of polished cross-sections of corrosion films.

Results and Discussion

Corrosion Film

X-ray diffraction analysis of the lead corrosion films detected $PbCO_3$ (cerussite) and $Pb_3(CO_3)_2(OH)_2$ (hydrocerussite). No other crystalline species were present. If chlorides, hydroxychlorides, or sulfates are in the corrosion film, they exist as amorphous species. The electron back-scatter SEM photomicrographs of the Albany and Newport corrosion films are shown in Figure 3. The Albany corrosion film (3a) exhibits extensive cracking, which would allow penetration of the film by precipitation. The cracking also suggests that the film has low integrity and may be susceptible to flaking. The Newport film (3b) shows more porosity, less cracking, and a better developed crystal phase than Albany. The Newport corrosion film may have more integrity than the Albany film.

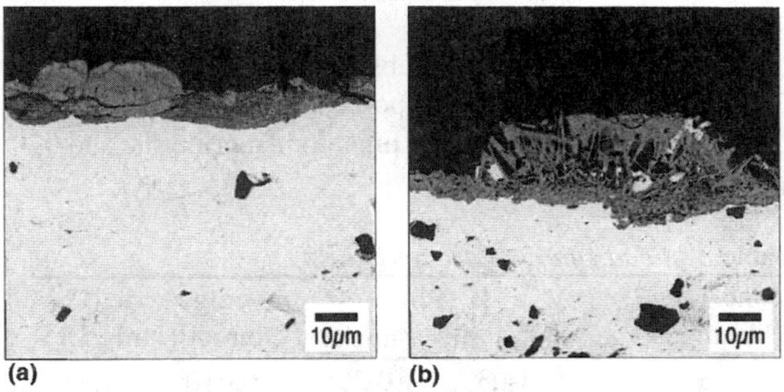

Figure 3 - *Electron back-scatter photomicrograph of Albany (a) and Newport (b) lead corrosion films.*

Runoff Chemistry

The range for lead ion concentration in the runoff was 0.7 to 3.7 mg/L Pb, significantly higher than the U.S. EPA action level of 0.015 mg/L for Pb in drinking water. Because both the Albany and Newport sites are not subject to acid rain (i.e. rainfall pH > 5.6), only hydrogen ions from dissociation of the weak acid H_2CO_3 (from dissolved CO_2) was available for dissolution of the alkaline lead compounds. As the lead compounds dissolve, hydrogen ion is consumed and bicarbonate ion produced, raising the pH level with a net increase of bicarbonate ion concentration in the runoff (Equation 5). The runoff solution that drains from the panel has a pH typically one unit higher than the incident rainfall.

$$Pb_3(CO_3)_2(OH)_2 + 4H_2CO_3 = 3Pb^{+2} + 6HCO_3^- + 2H_2O \quad \textbf{(5)}$$

Cumulative corrosion film dissolution curves for Albany and Newport are shown in Figure 4. The curves show periodicity caused by the variation between the Oregon rainy season in the winter and the dry season occurring in late summer. Even with these variations, it is apparent that the curves vary around a straight line over the collection period. The data were fitted by least squares to straight lines with slopes (dR/dt from Equation 2) of 19.6 and 14.3 mmol Pb/m^{2}y for Newport and Albany (Table 3) respectively, and R^2 values of 0.96 and 0.98. These slopes are the lead precipitation runoff rates and equivalent to the steady state runoff rate in Equation 3. Runoff rates in Table 3 expressed in terms of μm of lead removed per year are 0.36 and 0.26 μm Pb/yr for Newport and Albany. These values are comparable to reported corrosion rate values of 0.1 to 2.2 μm/y for a marine site and 0.4 to 1.9 μm/y for a rural site [*10*].

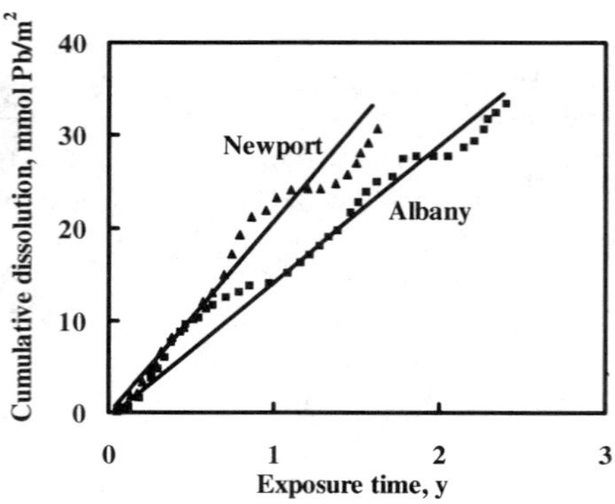

Figure 4 - *Cumulative lead corrosion film dissolution vs exposure time.*

Table 3 - *Lead runoff characteristics.*

Exposure site	Site type	conc. range (Pb) mg/L	β (Pb) mmol /m^{2}y	(μm/y)	average conc.(Pb) mmol/L	(mg/L)
Albany OR	Unpolluted	0.7 – 3.7	14.3	(0.26)	0.010	(2.1)
Newport OR	Marine	1.4 – 3.4	19.6	(0.36)	0.010	(2.1)

Lead runoff can also be viewed as a function of precipitation volume (Figure 5a). After an initial induction period, plots for both Newport and Albany are linear and have slopes corresponding to the steady-state concentration of Pb in runoff, 0.010 mmol Pb/L (2.1 mg Pb/L) with an R^2 value of 1.00 for both sites (Table 3). The concentrations for Albany and Newport are identical because lead in the runoff is due entirely to weak acid (carbonic acid) dissolution of the corrosion film. Since the carbonic acid concentrations in the precipitation at these two unpolluted sites are nearly the same, lead concentration in the runoff should be the same. The observed linear behavior shows no sign of change at the end of the study period (2.3 years) and may persist for much longer periods. During the induction period, differences in Pb concentration reflect the differing environmental conditions existing at each site while the corrosion film was first developing. In previous work [*9*], it was noted that a substantial deposition of basic particulates occurred on the panel at the Newport site, but not at Albany. These basic particulates can compete with the corrosion film for hydrogen ions produced by carbonic acid. The net effect is a higher rainfall pH and a reduction in the amount of corrosion film dissolved at Newport during the induction period. However, when the corrosion film is well developed, competition from these basic particulates becomes much less significant.

A range of lead concentrations of 0.7 to 3.7 mg Pb/L was observed during the individual collection periods. Higher rainfall rates typically occur during the late fall and winter at Albany and Newport and give slightly lower Pb concentrations in the runoff than the average runoff concentration of 2.1 mg Pb/L (Figure 5b). During a "hard rain" many of the raindrops never come into contact with the corrosion film, because a sheet of water covers the surface. This tends to dilute the Pb dissolved by water in contact with the corrosion film. Slower rates of rainfall, occurring in the summer and early fall at Albany and Newport, tend to have higher concentrations of Pb in the runoff than the average. When this occurs, rainfall has a longer residence time on the panel allowing more opportunity for Pb ions to dissolve in the runoff.

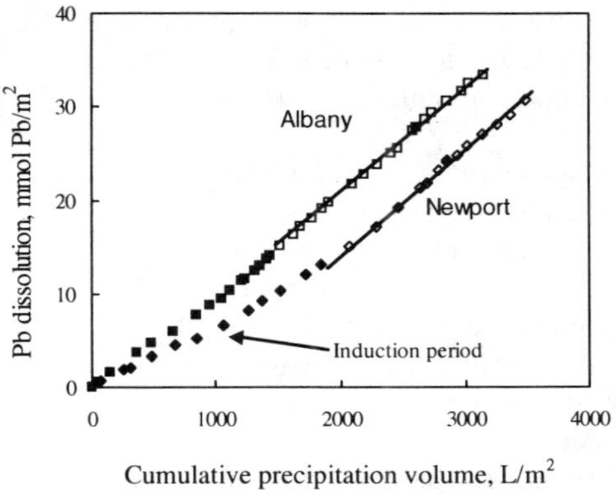

Figure 5a - *Cumulative lead corrosion film dissolution vs precipitation volume.*

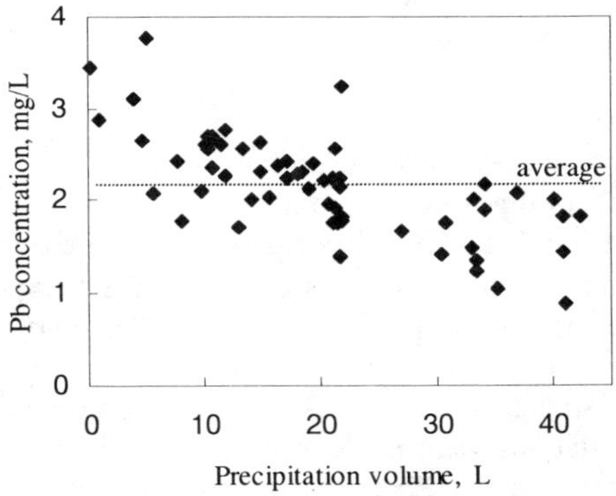

Figure 5b - *Pb concentration in the runoff as a function of precipitation volume for each collection period at Albany and Newport.*

Dry Deposition

Much information about the local environment can be obtained by analyzing runoff chemistry. For example, atmospheric particles impact the collection panels and are found in the runoff. Their concentrations can give important clues about the environment. A unique event occurred on the runoff panels at Newport, OR. Zinc was detected in runoff from the lead panel and in the incident precipitation collected by the wet-dry collector for part of a year during the study. Analysis of the data indicated that small amounts of zinc had been dry deposited on the panels.

This zinc deposition was coincident with the zinc thermal spraying of the south approach of the Yaquina Bay Bridge. The thermal spray zinc serves as the anode in an impressed current cathodic protection system for the bridge [*11*]. Even though the thermal spraying took place in a containment structure, enough zinc escaped to have a noticeable impact on the lead runoff during the thermal spraying (Figure 6). As the thermal spraying approached the collection site, zinc levels rose in the runoff from the lead panel until thermal spraying ceased.

Lead runoff was suppressed by the zinc metal particles and zinc corrosion products in the Pb corrosion film dissolving during precipitation events in preference to the less soluble hydrocerrusite during this period (Figure 6). Because zinc was partially incorporated in the Pb corrosion film there was a lag time until zinc levels in the runoff were reduced to normal background levels. Lead runoff resumed the normal linear behavior once the zinc returned to background levels after the thermal spraying was completed. Because of the effect of the zinc, data points for lead runoff and dry deposition were omitted from Figures 4, 5, and 8a for the collection period in which zinc was present in the Newport runoff. The lead runoff rate, β, was the same before the appearance of zinc in the runoff and after the zinc was flushed from the lead corrosion product surface.

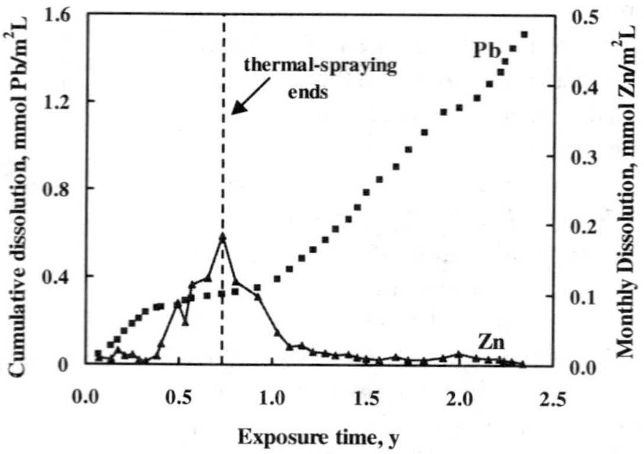

Figure 6 - *Uncorrected cumulative lead dissolution from Newport lead panel showing effect of zinc found in the runoff.*

Using a geochemical modeling program from the U.S. Geological Survey [*12*], calculations were made to produce a plot of solubilities versus pH of compounds likely present on the Pb panels (Figure 7). On top of the solubility curves is overlaid the concentration ranges of Pb found in the runoff solutions (cross-hatched area). All compounds with solubilities above the region of Pb runoff concentrations may contribute to the Pb runoff. Because the Pb concentrations of the runoff directly overlap the solubility curve of hydrocerrusite at the pH of the runoff solution and because hydrocerrusite was detected in the corrosion film by X-ray diffraction, it

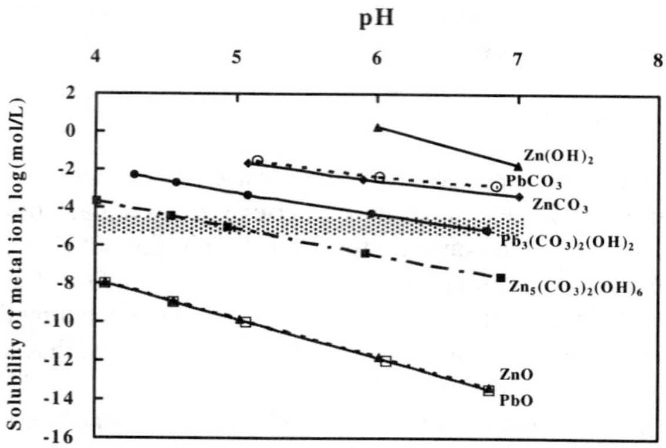

Figure 7 - *Solubility of minerals in lead corrosion film.*

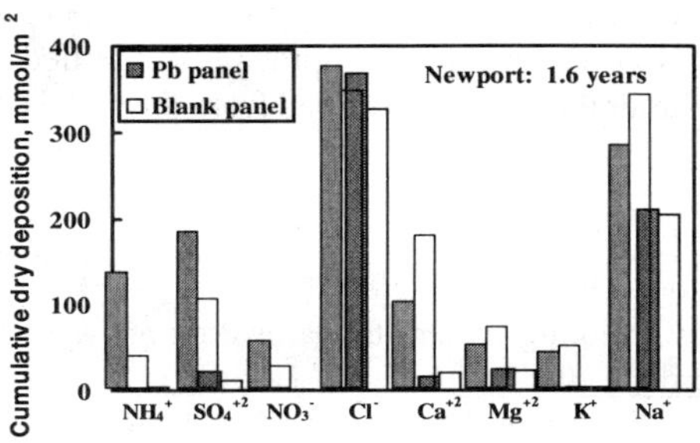

Figure 8(a) - *Cumulative dry deposition on Newport panel compared with blank panel.*

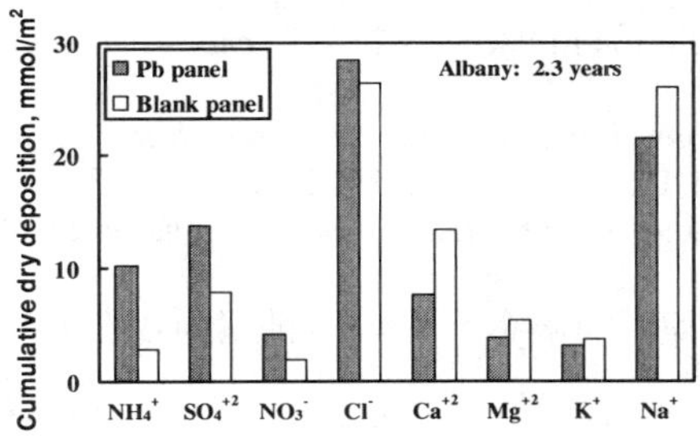

Figure 8(b) - *Cumulative dry deposition on Albany panel compared with blank panel.*

is likely that hydrocerrusite solubility in the corrosion film controls the final concentration of Pb in the runoff. Because the solubility of $PbCl_2$ (cotunnite) is orders of magnitude greater than hydrocerrusite [*13*], any $PbCl_2$ formed in the high chloride environment at Newport is removed completely during each runoff event. This prevents the accumulation of lead chloride in the corrosion film, verified by the X-ray diffraction data for the Newport lead panel.

Other ions present in the runoff were analyzed to determine the rate of dry deposition of those ions to the lead panel. Figures 8(a) and 8(b) show the relative rate of dry deposition of NH_4^+, SO_4^{-2}, NO_3, Cl^-, Ca^{+2}, Mg^{+2}, K^+, and Na^+ to the lead panel at Albany and Newport compared with deposition to Lexan plastic panels (blank). Dry deposition of NaCl dominates at the Newport site, and the results can be used to determine the rate at which NaCl is deposited on surfaces at marine sites. At the rural inland site (Albany), NaCl also predominates but is present as a much smaller percentage of the total dry deposited species. Differences in dry deposition between the Pb panel and the blank indicate that the Pb corrosion film serves as an "accumulator" for some ionic species (NH_4^+, SO_4^{-2}, NO_3^-), tending to concentrate them in the film until they are washed away in the next rainfall event.

Conclusions

Precipitation runoff from Pb surfaces at both the unpolluted marine and rural sites behaves in a predictable way and is based on the precipitation volume and the Pb corrosion product solubility. X-ray diffraction studies of the corrosion film indicate it is composed of cerrusite (lead carbonate) and hydrocerrusite (lead hydroxy carbonate). As indicated by the plot of solubility of hydrocerrusite versus pH, the final Pb concentration in the runoff is determined by the solubility of hydrocerrusite, at least at sites where lead carbonate species predominate in the corrosion film. Environmentally significant concentrations of Pb are found in the runoff (0.7- 3.7 mg/L) from Pb panels at both Albany and Newport.

The average Pb concentration in the runoff once the corrosion film was well developed was 0.010 mmol Pb/L (2.1 mg Pb/L) for both the unpolluted marine and rural sites. The steady-state Pb runoff rates were 14.3 mmol Pb/m^2y (0.26 μm Pb/y) for Albany and 19.6

mmol Pb/m^2y (0.36 μm Pb/y) for Newport. Lead runoff was suppressed from the Newport panel during a period when zinc dust from an adjacent zinc thermal spray project settled on the lead panel surface. Lead runoff continued after the zinc was flushed from the surface at the same rate as before the zinc appeared.

References

[1] U.S. Environmental Protection Agency Drinking Water Standards Program website, http://www.epa.gov/safewater/standards/leadfs.html

[2] Holmes, J.F., "Lead and Lead Alloys, Corrosion Resistance- Part 1," *Corrosion Technology*, Vol. 10, 1963, pp. 11-13.

[3] Boller, M., "Tracking Heavy Metals Reveals Sustainability Deficits of Urban Drainage Systems," *Water Science Technology*, Vol. 35, No. 9, 1997, pp. 77-87.

[4] Gromaire-Mertz, M.C., Garnaud, S., Gonzales, A., and Chebbo, G., "Characterization of Urban Runoff Pollution in Paris," *Water Science Technology*, Vol. 39, No. 2, 1999, pp. 1-8.

[5] Travis, B.R.,"Lead Coatings," *The Encyclopedia of Engineering Materials and Processes*, H.R. Clauser, Ed. Reinhold, Eds., New York, 1963, pp. 370-373.

[6] DuRose, A.H.," Atmospheric Exposure of Electroplated Lead Coatings on Steel," *ASTM STP 197*, 1957, pp. 97-106.

[7] Tranter, G.C, "Patination of Lead- An Infrared Spectroscopic Study," *British Corrosion Journal*, Vol. 2, No. 4, 1976.

[8] Cramer, S.D., Matthes, S.A., Holcomb, G.R., Covino, B.S. Jr., and Bullard, S.J.,"Precipitation Runoff and Atmospheric Corrosion," *Paper No. 00452, CORROSION/ 2000*, NACE International, Houston TX, 2000, 16pp.

[9] National Acid Deposition Program, *NADP/NTN Annual Data Summary: Precipitation Chemistry in the United States. 1994,* Natural Resource Ecology Laboratory, Colorado State University, Fort Collins CO, 1996, 256 pp.

[10] Leygraf, C., and.Graedel, T., "The Atmospheric Corrosion of Lead," *Atmospheric Corrosion*, The Electrochemical Society, John Wiley and Sons, 2000, p 296.

[11] Covino, B.S. Jr., Cramer, S.D., Bullard, S.J., Holcomb, G.R., Collins, W.K., McGill, G.E., "Thermal Spray Anodes for Impressed Current Cathodic Protection of Reinforced Concrete," *Materials Performance*, Vol.38, No.1, 1999, pp. 27-33.

[12] Parkhurst, D.L., "User's Guide to PHREEQC – A Computer Program for Speciation, Reaction-path, Advective-Transport, and Inverse Geochemical Calculations," *Water Resourses Investigations Report 95-4227*, US Geological Survey, Lakewood CO, 1995, 143 pp.

[13] Graedel,T.E., "Chemical Mechanisms for the Atmospheric Corrosion of Lead," *Journal of the Electrochemical Society*, Vol. 141, No. 4, 1994, pp. 922 - 927.

LONG-TERM OUTDOOR CORROSION PERFORMANCE OF ENGINEERING MATERIALS

Edward L. Hibner[1]

Evaluation of Nickel-Alloy Panels from the 20-Year ASTM G01.04 Atmospheric Test Program Completed in 1996

Reference: Hibner, E. L., "**Evaluation of Nickel-Alloy Panels from the 20-Year ASTM G01.04 Atmospheric Test Program Completed in 1996,**" *Outdoor Atmospheric Corrosion, ASTM STP 1421*, H. E. Townsend, Ed. American Society for Testing and Materials International, West Conshohocken, PA, 2002.

Abstract: This paper is an evaluation of 20-year atmospheric exposure test panels of Nickel 200 (UNS N02200), MONEL® alloy 400 (UNS N04400), INCONEL® alloy 600 (UNS N06600), INCONEL alloy 625 (UNS N06625, INCOLOY® alloy 800 (UNS N08800) and INCOLOY alloy 825 (UNS N08825)[2] from the ASTM Committee G01.04 Atmospheric Test Program, started in the United States of America's bicentennial year of 1976 and completed in 1996. The 20-year exposure test panels were evaluated to determine corrosion rates, pit depth data, and mechanical properties. The results are reported for each alloy as average values of three individual test panels exposed at each of five test sites. The test sites are ranked as to severity.

Keywords: nickel-alloys, atmospheric corrosion, corrosion rates, pit depths, mechanical properties

Introduction

Atmospheric corrosion has been a significant cause of metal degradation ever since metals have been used. Nickel alloys have been used in outdoor applications for many years. For example, alloy 400 is prominent on Washington D.C. rooftops: The Smithsonian Institute's Museum of Natural History, the National Gallery of Art and the Washington National Cathedral. Each building's rooftops are partially covered with 30 to 60 year-old alloy 400 materials.

The nickel-alloy atmospheric test panels from the ASTM Committee G01.04 Atmospheric Test Program were exposed for 20 years, with removal of some test panels in 1986 for evaluation after the initial ten-years of exposure. The exposure sites were: Kure Beach, NC, ocean front (25 m) test lot (East Coast Marine), Kearney, NJ (Industrial), Point Reyes, CA (West Coast Marine), State College, PA (Rural), and Panama Canal Zone (Tropical). Corrosion and mechanical property results are determined for the 20-year exposure nickel-alloy panels.

---

[1] Sr. Metallurgist, Research and Development, Special Metals Corporation, 3200 Riverside Dr., Huntington, WV 25705

[2] MONEL®, INCOLOY® and INCONEL® are Registered Trademarks of the Special Metals Family of Companies.

Panels tested at each of the five test sites exhibited excellent corrosion resistance with rates of <0.001mm/y (<0.04 mpy) and pit depth averages of <0.046mm (<0.0018 in.). In each of five test sites, panels exposed for 20 years exhibited excellent mechanical properties when comparisons are made to unexposed control panels. The ultimate tensile strength decreased by ≤ 2.1% (≤ 0.01% per year) and % elongation decreased by ≤ 7.8% (≤ 0.4% per year).

Experimental

Corrosion Evaluation

All exposed tests panels were examined under a light microscope at 20X magnification for pitting. Pit depths were measured with a depth gauge in accordance with ASTM G46, Standard Guide for Examination and Evaluation of Pitting Corrosion, *Annual Book of ASTM Standards*, Vol 03.02. For each alloy, comparisons were made to unexposed control panels. The ASTM Committee G01.04 Atmospheric Test Program called for cutting 2 tensile specimens from each of three 4 in. x 8 in. (102 mm x 203 mm) test panels to determine loss of mechanical properties. The values incorporated in Table 3 are averages of 6 test specimens, cut two each from the triplicate samples of each alloy.

Mass loss corrosion rates were determined in accordance with ASTM G1, Standard Practice for Preparing, Cleaning, and Evaluating Corrosion Test Specimens, *Annual Book of ASTM Standards*, Vol 03.02.

The test data for the panels removed after the initial 10 years of exposure is on file at Special Metals Corporation in Huntington, WV, USA. The 10 year data was not analyzed for this paper. Table 1 lists the nominal chemical composition of evaluated nickel alloys.

Table 1 – Nominal Composition of Evaluated Nickel Alloys

Alloy:	200 N02200	400 N04400	600 N06600	625 N06625	800 N08800	825 N08825
Ni	99.5	66.5	76.0	61.0	32.5	42.5
Cu	-	31.5	-	-	0.4	2.2
Cr	-	-	15.5	21.5	21.0	21.5
Mo	-	-	-	9.0	-	3.0
Fe	-	-	8.0	2.5	46.0	30.0
Al	-	-	-	0.2	0.4	0.1
Ti	-	-	-	0.2	0.4	0.9
Nb	-	-	-	3.6	-	-

Mechanical Testing

Room temperature tensile tests were conducted in accordance with ASTM E8, Standard Test Method for Tension Testing of Metallic Materials, *Annual Book of ASTM Standards*, Vol 03.01. The % loss in both ultimate tensile strength (UTS) and % elongation (%El) was calculated by dividing the parameter from the 20-year exposed panels by the same parameter determined from an unexposed panel of the alloy.

Examples:

$$\% \text{ loss in UTS} = \frac{\text{UTS exposed}}{\text{UTS unexposed}} \times 100 \quad (1)$$

$$\% \text{ loss in El} = \frac{\%\text{El exposed}}{\%\text{El unexposed}} \times 100 \quad (2)$$

Results and Discussion

Corrosion Data

Table 2 displays average 20-year exposure mass loss, corrosion rates and pit depths for nickel-alloys UNS N02200, N04400, N06600, N06625, N08800 and N08825. Panels tested at each of the five test sites exhibited excellent corrosion resistance with rates of <0.001mm/y (<0.04 mpy) and pit depth averages of <0.046mm (<0.0018 in.).

Mechanical Properties

Table 3 displays the ultimate tensile strength and % elongation, along with % loss in these parameters relative to unexposed panels. Panels exposed for 20 years in each of five test sites exhibited excellent mechanical properties when comparisons were made to control panels not exposed to the environment. The ultimate tensile strength decreased by ≤ 2.1% (≤ 0.01% per year) and % elongation decreased by ≤ 7.8% (≤ 0.4% per year).

Table 2 – *ASTM 1976 Atmospheric Test Program, 20 Years Exposure Results, Corrosion Rates and Pit Depths*

Location	Alloy	Average Mass Loss, mg/sq dm	Average Corrosion Rate		Average of 4 Deepest Pits, mm
			mdd	mm/y	
Kure Beach, NC (East Coast Marine)	N02200	455.1	0.06	<0.001	0.0463
	N04400	444.6	0.06	<0.001	0.0081
	N06600	23.95	0.0035	<0.001	0.0232
	N06625	2.6	0.00033	<0.001	0.0076
	N08800	26.4	0.0036	<0.001	0.0211
	N08825	20.36	0.003	<0.001	0.0169
Kearney, NJ (Industrial)	N02200	698.2	0.1	<0.001	0.0230
	N04400	652.5	0.09	<0.001	0.0178
	N06600	0.1	0.00002	<0.001	0.0091
	N06625	0	0	<0.001	0.0005
	N08800	0.31	0.00005	<0.001	0.0120
	N08825	0	0	<0.001	0
Point Reyes, CA (West Coast Marine)	N02200	87.07	0.01	<0.001	0.0184
	N04400	118.7	0.017	<0.001	0.0124
	N06600	5.27	0.0007	<0.001	0.0111
	N06625	0.97	0.00013	<0.001	0
	N08800	5.87	0.0007	<0.001	0.0098
	N08825	5.3	0.0006	<0.001	0.0046
State College, PA (Rural)	N02200	178.8	0.02	<0.001	0.00523
	N04400	211.9	0.03	<0.001	0.0066
	N06600	0.1	0.0004	<0.001	0.0048
	N06625	1.63	0.0002	<0.001	0.0034
	N08800	0.007	0	<0.001	0.0001
	N08825	2.5	0.0004	<0.001	0.0099
Panama Canal Zone (Tropical)	N02200	248.4	0.03	<0.001	0.043
	N04400	234.9	0.03	<0.001	0.0115
	N06600	7.4	0.001	<0.001	0.032
	N06625	0.6	0.00008	<0.001	0.0036
	N08800	10.13	0.001	<0.001	0.0096
	N08825	4.2	0.0006	<0.001	0.0013

Table 3 - *ASTM 1976 Atmospheric Test Program, 20 Years Exposure Results, Mechanical Properties*

Alloy	Test Sites	Ultimate Tensile Strength, MPa- Average		Elongation % in 50.8mm- Average	
		20 years	% loss	20 years	% loss
N02200	Initial	478.8		36.0	
	Kure Beach, N.C. (East Coast Marine)	477.2	0.3	35.1	2.5
	Kearney, N.J. (Industrial)	477.3	0.3	36.8	0
	Point Reyes, Ca. (West Coast Marine)	475.9	0.6	35.7	0.8
	State College, Pa. (Rural)	478.8	0	36.2	0
	Panama Canal Zone (Tropical)	476.2	0.5	35.6	1.1
N04400	Initial	536.6		40.0	
	Kure Beach, N.C. (East Coast Marine)	536.6	0	39.3	1.8
	Kearney, N.J. (Industrial)	534.7	1.9	38.7	3.3
	Point Reyes, Ca. (West Coast Marine)	536.1	0.09	38.6	3.5
	State College, Pa. (Rural)	536.2	0.07	38.8	3.0
	Panama Canal Zone (Tropical)	538.7	0	39.1	2.3
N06600	Initial	664.9		40.0	
	Kure Beach, N.C. (East Coast Marine)	666.6	0	36.9	7.8
	Kearney, N.J. (Industrial)	659.3	0.8	39.4	1.5
	Point Reyes, Ca. (West Coast Marine)	662.5	0.4	39.1	2.3
	State College, Pa. (Rural)	660.4	0.7	38.6	3.5
	Panama Canal Zone (Tropical)	675.5	0	37.4	6.5
N06625	Initial	858.0		54.0	
	Kure Beach, N.C. (East Coast Marine)	861.7	0	53.0	1.9
	Kearney, N.J. (Industrial)	858.3	0	51.0	5.6
	Point Reyes, Ca. (West Coast Marine)	862.2	0	52.5	2.8
	State College, Pa. (Rural)	864.3	0	52.8	2.2
	Panama Canal Zone (Tropical)	859.4	0	53.3	1.3
N08800	Initial	594.4		39.0	
	Kure Beach, N.C. (East Coast Marine)	593.4	0.2	38.3	1.8
	Kearney, N.J. (Industrial)	592.2	0.4	37.7	3.3
	Point Reyes, Ca. (West Coast Marine)	596.4	0	37.9	2.8
	State College, Pa. (Rural)	597.7	0	37.9	2.8
	Panama Canal Zone (Tropical)	594.0	0.1	38.1	2.3
N08825	Initial	782.7		32.0	
	Kure Beach, N.C. (East Coast Marine)	785.7	0	30.0	6.3
	Kearney, N.J. (Industrial)	779.5	0.4	30.4	5.0
	Point Reyes, Ca. (West Coast Marine)	788.6	0	31.2	2.5
	State College, Pa. (Rural)	789.8	0	30.7	4.0
	Panama Canal Zone (Tropical)	789.5	0	31.1	2.8

Severity of Exposure Sites

As a measure of the severity of the exposure sites, average mass losses were determined for the six nickel-alloys together for each site.

The exposure site severity ranked somewhat differently for nickel and nickel-copper alloys, than for Ni-Cr-Mo and Ni-Cr-Fe alloys.

The exposure sites ranked by decreasing severity to alloys 200 and 400 as shown above for the entire group of alloys,

Kearney, NJ > Kure Beach, NC > Panama Canal Zone > State College, PA > Point Reyes, CA.

These were plotted in the bar graph shown in Figure 1.

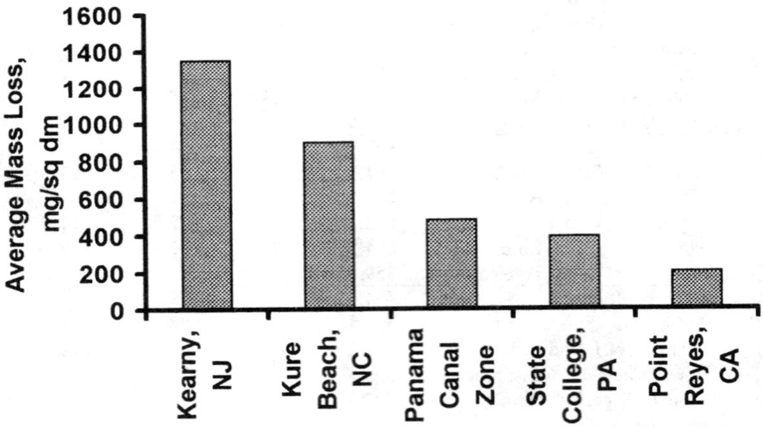

Figure 1 - *Average 20 Year Mass Loss for Nickel - Alloys 200 and 400 vs. Test Location*

The exposure sites ranked by decreasing severity to alloys 600, 625, 800 and 825 as shown above for the entire group of alloys,

Kure Beach, NC > Panama Canal Zone > Point Reyes, CA. > State College, PA > Kearny, NJ

These were plotted in the bar graph shown in Figure 2.

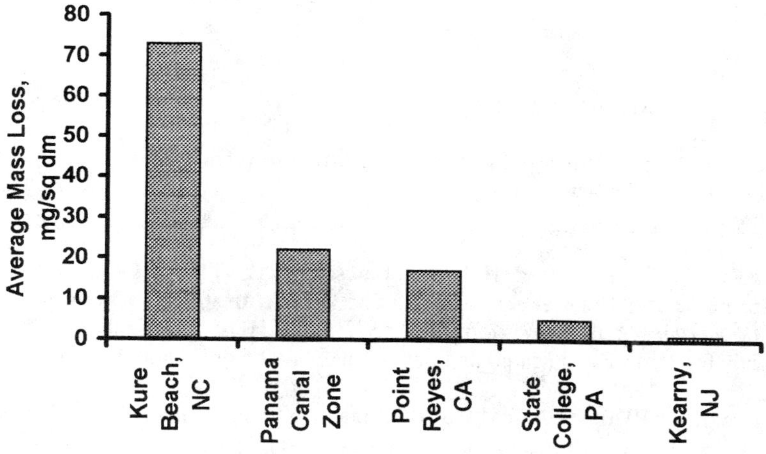

Figure 2 - *Average 20 Year Mass Loss for Nickel - Alloys 600, 625, 800 and 825 vs. Test Location*

Summary

The 20-year atmospheric exposure test panels of nickel-alloys UNS N02200, N04400, N06600, N06626, N08800 and N08825, tested at each of the five test sites throughout the United States and Panama, exhibited excellent corrosion resistance with rates of <0.001mm/y (<0.04 mpy) and pit depth averages of <0.046mm (<0.0018 in.).

The exposed panels exhibited little or no loss in mechanical properties when comparisons were made to control panels not exposed to the environment. The ultimate tensile strength decreased by ≤ 2.1% (≤ 0.01% per year) and % elongation decreased by ≤ 7.8% (≤ 0.4% per year).

The exposure sites ranked in severity according to alloy groups. For alloys 200 and 400 the exposure sites ranked by decreasing severity as follows, Kearny, NJ > Kure Beach, NC > Panama Canal Zone > State College, PA > Point Reyes, CA. For alloys 600, 625, 800 and 825 the exposure sites ranked by decreasing severity as follows, Kure Beach, NC > Panama Canal Zone > Point Reyes, CA. > State College, PA > Kearny, NJ.

Acknowledgement

The author would like to acknowledge the significant contribution of James R. Adams, Technician at Special Metals Corporation, in examining and evaluating the nickel-alloy atmospheric exposure panels.

Herbert E. Townsend[1] and Herbert H. Lawson[2]

Twenty-One Year Results for Metallic-Coated Steel Sheet in the ASTM 1976 Atmospheric Corrosion Tests

---

REFERENCE: Townsend, H. E., and Lawson, H. H., **"Twenty-One Year Results for Metallic-Coated Steel Sheet in the ASTM 1976 Atmospheric Corrosion Tests,"** *Outdoor Atmospheric Corrosion, ASTM STP 1421,* H. E. Townsend, Ed., American Society for Testing and Materials International, West Conshohocken, PA, 2002.

Abstract: In 1976, the ASTM G01.04 Subcommittee on Atmospheric Corrosion initiated a long-term atmospheric corrosion test program at five outdoor locations: Kure Beach, NC (25-meter East-coast marine), Newark-Kearney, NJ (industrial), State College, PA (rural), Point Reyes, CA (West-coast marine), and Fort Sherman, Panama (tropical marine). Among the forty-one materials exposed were four coated steel sheet products: G60 galvanized, G90 galvanized, 55% aluminum-zinc-coated, and Type 2 aluminum-coated. Results are presented for the performance of the coated sheet steels after 21 years. These results indicate the following qualitative order of corrosivity for five sites:

Panama ~ Kure Beach > Newark-Kearny > State College > Point Reyes

Based on changes in the corrosion of 55% aluminum-zinc-coated, an increase in the aggressiveness at Panama took place sometime during the 10th to 21st year of exposure. Future testing at these sites could be better quantified by short-term (1-2 years) exposures of standard materials. A loss of significant data from 2-, 5- and 10-year removals and initial weights for many of the test materials highlights the importance of safeguards to ensure the integrity of future long-term ASTM tests.

In terms of relative performance, the results are consistent with the following order of corrosion resistance (most-to-least).

T2 100 ~ AZ55 > G90 > G60

Keywords: atmospheric, corrosion, marine, rural, industrial, steel, galvanized, aluminum-coated, 55% aluminum-zinc alloy-coated

Introduction

In 1976, the ASTM G01.04 Subcommittee on Atmospheric Corrosion initiated a long-term atmospheric corrosion test program. A complete description of the program was

---

[1]Senior Research Consultant, Bethlehem Steel Corporation, Homer Research Laboratories, Bethlehem, PA, 18016.

[2]President, Lawson Consultants, 4821 Manchester Road, Middletown, OH, 45042.

given in 1978 by Ailor [*1*]. A total of forty-one materials were put into test at five outdoor locations: Kure Beach, NC (25-meter east-coast marine), Newark-Kearney, NJ (industrial), State College, PA (rural), Point Reyes, CA (west-coast marine), and Fort Sherman, Panama (tropical marine).

Materials in the program included stainless steels; alloys of copper, lead, nickel, titanium, zinc, and aluminum; and metallic-coated steel sheet products. The coated steels were G60 galvanized, G90 galvanized, AZ55 55% Al-Zn alloy-coated, and T2 100 aluminum-coated. This report documents the performance of the four metallic-coated steels after 21 years of exposure.

Materials

Test materials (Table 1) were produced on commercial hot-dip coating lines on which coils of cold rolled steel sheet are unwound, annealed, coated, and rewound in a continuous operation. A detailed description of the continuous hot-dip coating process is given elsewhere [*2*].

Table 1 - *Metallic-coated steel sheet test materials.*

Material Designation	Coating Metal	Coating Mass, g/m^2 (triple-spot total for both sides)	Nominal Coating Thickness, µm (per side)	Sheet Thickness, mm	Supplier
G60 Galvanized (ASTM A 653)	Zinc	180	13	1.2	Inland Steel Company
G90 Galvanized (ASTM A653)	Zinc	275	19	1.0	Inland Steel Company
AZ55 55%Al-Zn Coated (ASTM A 792)	55%Al-Zn Alloy	165	22	0.5	Bethlehem Steel Corporation
T2 100 Aluminum Coated (ASTM A 463)	Aluminum	300	48	0.8	AK Steel Corporation

G60 and G90 galvanized are zinc-coated sheet with two coating thicknesses and were produced by Inland Steel Company. According to the prevailing practice of that time, these coatings contained small amounts (0.1-0.2%) of added aluminum to improve formability. Similar levels of lead were also added to impart a spangled surface. It should be noted that today, most galvanized coatings are produced without added lead for environmental reasons.

AZ55 is steel sheet with a 55% Al-Zn alloy coating. It was produced by Bethlehem Steel under the trademark Galvalume[3]. Material from the same coil was also tested in

[3]Galvalume is a registered trademark of BIEC International for steel sheet with a 55% Al-Zn coating.

other rural, industrial, and marine environments, and the results are reported elsewhere [*3*]. Moreover, additional material from this coil will be included in the new atmospheric test program currently being organized by ASTM G01.04.

T2 100 aluminum-coated was produced by AK Steel Corporation. Results for other outdoor tests of this material have also been reported separately [*4*].

Atmospheric Tests

Sheet materials were sheared into 100- by 200-mm coupons, and a coded series of holes were drilled or punched for identification. The coupons were cleaned and weighed, and placed on racks inclined at 30 degrees from the horizontal at the five locations (Table 2). These sites are intended to provide a broad range of environmental and geographical conditions.

Table 2 - *Atmospheric test locations.*

Test Location	Description of Site
Fort Sherman, Panama	Tropical Marine, 1 km from Caribbean Sea at entrance to Panama Canal
Kure Beach, NC	Severe Marine, 25 Meters from Atlantic Ocean
Newark-Kearney, NJ	Industrial
State College, PA	Rural with Acid Rain
Point Reyes, CA	Mild Marine, on Hills Overlooking Pacific Ocean

The skyward surfaces of the coupons faced South at all test sites, except for the severe marine lot where they faced eastward to the Atlantic Ocean. Unless otherwise noted, the tests were conducted according to ASTM procedures, namely, Practice for Conducting Atmospheric Tests on Metals (G 50), and Practice for Preparing, Cleaning, and Evaluating Corrosion Test Specimens (G 1).

It was initially intended that triplicate coupons would be removed at intervals of 2, 5, 10, and 20 years. As a result of many changes that have occurred within ASTM and the material suppliers over the course of two decades, it is no longer possible to retrieve initial weights of the aluminum-coated coupons, losses for the 2-, 5-, and 10-year removals of the Al T2 100 and G60 and G90 coupons, and for the 5-year removals of the AZ55 coupons.

Corrosion losses after 21 years were determined in the case of those materials for which initial weights were available, namely, G60, G90, and AZ55. To avoid errors due to substrate corrosion, only specimens with less than 5% surface rust were used for determination of coating loss. Corrosion products were removed by cleaning in chromic acid according to ASTM G 1, Method C.9.3 [*6*], weighing to determine mass loss, and calculating the coating thickness loss by use of Equation 1.

$$\Delta t = \Delta m / Ad \quad (1)$$

where

Δm = mass loss
Δt = thickness loss
A = area of the test coupon
d = coating density

In the case of single-component zinc coatings, G60 and G90, the density was taken as that of zinc (7.14 g/cm^2). In the case of multi-component 55% Al-Zn alloy coating, it is known that the zinc-rich fraction of the binary alloy corrodes preferentially [*5*]. For this reason, the composition and density of the corroded portion are not the same as that of the coating alloy. This was accounted for by using the actual density of the corroded coating, which is determined as follows. After removal of corrosion products and re-weighing, the remaining uncorroded coating was stripped in hydrochloric acid, and the stripping solution was chemically analyzed. The panel was then re-weighed to determine the amount of uncorroded coating. A simple mass balance was then employed to calculate the composition of the corroded part of the coating and its density. These compositions ranged from a low of 33.4% Al at Point Reyes to a high of 53.1% Al at Panama, corresponding respectively to densities of 4.81 and 3.82 g/m^2.

Results

Corrosion losses after 21 years were determined for those panels with rust on less than 5% of the surface, and for which initial weights were available (Table 3).

Table 3 - *Corrosion losses for galvanized and 55% aluminum-zinc alloy-coated specimens with less than 5% rust after 21 years.*

		Corrosion Loss, μm		
Site	Material	Average	Replicates	Standard Deviation
	G60	11.7	1	-
Point Reyes	G90	13.2	3	0.5
	AZ55	4.3	3	0.5
State College	AZ55	6.1	3	0.5
Newark-Kearney	AZ55	9.9	3	0.5
Kure Beach	AZ55	19.8	1	-

In the case of 55% Al-Zn coated, these results indicate an order of site corrosivity (most corrosive to least corrosive) as follows:

Panama ~ Kure Beach > Newark-Kearny > State College > Point Reyes

The relatively low corrosivity of the Point Reyes marine site is probably the result of decreased transport of salt particles to the hilltop location. This order is slightly different than that reported in 1955 for 2-year tests of rolled zinc [*6*].

Panama > Kure Beach > Newark-Kearny > Point Reyes > State College

Average percent area of surface rust for the three replicates (Table 4) for galvanized suggests the same order of site corrosivity, except that the positions of Panama and Kure Beach are reversed.

Table 4 - *Percent of surface area with red rust.*

Site	Material	Average percent surface with red rust	
		Skyward surface	Groundward surface
Point Reyes	G60	15	0
	G90	1	0
	AZ55	0	0
	T2 100	2	3
State College	G60	93	56
	G90	72	12
	AZ55	0	0
	T2 100	0	0
Newark-Kearney	G60	100	67
	G90	100	30
	AZ55	0	0
	T2 100	0	trace
Panama	G60	100	75
	G90	100	60
	AZ55	5	35
	T2 100	5	65
Kure Beach	G60	*	*
	G90	100	78
	AZ55	12	10
	T2 100	trace	trace

Several characteristics of the coated sheets are apparent from the degree of rusting (Table 4). In the case of galvanized sheet, the amount of rust is generally greater on G60 than on G90. This result is consistent with the well-known increase in galvanized coating life with increasing coating weight.

Also, rusting of both G60 and G90 galvanized is generally greater on skyward surfaces, in agreement with previously reported results [*3*]. This contrasts with the behavior of T2 100 aluminum-coated, which is more prone to rusting on groundward surfaces, or of AZ55 aluminum-zinc alloy-coated, which exhibits more groundward rust at Panama, and is more uniform at Kure Beach.

It is also clear from Table 4 that AZ55 and T2 100 are significantly more durable than G60 or G90, confirming the results of previously reported tests [*3,6*]. The absence of G60 test coupons after 21 years at the Kure Beach site is probably the result of excessive corrosion.

Overall, these results are consistent with the following order of corrosion resistance (most-to-least).

$$\text{T2 100} \sim \text{AZ55} > \text{G90} > \text{G60}$$

Thickness-losses of AZ55 after 2, 10, and 20 years are plotted in Figure 1. The results at Kure Beach are about 20% greater than those found in other tests of the same material at this site [*3*]. Given that the relation between corrosivity and distance from the ocean is very steep at these distances, this difference may result simply from different sample positions within the exposure site.

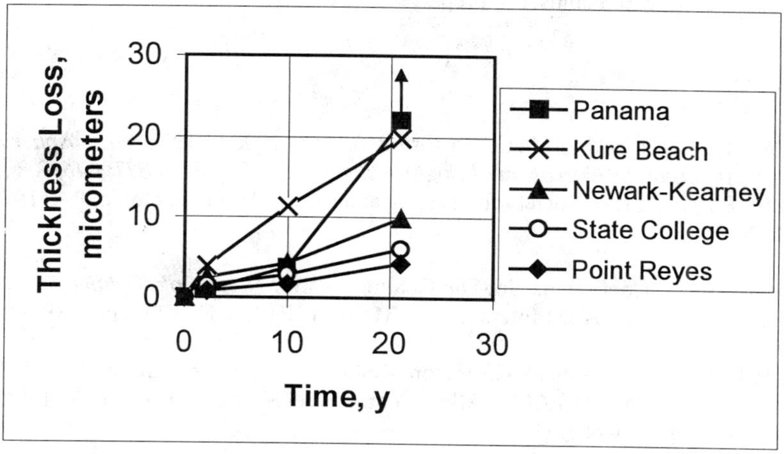

Figure 1 - *Thickness losses for 55% aluminum-zinc alloy-coated steel sheet.*

An unexplained increase in the corrosivity of the Panama site appears to have occurred after the tenth year of exposure. Responsible personnel at this site sugested that salt deposition rates may have increased during this period because high vegetation between the sea and the test site was cut down.

Summary

An evaluation of metallic-coated steel sheet after 21 years of exposure indicates the following qualitative order of corrosivity for five sites:

Panama ~ Kure Beach > Newark-Kearny > State College > Point Reyes

Uncertainty in the relative aggressiveness at Panama and Kure Beach may be related to a large increase in corrosivity at Panama that took place sometime during the 10th to 21st year of exposure.

Future testing at these sites could be better quantified by short-term (1-2 years) exposures of standard materials.

The loss of significant data from 2-, 5-, and 10-year removals and initial weights for many of the test materials highlights the importance of safeguards to ensure the integrity of future long-term ASTM tests.

In terms of relative performance, the results are consistent with the following order of corrosion resistance (most-to-least).

T2 100 ~ AZ55 > G90 > G60

Acknowledgments

The authors are indebted to H. J. Cleary for preparing the AZ55 coupons, to E. J. McInerney and J. D. Hoffman for evaluating test materials after exposure, and to J. J. Bachman for assisting with manuscript preparation.

References

[1] Ailor, W. H., "ASTM Atmospheric Corrosion Testing: 1906 to 1976," *Atmospheric Factors Affecting the Corrosion of Engineering Materials, ASTM STP 646,* S. K. Coburn, Ed., American Society for Testing and Materials, Philadelphia, PA, 1978, pp. 129-151.

[2] Townsend, H. E., "Continuous Hot-Dip Coatings," *ASM Handbook, Volume 5, Surface Engineering,* ASM International, Materials Park, OH, 1994, pp. 339-348.

[3] Townsend, H. E., "Atmospheric Corrosion of Skyward- and Groundward-Exposed Surfaces of Zinc and 55% Al-Zn Alloy-Coated Steel Sheet," *Corrosion,* Vol. 54, No. 7, 1998, pp. 561-565.

[4] Lawson, H. H., "Atmospheric Corrosion Testing of Metallic-Coated Steels - Forty Years Testing Experience," *Critical Issues in Reducing the Corrosion of Steels (Proceedings of the USA-Japan Seminar on Corrosion)*, H. Leidheiser, Jr. and S. Haruyama, Eds., NACE, Houston, TX, 1986, pp. 201-213.

[5] Zoccola, J. C., Townsend, H. E., Borzillo, A. R., and Horton, J. B., "Atmospheric

Corrosion Behavior of Aluminum-Zinc Alloy-Coated Steel," *Atmospheric Factors Affecting the corrosion of Engineering Metals, STP 646,* S. K. Coburn, Ed., American Society for Testing and Materials, Philadelphia, PA, 1978, pp. 165-184.

[6] Anderson, E. A., "The Atmospheric Corrosion of Rolled Zinc," *Symposium on Atmospheric Corrosion of Nonferrous Metals, STP 175,* American Society for Testing and Materials, Philadelphia, PA, 1955, pp. 126-134.

Herbert E. Townsend[1]

Estimating the Atmospheric Corrosion Resistance of Weathering Steels

---

Reference: Townsend, H. E., "**Estimating the Atmospheric Corrosion Resistance of Weathering Steels,**" *Outdoor Atmospheric Corrosion, ASTM STP 1421*, H. E. Townsend, Ed., American Society for Testing and Materials International, West Conshohocken, PA, 2002.

Abstract: Although important properties such as strength, toughness, and weldability can be easily measured in the laboratory, there is no generally accepted laboratory test for determining the atmospheric corrosion resistance of low-alloy weathering steels. Because many years of outdoor testing are needed to develop corrosion data, there is a need for reliable methods of estimating corrosion performance. The relative corrosion resistance of weathering steels can be estimated from the corrosion index calculated from composition by use of the ASTM Standard Guide for Estimating the Atmospheric Corrosion Resistance of Low-Alloy Steels (G 101). This paper describes a new, alternate method for estimating corrosion resistance from composition that has been recently added to G 101. The new method involves calculation of a corrosion index based on historical data recently published by Bethlehem Steel. It overcomes several limitations of the original G 101 corrosion index. Specifically, more elements are considered. In addition to the five elements (copper, nickel, phosphorus, chromium, and silicon) considered in G 101, the new method also takes account of the effects of several other elements, including carbon, molybdenum, sulfur, and tin. Because the method is based on data from three test sites, as compared to only one site for G 101, the results should be applicable to a wider range of environmental conditions. Also, the ranges of the elements are generally larger than those of G 101. Finally, the effects of the elements increase monotonically. As a result, the new method is free of anomalies resulting from quadratic terms, such as a maximum effect of copper at about 0.25% that is predicted by G 101. The overall absence of quadratic terms allows for more reliable extrapolation beyond the ranges of the original data. To facilitate computation of the G 101 corrosion indices, a calculator is available on the ASTM website at http://www.astm.org/COMMIT/G01_G101Calculator.xls.

Keywords: atmospheric, corrosion, weathering, steel, composition, corrosion index

---

[1] Senior Research Consultant, Bethlehem Steel Corporation, Homer Research Laboratories, Bethlehem, PA, 18016.

Introduction

Weathering steels are low-alloy structural steels that develop an adherent, protective rust layer during outdoor exposure. As a result of corrosion protection from this rust layer, weathering steels, unlike ordinary plain carbon steels, are generally used without painting. This avoids costs associated with: (1) initial painting, (2) periodic repainting, (3) containment and disposal of blasting debris during repainting, and (4) interruptions in service during repainting. The largest applications of unpainted weathering steels are highway bridges, utility towers, light poles, and highway guiderails.

Protective rust forms on weathering steels as a result of certain alloying elements, generally totaling less than 3%, which include chromium, copper, nickel, silicon, and phosphorus. These elements also add strength to weathering steels, thereby enabling further cost and weight savings compared to ordinary structural steels. A typical weathering steel is described in the ASTM Specification for High-Strength, Low-Alloy Structural Steel with 50 ksi (345 MPa) Minimum Yield Point to 4 in. (100 mm) Thick (A 588).

Current efforts aimed at developing new and improved grades of weathering steel must take corrosion behavior into account. Important properties such as strength, toughness, and weldability can be measured in the laboratory, but unfortunately there is no generally accepted laboratory test for determining the atmospheric corrosion resistance of low-alloy weathering steels. Because many years of outdoor testing are needed to develop corrosion data, reliable means to estimate corrosion performance are needed to minimize development time for new grades of weathering steel. The ASTM Guide for Estimating the Atmospheric Corrosion Resistance of Low-Alloy Steels (G 101) was created to address this need.

As originally conceived, G 101 provided two general methods for estimating the corrosion resistance of weathering steel. The first method (6.2) provides for the extrapolation of short-term exposure data to longer times by use of regression analysis to fit an exponential equation.

The second G 101 method (6.3) allows calculation of a relative corrosion index from composition based on historical data. Until recently, there was only one method of calculating a corrosion index. This method was based on the effects of composition as reported by Larrabee and Coburn [*1*] and analyzed by Legault and Leckie [*2*]. The purpose of this paper is to describe a second method which has been added to G 101 based on recently published data [*3*].

USX Data and the Legault-Leckie Analysis

The development of weathering steels benefited from numerous studies on the effects of alloying elements on the atmospheric corrosion resistance of steels. The best known is that of Larrabee and Coburn of USX, which was published in 1961 [*1*]. This study gives the results of 15.5-year exposures of 273 low-alloy steel compositions in three environments, namely, industrial (Newark-Kearney, NJ), rural (State College, PA), and marine (250-meters from the ocean at Kure Beach, NC). These results helped to define the beneficial effects of chromium (Cr), copper (Cu), nickel (Ni), phosphorus (P), and silicon (Si).

In 1974, Legault and Leckie of Inland Steel published a statistical analysis of the Larrabee-Coburn data [*2*]. The results were reported in the form of regression equations relating the 15.5-year corrosion loss to steel composition for each of the three environments.

The Original G 101 Corrosion Index, C1

The Legault-Leckie analysis served as the basis of the original G 101 method for calculating a relative corrosion index. Their equation for the 15.5-year corrosion loss at the industrial site is given as

$$M = 10 - 26.01\%Cu - 3.88\%Ni - 1.2\%Cr - 1.49\%Si - 17.28\%P + 7.29\%Cu\%Ni + 9.1\%Ni\%P + 33.39\%Cu^2 \quad (1)$$

where

M = thickness loss in mils. [2]

According to Equation 1, iron with no alloying elements has a corrosion loss of 10 mils (250 micrometers) after 15.5 years in this environment. The terms following the constant 10 predict the reduction in loss brought about by the addition of alloying elements. Taken together with the sign reversed, these terms comprise the original G 101 corrosion index,

$$C1 = 26.01\%Cu + 3.88\%Ni + 1.2\%Cr + 1.49\%Si + 17.28\%P - 7.29\%Cu\%Ni - 9.1\%NiP - 33.39\%Cu^2 \quad (2)$$

where

$C1$ = G101 original corrosion index.

Thus, C1 is numerically equal to 10 minus the thickness loss calculated from the Legault-Leckie equation for the industrial site. It is a dimensionless quantity on a scale of approximately 0 to 10 that increases with corrosion resistance. Weathering steels are generally expected to have a value of 6 or more. C1 values calculated for a variety of low-alloy steels by use of Equation 2 are reasonable only for steels within the specified range of composition (Table 1). Beyond these composition limits anomalies such as the negative indices for high-copper alloys, of values greater than 10 for the high chromium alloy, are apparent.

---

[2]The mil (0.001 inch = 25.4 micrometer) unit is given above to be consistent with the original references.

Table 1 - *Ranges and corrosion indices (C1) for various steels according to the original ASTM G101 corrosion index.*

Element w/o	Range Max.	Pure Fe	Typical A36	A36 + 0.2% Cu	Min. A588	Typical A588	Alloy 1	Max. A588	Alloy 2	Alloy 3	Alloy 4	Alloy 5
P	0.120	0.000	0.012	0.012	0.005	0.012	0.007	0.040	0.007	0.004	0.010	0.017
Si	0.640	0.000	0.220	0.220	0.300	0.360	0.290	0.650	0.410	0.200	0.250	0.560
Ni	1.100	0.000	0.019	0.019	0.050	0.310	0.970	0.400	0.920	2.970	0.750	0.500
Cr	1.300	0.000	0.027	0.027	0.400	0.530	0.018	0.650	0.060	0.025	0.500	11.860
Cu	0.510	0.000	0.018	0.200	0.250	0.300	0.940	0.400	1.480	0.350	1.000	0.020
C1		0.00	1.09	4.48	5.53	6.67	-7.42	7.74	-40.25	9.25	-8.86	17.66

Even within its composition limits, C1 has three notable limitations. First, it involves only the five elements included in the Larrabee-Coburn data, and thus it overlooks the effects of other potentially beneficial elements, such as molybdenum, and of harmful elements, such as sulfur.

Secondly, C1 is based on corrosion data from a single location and thus may not be broadly applicable to all environments.

Thirdly, C1 predicts a minimum in the corrosion resistance of pure iron at about 0.39% copper, and for A 588 at about 0.25% copper. This anomaly stems from the quadratic and cross terms for copper in the Legault-Leckie regression equation, and it leads to negative values of C1 for steels high in copper.

Bethlehem Steel Data

During the period 1934 to 1938, Bethlehem Steel Corporation initiated a series of outdoor corrosion tests involving several hundred low-alloy steel compositions. The results of these tests were recently published [*3*]. The tests were conducted in the industrial atmospheres of Bethlehem, PA, Pittsburgh, PA, and Columbus, OH, for times up to 16 years. Their purpose was to define the effects of various alloying elements and impurities on the corrosion resistance of steel. Besides the elements Cr, Cu, Ni, P, and Si, these steels also contained systematic variations in the elements sulfur (S), carbon (C), manganese (Mn), arsenic (As), molybdenum (Mo), tin (Sn), vanadium (V), tungsten (W), aluminum (Al), and cobalt (Co).

Data from these tests were statistically analyzed, and a set of equations were derived that enable calculation of corrosion behavior as a function of the above elements [*3*]. This was done by fitting the results for each alloy to the power law relationship,

$$L = AT^B \tag{3}$$

where

L = corrosion loss in micrometers,
T = time of exposure in months, and
A and B = constants.

The constant A in Equation 3 is numerically equal to the corrosion loss during the initial month of exposure. Accordingly, it is a measure of the first reaction of the bare steel surface prior to the formation of a rust layer, and is related mainly to the corrosivity of the environment.

The B constant indicates how the corrosion varies with time, and thus it is a measure of the protective qualities of the rust layer that develops on the steel surface. Smaller B values indicate a more protective rust, which slows corrosion with increasing time of exposure.

At each exposure site, constants A and B were calculated for all steels. The constants were then fitted to equations expressed as linear combinations of the alloying elements (Equations 4 and 5)

$$A = a_o + \sum a_i x_i, \tag{4}$$

where

a_o = constant,
a_i = coefficient for element I, and
x_i = concentration of element i in weight percent

$$B = b_o + \sum b_i x_i, \tag{5}$$

where

b_o = constant, and
b_i = constant for element i.

The coefficients resulting from regression analyses of data from the three test sites are summarized in Table 2.

The value of each element in contributing to the development of protective rust can be most easily determined by inspection of the b_i values (Table 2). These show that P, Si, Cr, C, Cu, Ni, Sn, and Mo are beneficial to corrosion resistance. S is shown to have a very large adverse effect. V, Mn, Al, Co, As, and W have little, if any, effect on corrosion. Note also that the maximum concentrations in the test materials are usually greater for the elements in these tests (cf., the range maxima given in Tables 1 and 3).

Table 2 - *Constants and coefficients for calculating rate constants A and B from composition by use of Equations 4 and 5 (from Ref 3).*

	A (microns)			B (T in months)		
n site	275 Bethlehem PA	227 Columbus OH	248 Pittsburgh PA	275 Bethlehem PA	227 Columbus OH	248 Pittsburgh PA
Constant	15.157	16.143	14.862	0.511	0.539	0.604
Carbon	6.310	*	3.350	-0.102	-0.103	-0.046
Manganese	*	-2.170	-2.370	-0.097	-0.019	0.042
Phosphorus	-1.770	-10.250	-5.120	-0.592	-0.333	-0.546
Sulfur	-27.200	-15.970	*	2.408	0.908	1.004
Silicon	6.50	2.96	1.38	-0.20	-0.16	-0.13
Nickel	1.970	-1.380	1.180	-0.080	-0.029	-0.088
Chromium	*	2.560	2.370	-0.103	-0.095	-0.174
Copper	*	0.990	-1.970	-0.072	-0.067	-0.068
Aluminum	*	1.580	5.520	*	*	-0.087
Vanadium	*	6.110	*	*	-0.193	*
Cobalt	1.580	-1.770	-2.560	-0.063	-0.053	0.044
Arsenic	3.150	-6.110	-7.690	-0.157	*	0.097
Molybdenum	*	*	-2.960	-0.078	-0.038	*
Tin	-3.740	-7.490	-9.860	-0.151	-0.038	*
Tungsten	*	-5.520	*	-0.148	*	*

* indicates coefficient has greater than 50% probability of chance occurrence

Derivation of a New, Alternative Corrosion Index, C2

For purposes of developing a corrosion index from the Bethlehem results, the following procedure was employed with the results of each step shown in Table 3.

1. For each of the three test sites, A and B values for pure, unalloyed iron are calculated by use of the regression constants given in Table 2, and Equations 4 and 5.

2. The times for pure iron to reach a 10-mil loss at the three sites are then calculated from Equation 6, which is obtained by solving Equation 3 for T, and taking L as 250 micrometers

$$T = (250/A)^{1/B} \tag{6}$$

3. For a given alloy steel, A and B values at each site are calculated from the regression constants and coefficients in Table 2, and Equations 4 and 5.
4. The losses of the steel at each site are calculated from Equation 3 at the times required for pure iron to lose 250 micrometers as determined in Step 2.
5. The differences between the 250 micrometers losses for pure iron and the calculated losses for the steel at each site are averaged to give the new corrosion index.

Results for each step of the above calculations, and the resulting corrosion indices for a variety of steel compositions are summarized in Table 3. Unlike the original corrosion indices (C1 in Table 1), the new indices (C2 in Table 3) give a ranking for these steels in the range 0 to 10 (Figure 1).

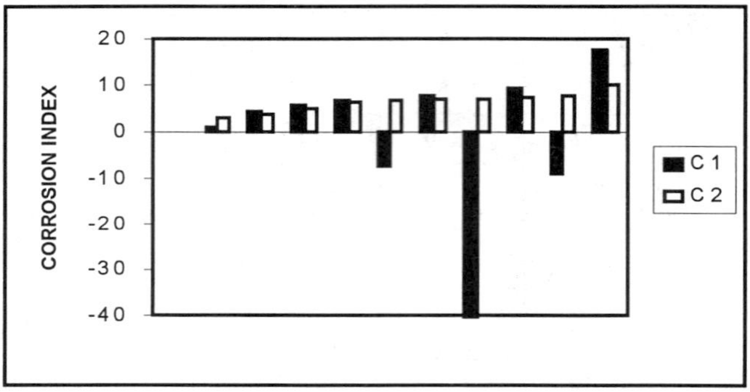

Figure 1 - *Comparison of the old (C1) and new (C2) G 101 corrosion indices for the steels included in Tables 1 and 3 (Order shown is same as that given in Tables 1 and 3).*

Summary

A new, alternative method for predicting the relative corrosion resistance of weathering steel from composition has been added to the ASTM Guide for Estimating the Atmospheric Corrosion Resistance of Low-Alloy Steels (G 101). Compared to the original method , the new method has the following advantages:

1. More elements are considered. In addition to the five elements (Cu, Ni, P, Cr, and Si), the new method also takes account of the effects of C, Mo, S, and Sn.
2. The new method is based on data from three test sites, so the results should be applicable to a wider range of environmental conditions.
3. The ranges of elements are generally larger.
4. Because the effects of the elements are monotonic, the new method is free of anomalies resulting from quadratic terms, such as a maximum effect of copper at about 0.25%. The absence of quadratic terms also allows for more reliable extrapolation beyond the ranges of the original data.

Table 3 - *Range and corrosion indices (C2) for various steels according to ASTM G 101 6.3.2*

Element w/o	Range Maximum	Pure Fe	Typical A36	A36 + 0.2% Cu	Min. A588	Typical A588	Alloy 1	Max. A588	Alloy 2	Alloy 3	Alloy 4	Alloy 5
C	1.500	0.000	0.160	0.160	0.060	0.100	0.060	0.190	0.032	0.091	0.060	0.020
Mn	1.500	0.000	1.010	1.010	0.800	1.180	1.090	1.250	0.480	0.580	1.000	1.360
P	0.300	0.000	0.012	0.012	0.005	0.012	0.007	0.040	0.007	0.004	0.010	0.020
S	0.300	0.000	0.013	0.013	0.001	0.011	0.002	0.050	0.006	0.001	0.002	0.003
Si	1.500	0.000	0.220	0.220	0.300	0.360	0.290	0.650	0.410	0.200	0.250	0.620
Ni	1.100	0.000	0.019	0.019	0.050	0.310	0.970	0.400	0.920	2.970	0.750	0.470
Cr	1.100	0.000	0.027	0.027	0.400	0.530	0.018	0.650	0.060	0.025	0.500	11.000
Cu	1.500	0.000	0.018	0.200	0.250	0.300	0.940	0.400	1.480	0.350	1.000	0.000
Al	1.500	0.000	0.051	0.051	0.010	0.020	0.000	0.000	0.029	0.000	0.030	0.000
V	0.500	0.000	0.003	0.003	0.020	0.040	0.000	0.100	0.009	0.000	0.060	0.000
Co	1.500	0.000	0.000	0.000	0.000	0.000	0.000	0.000	0.010	0.000	0.000	0.000
As	0.500	0.000	0.000	0.000	0.000	0.002	0.000	0.000	0.000	0.000	0.000	0.000
Mo	0.500	0.000	0.004	0.004	0.000	0.005	0.004	0.000	0.048	0.006	0.500	0.300
Sn	0.500	0.000	0.003	0.003	0.000	0.002	0.000	0.000	0.012	0.000	0.000	0.000
W	0.500	0.000	0.000	0.000	0.000	0.000	0.000	0.000	0.000	0.000	0.000	0.000
B	Bethlehem	0.511	0.369	0.356	0.300	0.226	0.193	0.198	0.194	0.139	0.128	-0.948
	Columbus	0.539	0.472	0.460	0.410	0.369	0.374	0.308	0.330	0.375	0.307	-0.659
	Pittsburgh	0.604	0.604	0.592	0.502	0.465	0.454	0.416	0.376	0.307	0.376	-1.390
A	Bethlehem	15.157	17.339	17.303	17.523	18.420	19.118	20.033	19.410	22.803	18.613	20.347
	Columbus	16.143	14.437	14.618	16.580	15.840	14.179	16.302	16.458	11.745	15.845	42.053
	Pittsburgh	14.860	13.559	13.200	14.064	13.827	12.172	14.256	12.534	16.907	11.588	38.203
Years to 10-mil loss	Bethlehem	20.800										
	Columbus	13.820										
	Pittsburgh	9.180										
20.8-yr mils	Bethlehem	10.000	5.230	4.850	3.620	2.530	2.180	2.350	2.230	1.930	1.480	0.000
13.82-yr mils	Columbus	10.000	6.340	6.030	5.320	4.120	3.770	3.090	3.500	3.150	2.990	0.060
9.18-yr mils	Pittsburgh	10.000	9.140	8.400	5.860	4.840	4.050	3.960	2.890	2.820	2.670	0.000
Site Index	Bethlehem	0.000	4.770	5.150	6.380	7.470	7.820	7.650	7.770	8.070	8.520	10.000
	Columbus	0.000	3.660	3.970	4.680	5.880	6.230	6.910	6.500	6.850	7.010	9.940
	Pittsburgh	0.000	0.860	1.600	4.140	5.160	5.950	6.040	7.110	7.180	7.330	10.000
C2		0.00	3.09	3.57	5.07	6.17	6.66	6.86	7.13	7.37	7.62	9.98

To facilitate computation of the G 101 corrosion indices, a calculator is available on the ASTM website at http://www.astm.org/COMMIT/G01_G101Calculator.xls.

Acknowledgments

Sincere appreciation for many helpful suggestions is due to Alex Wilson of Bethlehem Steel Corporation and Sheldon Dean of Air Products and Chemicals, Inc.

References

[1] Larrabee, C. P. and Coburn, S. K., "The Atmospheric Corrosion of Steels as Influenced by Changes in Chemical Composition," *Proceedings of the First International Congress on Metallic Corrosion*, Butterworths, London, 1961, pp. 276-285.

[2] Legault, R. A. and Leckie, H. P., "Effect of Alloy Composition on the Atmospheric Corrosion Behavior of Steels Based on a Statistical Analysis of the Larrabee-Coburn Data Set," *Corrosion in Natural Environments, ASTM STP 558*, American Society for Testing and Materials, Philadelphia, PA, 1974, pp. 34-347.

[3] Townsend, H. E., "Effects of Alloying Elements on the Corrosion Resistance of Steel in Industrial Environments," *Corrosion*, Vol. 57, No. 6, 2001, pp. 497-501.

Myron L. Hoitomt[1]

Performance of Weathering Steel Tubular Structures

Reference: Hoitomt, M. L., "**Performance of Weathering Steel Tubular Structures,**" *Outdoor Atmospheric Corrosion, ASTM STP 1421,* H. E. Townsend, Ed., American Society for Testing and Materials International, West Conshohocken, PA, 2002.

Abstract: An extensive study was conducted to determine the performance of weathering steel tubular utility structures. Utilization of weathering steel started more than 30 years ago and has slowly grown as some utilities realized it was a viable alternative to conventional protective coatings such as hot-dip galvanizing and paint. Unfortunately, weathering steel has not been widely accepted by potential users. Their reluctance is a result of corrosion problems identified in other types of steel structures. Those problems, which have since been resolved by design changes, included: lattice (angle-iron) towers containing pack-out in lap-joints; excessive rusting of bridge members detailed so joints held runoff-water and decaying refuse such as bird nests; rust stains on concrete supporting steel members of buildings and bridges. In comparison, tubular weathering steel utility structures have performed even better than anticipated. The corrosion data obtained from 23 sites correlate well with published laboratory test data.

Keywords: weathering steel, corrosion, performance, tubular, utility structures, pack-out, galvanic corrosion

Introduction

A study of 47 structures at 23 sites was undertaken to provide data/information that would answer those questions being asked about the service life of weathering steel tubular structures in a variety of environments. This data was needed because the utilization of tubular weathering steel was impeded by the numerous reports of corrosion problems on other types of weathering steel structures. The problem structures included: pack-out between overlapping angle-irons legs of lattice transmission towers; excessive corrosion of below–deck bridge members; galvanic corrosion resulting when dissimilar metals are in direct contact; and rust staining of supporting concrete in a weathering steel structure. Other concerns included issues such as: exposure to sea-coast environments; prolonged periods of rain/fog and/or high rainfall; heavy snow; industrial pollution; acid rain; contact with soil; agricultural fertilizers and herbicides, etc.

[1] Consultant, Hoitomt Consulting Services, N2264 U. S. Hwy 63, Hager City, WI, 54014.

The use of weathering steel for the manufacture of tubular structures (poles) to support electricity transmission and distribution lines has been proven to be an ideal application. The electric power line is by design free of other structures or trees that could shelter the structure surface and prevent the drying cycle, which is required to develop the protective iron oxide (rust) coating. The practicality of weathering steel for utility structures is inherently enhanced by the vertical installation and basic design characteristics that eliminate water and debris traps.

History

Weathering steel has been used in the manufacture of approximately one hundred thousand tubular steel structures, starting about 30 years ago. Those structures have been used for the transmission and distribution of electric power, lighting of athletic fields and parking lots, highway interchange lighting, wind-power generation, and recently cellular communication. Those weathering steel structures have been installed in environments that range from the tropical climate of Florida and Puerto Rico to frigid Alaska, from arid Southwest Texas to rainy/damp Puget Sound. They have been installed near saltwater, large industrial complexes, and chemical plants. The composition of those structures has been primarily that of ASTM A588 Specification for High-Strength Low-Alloy Structural Steel with 50 ksi (345Mpa) Minimum Yield Point to 4 in. (100 mm) Thick with minor modifications for increased strength (60 or 65 ksi (415 or 450 MPa) yield strength). ASTM A871 Specification for High-Strength Low-Alloy Structural Plate With Atmospheric Corrosion Resistance is now the standard covering the increased strength material, or ASTM A852 Specification for Quenched and Tempered Low-Alloy Structural Steel Plate with 70 ksi (485Mpa) Minimum Yield Strength to 4 in. (100 mm) Thick.

The tubular weathering steel structures have been designed with tube plate thickness ranging from 3/16" (5 mm) to 1-1/4" (32 mm) and the base plates ranging from 1-1/2" (38 mm) to 8" (203 mm). Annual use now is conservatively estimated at thirty thousand tons. The shaft plate is ordered to ASTM A6 tolerances, which allow for thickness up to 0.032" (810 μm) greater than nominal and 0.010" (250 μm) less than nominal. Ultrasonic thickness measurements in spots on which rust was removed, revealed that only a few out of the hundreds of measurements taken were less than the specified nominal thickness. No thickness measured was thinner than nominal minus 0.010" (250 μm), except of those areas in direct contact with the soil, which were noted on the data sheets. Throughout this paper, any thickness losses noted are relative to the actual thickness of the steel, as measured with the oxide removed. These losses should not be confused with variance from nominal thickness associated with tolerance.

Inspection - Locations and Procedures

Locations

Locations for examination of weathering steel structures were selected based on variation in environment, approval of the structure owner, accessibility, and length of exposure. Those locations were as follows:

1. Rathdrum, Idaho – Three structures in a rural environment: Two in a grassy valley, and one on top of the surrounding mountain.

2. Tacoma, Washington – Three structures: Urban industrial environment with one in a crushed-rock base substation, one in cut grass at edge of substation three miles from Puget Sound, and one in a weed and berry bush covered area near the harbor.

3. Glenallen, Alaska - One structure: Wooded brushy tundra.

4. Valdez, Alaska – One structure: Near top of mountain pass in rock terrain.

5. Kodiak, Alaska (Coast)– Two structures at urban industrial substation near the harbor with high (60" (1.5 m) annual) rainfall.

6. Kodiak, Alaska (Mountain) – Two structures in grassy meadows at the top (1200' (365 m)) of the mountain with high [10' (3 m) to 15' (4.6 m)] annual snowfall.

7. Anchorage, Alaska (Rural) – Three structures in forested brushy tundra at 400' to 600' (120 to 180 m) elevation northeast of city.

8. Anchorage, Alaska – One structure in mowed grass near a major highway in the city.

9. Hager City, Wisconsin – Two structures: Rural environment in blacktop parking lot, located 1/2 mile (800 m) from the Mississippi River.

10. Houston, Texas – One structure: Semi-urban grassy area near parking lot.

11. Austin, Texas - Four structures: Three in mowed grass near city streets, and one in ornamental shrub row near a city street.

12. Harrisburg, Pennsylvania - Two structures: Rural environment with tall grass, weeds and brush with dense forest on both sides of line.

13. Fairfax County, Virginia, Washington, DC – Two structures: Urban environment with one in thick tall brush and berry bushes close to I-495, and one in mowed grass close to I-495.

14. Katonah, New York - Three structures: Rural environment with tall grass, weeds and berry bushes in a wooded area near a large reservoir.

15. Martin County, Florida – Three structures: Rural environment with tall weeds and brush near sugar cane fields.

16. Eagan, Minnesota – Two structures: Urban environment with mowed grass between the street and sidewalk.

17. South St. Paul, Minnesota – One structure: Industrial environment with tall weeds and adjacent scrap yard, 300' (91m) from I-494 and 1/4 mile (400 m) from the Mississippi River.

18. Burkhart, Wisconsin – One structure: Rural environment with tall grass and brush close to County highway.

19. Turtle Lake, Wisconsin – One structure: Rural environment with tall grass and brush close to County highway.

20. Knoxville, Tennessee – One structure: Urban environment with grass or dirt in center of major highway intersections in heart of the city.

21. Nacogdoches, Texas -- Four structures: Semi-rural environment on concrete top of elevated sides of an open-air athletic stadium.

22. Lincolnton, Georgia – Three structures: Rural environment with tall grass and weeds in power line corridor with forest on each side.

23. Georgetown, South Carolina – One structure: Rural environment at power plant with cooling towers emitting water vapor adding to the natural humidity of this coastal region. Pulp and paper mills visible up-wind from this site.

Procedure

Examination was performed as a team concept following established guidelines. One person performed the tests and the other recorded the data. Specific information was developed and recorded for each structure. The information for three structures, which was typical for the 47 structures, is shown on the following tables (Tables #1, #2, and #3).

Table 1 - *Typical corrosion study data sheet*

• Location:	Tacoma, WA					
• Terrain:	Surrounded by weeds & berry bushes – in industrial environment ½ mile from Puget Sound.					
• Elevation:	288'					
• Utility:	City of Tacoma					
• Line Name:	Cowlitz Northeast Line					
• Structure Type:	Large angle on concrete/anchor-bolt foundation					
• Structure I.D.:	3-8A					
• Manufacturer:	Thomas & Betts					
• Years Exposure:	16					
• Annual Rain:	35.8"					
• Avg. Temperature:	50.4° F					
• Avg. High Temp.:	59.0° F					
• Avg. Low Temp.:	43.9° F					
• Oxide Appearance:	Uniform bottom to top, excellent appearance.					
• Oxide Color:	Dark-brown on south side, orange-brown on north side.					
• Oxide Thickness/Avg.:	7 to 16 mils/10.8					
• Metal Converted:	4.5 mils					
• Nominal Plate Thickness:	0.563"					
• Measured Thick.:	0.595	0.601	0.600	0.595	0.600	0.595
• Measured Thick. Ground:	0.588	0.591	0.591	0.582	0.584	0.585
• Oxide Thickness:	0.007	0.010	0.009	0.013	0.016	0.010
• Comments:	No visible pack-out in slip-joint. The performance of weathering steel on this structure was especially important as the environment of this region was considered to be marginal due to the lengthy period of wet weather in late fall and winter.					

Table 2 - *Typical corrosion study data sheet*

• Location:	Martin County, FL					
• Terrain:	Surrounded by weeds, brush & and sugar cane fields.					
• Elevation:	15'					
• Utility:	Florida Power & Light					
• Line Name:	Andytown-Martin Line					
• Structure Type:	Large H-Frame on concrete/anchor-bolt foundation					
• Structure I.D.:	2Z183					
• Manufacturer:	Thomas & Betts					
• Years Exposure:	20					
• Annual Rain:	60.76"					
• Avg. Temperature:	74.8° F					
• Avg. High Temp.:	82.9° F					
• Avg. Low Temp.:	66.7° F					
• Oxide Appearance:	Uniform bottom to top with some minor *rub marks near the base - overall excellent appearance.					
• Oxide Color:	Dark-brown.					
• Oxide Thickness/Avg.:	9 to 16 mils/12.0					
• Metal Converted:	5.0 mils					
• Nominal Plate Thickness:	0.375"					
• Measured Thick.:	0.392	0.392	0.390	0.387	0.383	0.392
• Measured Thick. Ground:	0.380	0.376	0.377	0.376	0.374	0.381
• Oxide Thickness:	0.012	0.016	0.013	0.011	0.009	0.011
• Comments:	*Owner was concerned about the possible removal of the oxide by wind blown vegetation, which results in visible rub-marks on the structure. Additional thickness measurements above the rub marks confirmed there was no loss of either oxide thickness or metal loss in the areas of rub.					

Table 3 - *Typical corrosion study data sheet*

- Location: Kodiak, AK
- Terrain: Graded substation near the sea, covered with crushed rock.
- Elevation: 60'
- Utility: Kodiak Electric Association
- Line Name: Terror Lake Transmission Line
- Structure Type: 4-leg Dead-End Structure on anchor bolt/concrete foundation.
- Structure I.D.: #103
- Manufacturer: Oregon Iron Works
- Years Exposure: 14
- Annual Rain: 63.63
- Avg. Temperature: 39.6° F
- Avg. High Temp.: 45.6° F
- Avg. Low Temp.: 33.8° F
- Oxide Appearance: Uniform dense oxide from bottom to top.
- Oxide Color: Bright orange.
- Oxide Thickness/Avg.: 5 to 8 mils/6.7
- Metal Converted: 2.8 mils
- Nominal Plate Thickness: 0.312"
- Measured Thick.: 0.312 0.313 0.313 0.311 0.309 0.311
- Measured Thick. Ground: 0.306 0.306 0.305 0.303 0.304 0.305
- Oxide Thickness: 0.006 0.007 0.008 0.008 0.005 0.006
- Comments: The Owner was concerned that the orange color meant the corrosion protective oxide was not forming, which would eventually result in failure. The surrounding crushed rock base didn't have any rust stains, nor could any be wiped from the structure, which verified that the protective oxide had developed and was mature.

The examination, evaluation and data recording process consisted of the following functions.

- Take pictures of structure and surroundings; close-up pictures of the plate surface, before and after removal of the protective oxide, were taken with a 35 mm camera equipped with a macro lens. Oxide was removed with a battery powered 2" (51 μm) sanding disc, exposing shiny metal in a circular spot of approximately 1" (25 μm).

- Take ultrasonic thickness measurements of plate with oxide, and with oxide removed. Most measurements were taken at approximately 5' (1.5 m) above the base plate or ground-line, as applicable or convenient for sanding and measurement. All measurements were started on the south side of the structure, rotating clock-wise in 60° increments around the structure. Thickness data was recorded in the same sequence. One Technician using the same calibrated equipment at each site conducted all tests. A Krautkramer ultrasonic thickness gage was used for all the tests. Calibration accuracy to 0.001" (25 μm) was verified on calibrated thickness blocks fabricated from weathering steel.

- Visual check for oxide condition, pitting, and color.

- Visual check of slip-joints for any evidence of pack-out.

- Visual inspection of galvanized anchor bolt nuts or any other accessible galvanized line hardware for galvanic corrosion.

- Visual inspection of ground-line conditions of embedded structures or structures with buried concrete foundations.

- Record all findings including, location, owner, structure type, structure identification, and any other unique/unusual conditions.

- Obtain and record key weather data for each general location. This data included average annual rainfall, average temperature, average high temperature, and average low temperature.

Findings

This study provides information that verifies the effectiveness of weathering steel to resist corrosion in different environments. It also provides a comparison to the effects of years of exposure. Visual examination coupled with thorough evaluation of all data was used to determine the condition of each structure. The following conditions were observed.

1. The protective oxide color varies from light orange-brown to bright orange, to light brown to dark brown-black. It is evident the environment is a key factor in color, with the direction also a contributing factor. The cool, wet seacoast environment of

Kodiak, Alaska, creates a bright orange oxide of only moderate thickness. The hot, wet environment of Houston, Texas, creates a dark brown thick oxide. The exceptional extreme temperature variation [-40° F to 100° F (-40° C to 38° C)] of Hager City, Wisconsin, results in a brown-black oxide of moderate thickness.

2. Structures designed with a base-plate form a shelf that stays wet for an extended period after a rain or heavy dew. The oxide in the band of 3 to 4 inches (75 to 100 mm) from the base upward is darker and slightly thicker on most structures. This is especially true in regions that experience extended periods of fog. The oxide thickness is also thicker on the north side, which stays wet longer than the sun sides of the structure.

3. Mature protective oxide is typically very hard and dense. It is not damaged or eroded by weeds and brush that contact it continuously, or is wind blown resulting in what appears as a polished surface.

4. Pitting has not been found to be a structural detriment. Only one location (Fairfax County, Virginia) contained pits of a size or quantity that was worth noting.

5. There was no evidence of deterioration of galvanized hardware or fasteners. This included anchor bolt nuts at the base-plate and bolt-on line hardware. This has been a source of concern of those referencing the steel producer's weathering steel product bulletins, which caution against designing structures with dissimilar metals.

6. Several structures contained slip-joints, which consist of the lower tapered tube sliding inside of the upper tapered tube, being held together by friction resulting from hoop-stress. There is no visible evidence of any pack-out. This has been a constant source of concern by users and potential users of weathering steel tubular structures. One structure was taken down and disassembled, providing positive verification that pack-out is not a factor in this type of joint.

7. Bare weathering steel should not be in contact with soil or excessive rotting vegetation. A 25 year-old site at Katonah, NY, contains excessive ground-line corrosion. This line was designed with concrete foundations installed approximately 12" (305 mm) below grade. This was done to accommodate covering the base and anchor bolts with soil backfill. The base and structure were coated with a protective coating to a height of approximately 15" (380 mm) Unfortunately, backfill or washed-in soil is in some cases 24" (610 mm) high, the result being metal loss of as much as 0.100" (2.5mm). Decaying vegetation just above the ground-line resulted in metal loss as much as 0.010" (.25 mm).

Note: Excessive ground-line corrosion typically starts at the ground-line and extends downward in a band 3 to 4 inches (75 to 100 mm) wide; corrosion below that depth is minimal, as a result of less oxygen.

8. The barrier coating on a few embedded structures revealed loosening at the top edge that is exposed to the above-ground environment. However, there is no excessive corrosion due to trapped water; the protective oxide develops as soon as the water

evaporates or the coating cracks allowing the water to escape. It should be noted that no loosening was observed on any structures on which the urethane barrier coating was feathered out to 0 mils.

9. One of the major concerns of first-time or potential users of weathering steel is the amount of metal converted to protective oxide in the oxidation process. The protective oxide thickness varies by environment (Table #4). The concern being the conversion process will continue until such time as the metal thickness is reduced so much as to jeopardize structural integrity (Figure #1). Ultrasonic thickness measurements taken in this study show the maximum oxide thickness in one spot to be 17 mils (430 μm), which equates to converted metal thickness of 7.1 mils (180 μm). Analysis of all the data obtained from the 23 sites revealed the 17-mil spot was part of the data obtained from the Washington DC, site, which tied with Houston for thickest oxide. Compare this data obtained at one of the most corrosive environments in the lower 48 states with the short summers and minimal rainfall of mainland Alaska that results in an environment that is almost void of corrosion. Structures exposed for 19 - 20 years contain a protective oxide film of only 1 - 4 mils (25 to 100 μm), with an average thickness of less than 3 mils (75 μm), which equates to converted metal thickness loss of approximately 60 μm.

Table 4 - *Location of sites, years of exposure, oxide thickness,*[1] *range and average*

Location:	Years:	Range:	Average:
• Idaho - Western region	19	4-7	5.4
• Washington – Tacoma	16	7-16	11.2
• Alaska – Glenallen	19	1-3	0.7
• Alaska – Valdez	19	1-4	3.0
• Alaska - Kodiak Harbor	14	5-8	6.7
• Alaska Kodiak Mountain	14	1-6	3.9
• Alaska – North/NE of Anchorage	13 – 20	2-5	2.8
• Alaska – Anchorage	16	4-7	5.3
• Wisconsin - Hager City	27	3-5	4.0
• Texas - Houston	17	11-16	13.5
• Texas – Austin	5 – 7	3-8	5.2
• Pennsylvania - North of Harrisburg	11	5-9	7.8
• Virginia - North of Washington, DC	15	8-17	13.7
• New York – Katonah	25	4-17	12.3
• Florida - Martin County	20	9-16	12.0
• Minnesota – Eagan (Minneapolis)	6	4-8	6.1
• Minnesota – St Paul	10	6-9	7.5
• Wisconsin – Burkhart	17	7-10	9.3
• Wisconsin – Turtle Lake	7	5-8	6.7
• Tennessee – Knoxville	15	7-8	7.5
• Texas – Nacogdoches	25	4-11	7.6
• Georgia – Lincolnton	24	13-14	13.4
• South Carolina – Georgetown	10	9-16	12.5

[1]The protective oxide development on weathering steel is dependent on the environment. It is evident the oxide development is much less in a cold-dry environment than in one that is hot-wet. The protective oxide thickness is a measure of how much metal has been consumed/converted in the oxide development process. The conversion formula used for this study is 1 mil (25.4 μm) of metal converted to 2.39 mils (61 μm) of protective oxide.

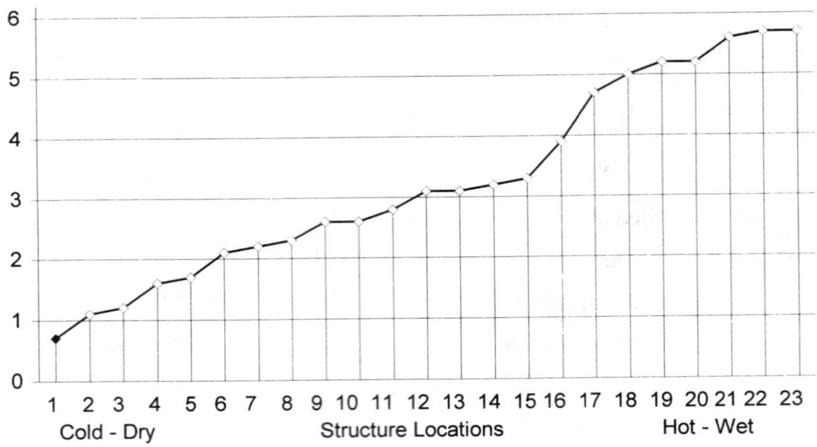

No.:	Location:	Metal Converted
1.	Alaska – Glenallen	0.7 mils (18 μm)
2.	Alaska – Valdez	1.1 mils (28 μm)
3.	Alaska - N/NE of Anchorage	1.2 mils (30 μm)
4.	Alaska - Kodiak Mountain	1.6 mils (41 μm)
5.	Wisconsin – Hager City	1.7 mils (43 μm)
6.	Texas – Austin	2.1 mils (53 μm)
7.	Alaska – Anchorage	2.2 mils (56 μm)
8.	Idaho – Western Part	2.3 mils (58 μm)
9.	Alaska - Kodiak Harbor	2.6 mils (66 μm)
10.	Minnesota – Minneapolis	2.6 mils (66 μm)
11.	Wisconsin – Turtle Lake	2.8 mils (71 μm)
12.	Minnesota - St Paul	3.1 mils (79 μm)
13.	Tennessee – Knoxville	3.1 mils (79 μm)
14.	Texas – Nacogdoches	3.2 mils (81 μm)
15.	Pennsylvania - N. Harrisburg	3.3 mils (84 μm)
16.	Wisconsin – Burkhart	3.9 mils (99 μm)
17.	Washington – Tacoma	4.7 mils (119 μm)
18.	Florida – Martin County	5.0 mils (127 μm)
19.	New York – Katonah	5.2 mils (132μm)
20.	South Carolina – Georgetown	5.2 mils (132 μm)
21.	Georgia – Lincolnton	5.6 mils (142 μm)
22.	Texas – Houston	5.7 mils (145 μm)
23.	Virginia - N. Washington, DC	5.7 mils (145 μm)

Figure 1 – *Metal converted as a function of location*

10. A Florida site provided data from three structures with respective exposure time of 2-years, 5-years and 20-years. The results reveal that the oxide develops very rapidly in the first two years and gradually slows to a point where additional growth is measured in microns (Figure #2). (Additional studies will need to be conducted to determine if the oxide development follows a similar pattern in more moderate environments, such as the Midwest.)

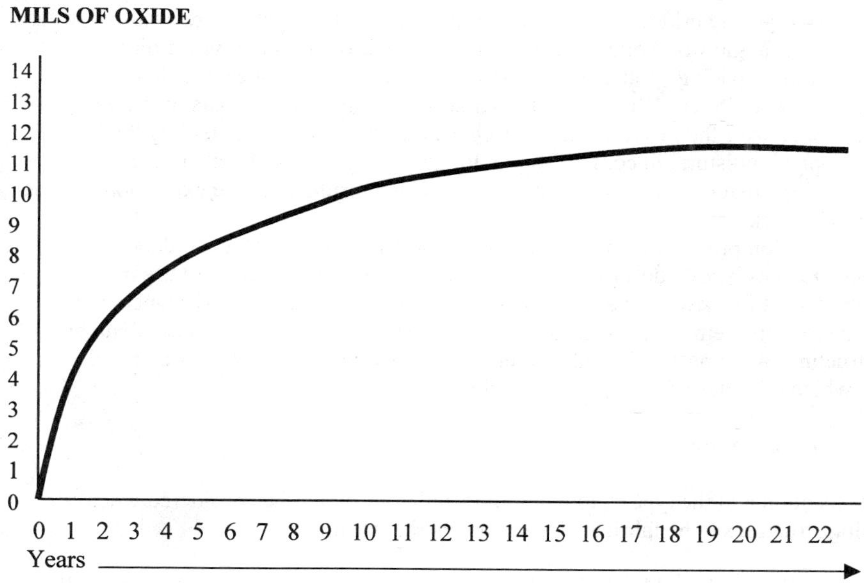

Figure 2 - *Oxide Growth as a Function of Time in Florida*[1]

[1]This figure represents oxide growth, which is 12 mils (305 μm) after 20 years exposure. The data used to develop this graph was obtained from structures in the same area that had exposure time of 2, 5, and 20 years respectively.

- 47% (5.7 mils (145 μm)) of oxide growth occurs in the first two years of exposure.
- 61% (7.3 mils (185 μm)) of oxide growth occurs in the first five years of exposure.
- The remaining 39% (4.7 mils (120 μm)) growth is spread over the next 13 years, with stability reached at approximately 18 years of exposure.
- Any additional oxide growth is estimated at less than .2 mil (5 μm) per year.

Comparison With Historical Weathering Results

The majority of structures included in this study were fabricated from steel purchased with chemical composition that provides a minimum corrosion index of 6.0. A formula for calculating the corrosion index of low-alloy steel is provided in ASTM G101 Guide for Estimating the Atmospheric Corrosion Resistance of Low-Alloy Steels. Comparison of the data obtained from this study, and other published data [1] [2] reveals good correlation, which adds credibility to the reliability of Guide G101 and the data obtained in this study. The present minimum corrosion index of 6.0, as required by ASTM A588, ASTM A852, and ASTM A871, provides assurance of corrosion resistance when any one or combination of these high-strength, low-alloy structural steels are used to fabricate tubular utility structures.

When potential users of weathering steel tubular structures question weathering steel's reliability, they often relate problems connected with weathering steel lattice structures or highway bridges. The primary problem with lattice structures was oxide pack-out in the angle-iron bolted connections. The pack-out problem was a direct result of improper bolt spacing. Bolt spacing standards were not available at the time the structures were designed. The typical tubular structure slip-joint consists of the upper section slipped over the lower section, which acts as an "umbrella" shielding the joint from entrapped moisture; in comparison, bolting two angle-irons together always provides a "lip" that can hold and absorb water, which results in oxide expansion and the potential for pack-out.

The corrosion problem with bridges was due to entrapped run-off and debris in designed crevices below the surface of the roadway. The recent increased use of weathering steel for new bridges is evidence that bridge engineers and designers now recognize that problem, and design accordingly to eliminate it. Tubular steel weathering steel structures are typically designed to eliminate crevices that can hold water and debris, which eliminates this type of corrosion problem.

Special Considerations

The location and the type of foundation/installation should be considered when designing structures to be fabricated from weathering steel.

1. The installation should be designed so soil or decaying vegetation at the ground-line does not contact bare steel. The below ground portion of an embedded structure should be protected with a high quality barrier coating that extends at least 2' above the ground-line, preventing damage from rotting vegetation. The structure installation process temporarily destroys the grass, weeds, and brush that could affect the development of the protective oxide. The temporary absence of those contaminants allows the protective oxide to develop adequate thickness, which inhibits corrosion resulting from the residue secreted by the re-growth of the surrounding plant life.

2. The barrier coating above the ground-line should be feathered back to a very thin edge, preferably 0 mils, eliminating the thick "ledge" that results from masking and applying the coating manufacturer's recommended minimum thickness [16 mils (406 μm)] to the masking line. This reduction in thickness will reduce the effect of the thermal expansion stress and eliminate or minimize the moisture holding ledge.

3. Grounding with copper wire should be done using special installation procedures. Bare copper wire should never be placed in the hole with the structure; it should be attached to a ground-rod that is driven in the ground several feet away from the structure to eliminate the potential for below-grade dissimilar metals corrosion.

4. Periodic inspection should be conducted at the base-plate of structures installed on foundations, and at the ground-line area of embedded structures. The inspection person should be looking for coating damage, buildup of decaying vegetation, and backfill or washed in soil that is in direct contact with the unprotected weathering steel. The contaminant should be cleaned from the structure and a protective coating applied to a height that will eliminate the potential for additional corrosion.

Summary

The utilization of weathering steel in the tubular utility industry has been impeded by corrosion problems observed in other types of applications. That impediment is increased when the potential first-time user reads the "cautionary" statements provided in the steel producers product bulletins. 90% of all potential first time user's questions about the life of weathering steel are based on those cautionary statements. Modification of the steel producers weathering steel bulletins would be a benefit to both the steel producer and the steel user.

The performance of weathering steel tubular structures has been excellent. The oxide growth and resultant metal loss is negligible after 25 years, even under the most corrosive conditions. Reduction of structural integrity due to corrosion is negligible and should not be a deterrent to the selection of weathering steel in the design of tubular utility structures by the owner's structural design engineer. However, the owner should always consider location and installation procedures of the weathering steel structure when developing the application. Product design should always include avoidance of water traps and direct contact of the steel with soil or damp decaying vegetation.

References

(1) Coburn, S. K., Komp, M. E., and Lore, S. C., "Atmospheric Corrosion Rates of Weathering Steels at Test Sites in the Eastern United States – Effect of Environment and Test Panel Orientation" Atmospheric Corrosion, ASTM STP 1239, W. W. Kirk and Herbert H. Lawson, Eds., American Society of Testing and Materials, Philadelphia, 1995.

(2) Townsend, H. E., "Effects of Silicon and Nickel Contents on the Atmospheric Corrosion Resistance of ASTM A588 Weathering Steel", Atmospheric Corrosion, ASTM STP 1239, W. W. Kirk and Herbert H. Lawson, Eds., American Society of Testing and Materials, Philadelphia, 1995.

Robert M. Kain[1] and Peter Wollenberg[2]

Atmospheric Corrosion and Weathering Behavior of Terne-Coated Stainless Steel Roofing

---

Reference: Kain, R. M. and Wollenberg, P., **"Atmospheric Corrosion and Weathering Behavior of Terne-Coated Stainless Steel Roofing,"** *Outdoor Atmospheric Corrosion, ASTM STP 1421*, H. E. Townsend, Ed., American Society for Testing and Materials International, West Conshohocken, PA, 2002.

Abstract: Terne (Pb-Sn) coated stainless steel has been utilized as a roofing material for a number of commercial, institutional and residential structures throughout the United States and elsewhere. As the base metal, stainless steel offers exceptional long-term resistance to atmospheric corrosion. While not requiring a protective coating, the application of terne coat is intended to provide the stainless steel with the aesthetic appeal of dull gray, aged metal roofing found on some period structures.

Within the recent ten years, several cases of reddish-brown discoloration of terne-coated stainless steel roofing have been documented. These undesirable changes in appearance have occurred in benign rural and more aggressive marine locations. A review of the corrosion literature provides historical evidence that terne coat, and lead, are more prone to corrosion in marine and rural sites than in more urban and industrial areas. This difference in behavior can be related to the absence of atmospheric pollutants, e.g., sulfur oxides, which otherwise enable terne coat and lead to develop a protective sulfate layer. X-ray diffraction analysis has determined that the reddish-brown discoloration found on roofing in some unpolluted atmospheres is lead oxide.

This paper reviews the occurrence and progression of discoloration on installed terne-coated stainless steel roofing, and that observed on test panels of the roofing material exposed at the LaQue Center for Corrosion Technology, Inc., marine atmospheric test sites at Kure Beach, NC. In addition to the issue of discoloration, mass loss and metallographic examination have quantified the loss of terne coat due to corrosion and weathering.

Keywords: roofing, terne coat, terne-coated stainless steel, atmospheric corrosion, and reddish-brown discoloration

Introduction

Overview of Metal Roofing

---

[1]Corrosion Consultant, LaQue Center for Corrosion Technology, Inc., 702 Causeway Drive, Wrightsville Beach, NC, 28480.

[2]President, Wollenberg Building Conservation, LLC, 2320 Hampton Avenue, St. Louis, MO, 63139.

Characteristically, metals used for roofing need to be sufficiently thin to facilitate fabrication and to minimize weight. Consequently, there is relatively little material allowance for corrosion. As such, the selected metal must have inherent corrosion resistance or be coated for additional corrosion protection. Both metallic and non-metallic coatings have been utilized, and these may offer decorative as well as protective value. Common examples of metal roofing intended to be used with or without painted topcoats include galvanized (zinc), galvalume (aluminum-zinc), terne (lead-tin) and tin-coated carbon steel. Other examples of metallic roofing include sheet lead and sheet zinc, copper, lead-coated copper, copper-nickel and nickel-copper alloys, aluminum and stainless steel.

Depending on the selected material of construction, metal roofing has been fabricated for use as overlapping corrugated sheets, flat solder seam layment and in standing seam construction. While introduced around the 1860s [*1*], the latter has gained more recent popularity for commercial, institutional and residential building projects. Associated trim, leaders and downspouts are often fabricated also from the selected metallic roofing materials.

Focus on Terne-Coated Stainless Steel

Austenitic stainless steels, for example, Type 301 (UNS S30100), Type 304 (UNS S30400) and Type 316 (UNS S31600) have excellent long-term atmospheric corrosion resistance and are widely used for architectural purposes. In rural and urban settings, little change in the visual appearance of unprotected stainless steel occurs during decades of exposure [*2*]. In more aggressive coastal atmospheres containing chlorides from sea spray or fog originating over seawater, rust staining and superficial pitting may develop. Despite these occurrences, extremely low rates of corrosion are the norm [*3*].

In addition to mill-finished products, clear coated, colorized and terne-coated stainless steel has been employed as a roofing material. The focus of this paper is on the corrosion and weathering resistance of unpainted, terne-coated stainless steel.

According to historical sources, tin was used in combination with iron sheets in central Europe as early as the fourteenth and fifteenth centuries [*1*]. Tin plate, produced by repetitive dipping of iron sheets in molten tin, ultimately became a roofing material and was produced extensively in Great Britain in the eighteenth and nineteenth centuries following the introduction of plate rolling machinery in 1728 [*4*]. Early examples of tin plate used in the United States, include Jefferson's Monticello (circa 1808) in Charlottesville, VA, Philadelphia's Arch Street Meeting House (circa 1804) and the Exchange Coffee House (circa 1808) in Boston, MA [*1*].

The use of lead for greater corrosion resistance in combination with tin was introduced in the early nineteenth century. Initially, tin plate was immersed in molten lead. Later, iron sheets (and steel sheets) were immersed in a molten bath comprising lead and 7% to 25% tin. The product of both processes was called terne, after the French word for dull [*3*]. Eventually, the bath composition was optimized to a tin content of 15% to 20%.

Terne-coated stainless steel roofing was developed and introduced in the United States in 1968. One of the earliest roofing applications was the Jefferson Market Courthouse in Greenwich Village, New York City [*5*].

In addition to corrosion resistance, fabricability and an attractive luster are commonly mentioned features of stainless steel. In some cases, however, the luster of stainless steel may not always be desirable. The use of terne coating on stainless steel is intended to reduce glare and provide the dull, matte look of period architecture. Reportedly, its selection for roofing on airport structures in Pittsburgh and Chicago was based on the non-reflective gray appearance and low maintenance features [*6*]. As part of the present investigation, 85° angle gloss measurements of groundward facing surfaces of terne-coated stainless steel and 2B finished Type 304 test panels exposed for nine years at the LaQue Center for Corrosion Technology, Inc.'s atmospheric test site at Kure Beach, NC, yielded values of 11 and 73, respectively. The higher gloss value is indicative of greater reflectivity or luster.

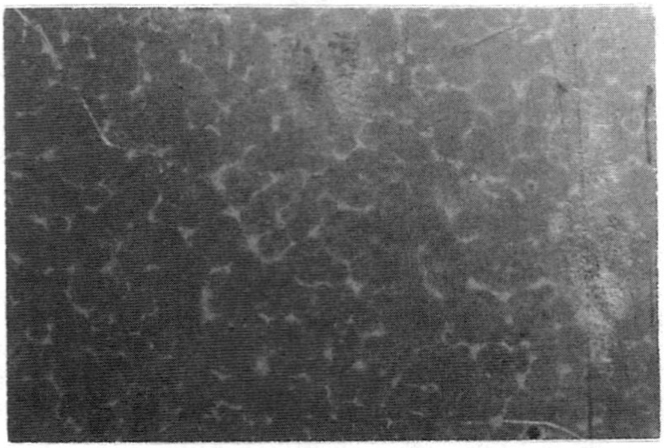

Figure 1 – *As-produced appearance of terne-coated stainless steel.*

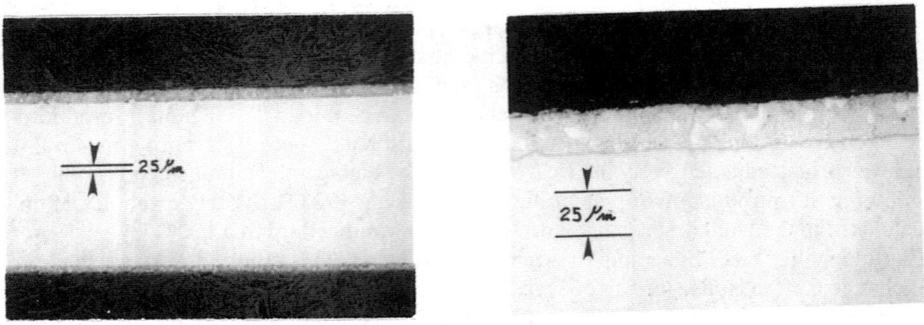

Figure 2 – *Photomicrographs of a crossed-sectioned sample of as-produced terne-coated stainless steel, shown above (left) full thickness @ ~50X and (right) terne coat layer @ ~250X.*

The Product Investigated

The subject material's processing involves immersion of chemically activated stainless steel in a molten bath of nominally 80% lead and 20% tin. The resulting, preexposure, brightly mottled appearance of a terne-coated stainless steel roofing panel is shown in Figure 1. Figure 2 shows the same material in cross-section; the general appearance of which is consistent with that described earlier by Ferrell and Gedge [*7*]. The above sample comprises a 26 gauge (0.46 mm or ~18 mils) core of stainless steel with about 25 μm (~1 mil)) or less terne coat on both sides. X-ray mapping of the section shown in Figure 2 (right) confirmed that the bright second phase particles is the tin-rich constituent. Specifications for this particular ~18 kg (40-lb.) terne coating reflect the weight of cover on both sides per double base box. A double base box is a unit of surface area equivalent to approximately 40 m^2 (62,720 in^2) of sheet metal [*8*].

The Observed Problem

While it is intended that upon weathering terne-coated stainless steel becomes dull gray, reddish-brown discoloration has been observed in some instances. Through product literature and personal endeavor, this investigation has identified a number of terne-coated roofing applications throughout the United States, including Alaska. Nationwide, however, this discoloration problem does not appear to be widespread. The authors are also aware of some similar discoloration problems with lead coated copper roofing [*9*]. Thus far, these occurrences have been associated with rural and coastal areas, as opposed to more urban or industrial regions. The following describes two case histories involving discoloration of terne-coated stainless steel, as well as collaborating test results from marine atmospheric exposures at the previously mentioned Kure Beach, NC, test sites. In addition to documenting the perceived discoloration problem, this investigation revealed significant degradation and coating loss. As will be discussed, results from earlier ASTM committee testing of terne-coated carbon steel indicated that terne coating offers poor resistance to atmospheric corrosion in rural and marine environments as contrasted with that in more industrial type atmospheres [*10*]. Corrosion literature attributes the good performance of terne coat, and lead, in more "polluted" atmospheres to the formation of protective lead sulfates. In contrast, lead oxides formed elsewhere are significantly less protective [*11*]. There is scant reporting of the aforementioned reddish-brown discoloration problem [*7*].

Present Investigation

Midwest Case History

The authors have investigated discoloration and deterioration of a terne-coated stainless steel roof on a building of historical importance in a rural midwest town. On-site monitoring at that location with wire-on-bolt sensors per ASTM Practice for Conducting Wire-on-Bolt Test for Atmospheric Galvanic Corrosion (G116-99) [*12*] resulted in aluminum wire mass losses of ≤0.1 percent. Mass losses of 0 to 1 percent are indicative of "negligible" atmospheric corrosivity typical of rural and suburban environments [*13*]. Reportedly, the roofing was installed around 1981 as a replacement for deteriorated galvanized steel that had been painted. The problems noted above were reportedly observed initially about ten years after installation and progressed to the state depicted, for example, in Figure 3. The general patterns of discoloration shown are consistent with those found on similar roofing exposed elsewhere at two seashore locations on the East coast. Moreover, as the pattern appears on sheets in different orientations, some influence of processing or coiling is suspected.

Figure 3 – *Extent of discoloration/discoloration documented on 18-year old terne-coated stainless steel roof in rural midwest location.*

Surface Analyses

In-house energy dispersive X-ray (EDS) analysis of the discolored material revealed strong energy peaks for lead (Pb), tin (Sn) and oxygen (O), along with a lesser peak for silicon (Si). The ratio of lead to tin varied from one analytical site to another. The instrumentation used was a Hitachi 3200N environmental scanning electron microscope with KEVEX-EDS operating at 20KV. The absence of an energy peak for iron (Fe) supported the belief that the widespread brown discoloration was not associated with corrosion of the stainless steel substrate.

Outsourced EDS (JEOL/TRACOR-NORTHERN 5500 instrumentation) analyses was also performed. Analysis of the "normal" gray appearing surface revealed a major level of tin with a minor level of lead. In contrast, a major level of lead and a minor level of tin were reported for the discolored areas. Very low levels of iron appeared in the spectra for both areas. X-ray diffraction (XRD) analysis was also performed to identify the compound associated with the brownish appearance. No evidence of carbonates was detected. The best matches were found for the compounds lead oxide (PbO_2) and tin oxide (SnO_2), along with Sn. Wavelength dispersive X-ray analysis showed only very low level of sulfur in one of two areas tested. Thus confirming the absence of protective lead sulfate.

Examination of the subject roofing with the aid of an optical microscope also revealed the presence of "black nodules". Of the various compounds of lead and tin, only a few are black or gray-black. These include the oxides Pb_2O and SnO and the sulfides PbS and SnS. In the absence of an apparent industrial or native source for sulfur/sulfide pollution at the subject rural sites, the sulfides can reasonably be eliminated. Considering that tin is five times more resistant to a rural atmosphere than in a marine one [*11*], SnO is suspected. Similar nodules have not been observed on terne-coated stainless steel exposed in the marine locations cited later.

Cross-sectional metallographic examination of a sample of affected roofing, as shown in Figure 4, revealed that the original lead-rich portion of the coating which remained on the skyward surface had become porous and shows little resemblance to the original structure shown previously in Figure 2.

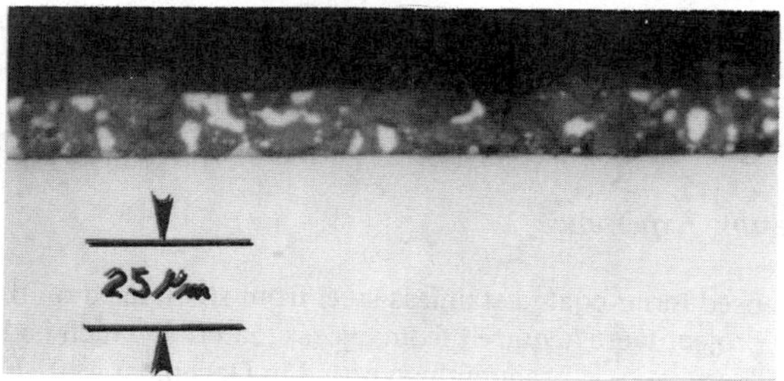

Figure 4 – *Photomicrograph of cross-sectioned sample of midwest roofing showing (dark)lead phase degradation of skyward surface during 18 years exposure, shown at ~50X.*

Figure 5 – *Discoloration found largely on standing seams and trim on coastal building roof after approximately nine years.*

Coastal Carolina Case History

Although solid copper roofing was considered initially, concerns about green copper runoff and staining of stuccoed building surfaces resulted in an alternative selection of terne-coated stainless steel roofing for a multi-story oceanfront condominium built on a South Carolina barrier island. Within one to two years of installation, a "red rust" color appeared on the standing seams at a few locations. Initially, it was suspected that breaks in the terne coat developed during fabrication, and these resulted in exposure of the 300 series substrate material along with normal rust staining. It soon became apparent that this was not the case, as the discoloration became more widespread in the ensuing eight years. When last inspected in July 1999, the discoloration was still largely confined to standing seams and rain gutters (Figure 5). Although not to the extent observed on the older midwest roof, some flat surface discoloration had developed as well. It is estimated that the affected structure's setback distance from the mean tide line is somewhere between that for the Kure Beach, NC, oceanfront and near-ocean atmospheric test lots.

The propensity for the pattern of discoloration observed on the vertical portion of the standing seam at this and the above midwest installation suggests the possible influence of mechanical working during the crimping operation.

Kure Beach, NC, Testing Experience

Samples of unexposed terne-coated stainless steel from stock used on the South Carolina condo roof project were acquired from excess material, which had been stored out of the weather. These were subsequently exposed in December 1991 in the oceanfront and near-ocean test sites at Kure Beach, NC. The exposure angle was 30° above horizontal. Wire-on-bolt corrosivity analyses consistently rate these sites as severe to very severe [*14*], and moderately severe to severe, respectively. Within a few months, specimens in the near-ocean test site began to develop discrete areas of reddish-brown discoloration akin to that observed at the South Carolina building site and that later found at the mid-West site. Samples cut from one of the test pieces exposed at Kure Beach were outsourced analyzed in 1995 by X-ray diffraction; the reddish-brown discoloration was identified as oxides of lead. Both the North Carolina and South Carolina sites mentioned above are in residential-resort locations, far removed from major sources of atmospheric pollution.

The occurrence of discoloration first appeared and progressed more rapidly for the specimens exposed to the less corrosive atmosphere found in the near-ocean test lot. Specimens in the more corrosive oceanfront test lot remained largely gray in appearance through the initial six years of exposure. An abrupt change in appearance of these specimens is coincident with a post-hurricane *Fran* beach renourishment project (1997), which increased the distance to the mean tide line [*14*]. In the ensuing time, the overall extent was similar at both locations. Figure 6, for example, shows the extensiveness of the discoloration following nine years exposure. In contrast, the groundward surfaces, which were the subject of the previously mentioned gloss measurements, remained almost entirely dull gray.

During the course of the ongoing Kure Beach exposures, the terne-coated stainless steel specimens were periodically removed to determine mass loss. Each time the specimens were weighed without any cleaning whatsoever. As shown in Figure 7, mass loss was more or less linear over the time span indicated. After 7.5 years exposure, for example, the average mass loss for duplicate specimens was 0.730 g (oceanfront) and 0.530 g (near-ocean). Assuming that most of the terne coat mass loss was from the boldly exposed and weathered skyward surface (70 cm^2), the average Kure Beach specimen mass loss (0.630 g) over 7.5 years extrapolates to 90 g/m^2 or 90 kg for a nominal 1000 m^2 roof. Roughly 80% of the total loss would be lead.

Figure 6 – *Comparison of terne-coated stainless steel specimens after nine years exposure in (left) oceanfront and (right) near-ocean test lots at Kure Beach, NC.*

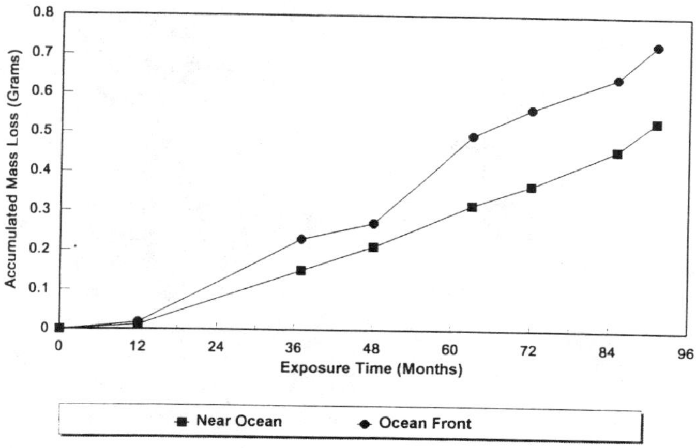

Figure 7 – *Mass loss results for terne-coated stainless steel roofing specimens exposed at the indicated Kure Beach, NC, test sites.*

Metallographic Examination – Photomicrographs in Figure 8 show cross-sectional views of the groundward and skyward surfaces of the terne-coated specimens exposed in the near-ocean lot for 7.5 years. The darker, outermost terne layers appear consistent with the "porous lead patina" previously described by others [*7*]. Elsewhere in the above cross-section, porosity affecting the skyward surface penetrated to the stainless steel substrate, as it did also in the previously discussed midwest case history.

The remaining portion of the specimens removed after 7.5 years were initially cleaned by alternate immersion in 30% acetic acid (at room temperature) and brushing. This removed the discoloration, but not the general "running" pattern. Ultimately, the remaining coating was stripped by immersion in 30% HNO_3 (at room temperature). This was done to enable detailed inspection of the substrate.

Figure 8 – *Example of lead phase degradation during 7.5 years exposure in near-ocean test lot at Kure Beach, NC, (left) skyward surface and (right) minimally affected groundward surface, both shown at ~250X.*

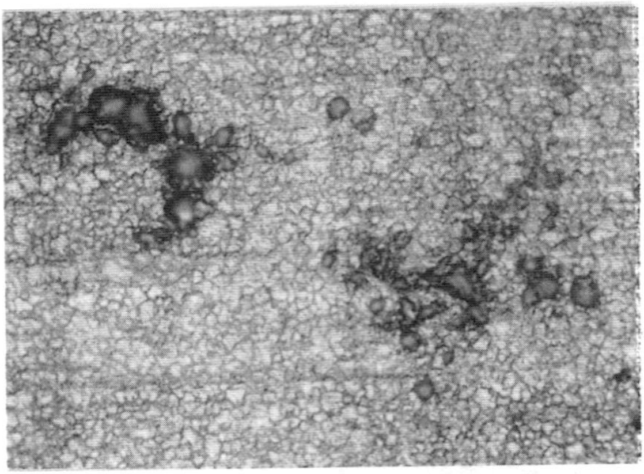

Figure 9 – *Photomicrograph showing general etching attributed to process etching and shallow pitting resulting from 7.5 years exposure in oceanfront test lot at Kure Beach, NC. (Original magnification 200X)*

Shallow intergranular attack of the stainless steel surface, and in cross-section (Figures 2 and 7), is likely the result of deliberate etching by hot hydrochloric acid used to prepare the stainless steel for immersion in the molten Pb/Sn bath [*7*]. Examination of the stripped surfaces revealed no evidence of micropitting following 7.5 years exposure in the near-ocean test lot. However, as documented in Figure 9, clusters of shallow pits were found on the skyward surface of the companion specimen exposed for the same period in the higher chloride containing oceanfront site. Based on more than 20 focusing microscope depth measurements, the pits were $\leq$25 µm ($\leq$1 mils) deep. This constitutes less than 5% of the substrate metal thickness. It may be conjectured that some shallow pits observed by present and previous investigators [*7*] are actually sites of lost surface grains associated with the above mentioned acid etching activation process.

Related Experience

20-Year Atmospheric Corrosion Testing of Terne-Coated Steel

Under the auspices of ASTM Committee A05.14, an atmospheric test program was initiated in 1960 to examine the performance of corrugated metallic-coated steel panels. The matrix of materials included galvanized steel from six types of continuous processes, pure aluminum coated steel, steel coated with an aluminum-silicon alloy, and **terne-coated steel**. Again, this was terne-coated carbon steel, not terne-coated stainless steel. The weight of the terne coating was reported at 397 g/m^2 (1.3 oz/ft^2) of coverage.

Exposure of the terne-coated material was limited to one panel per exposure site. The test sites selected by the ASTM committee included: State College, PA (rural), Newark/Kearny, NJ (industrial), Kure Beach, NC (marine, ~250 m from the ocean), Point Reyes, CA (marine, ~600 m from the ocean), and Brazos River-Freeport, TX (marine, ~1220 meters from the ocean). These same sites have been used in other ASTM atmospheric test programs both prior to and after the above 1960 one. Test panels originally exposed in Newark, NJ (near the airport) were relocated mid-way through this program to Kearny, NJ (near a power plant). The specimens were inspected annually by committee members and findings presented in the *Proceedings of the Annual ASTM Meeting*.

After one year of exposure, the only reported change in appearance of the terne-coated steel specimens was the development of "white spots" on the panel at Point Reyes. Interestingly, this was also observed on the more recent terne-coated stainless steel specimen exposed at Kure Beach.

The ensuing reports quantified the appearance of the panels in terms of pinhole rust, % rust and/or % rust stain. After approximately five years exposure, the panel in Newark exhibited only "trace rust," whereas the other four non-industrial sites ranged from 50% to 85% rust staining.

In the final 20-year analysis, Tonini [*10*] reported "the terne sheets very quickly showed significant rusting at all locations except Newark/Kearny, the most aggressive test site in the test. Despite this, the integrity of the sheets has been maintained without perforation." The above comment concerning aggressiveness stems from other tests aimed at characterizing the corrosivity of the various sites as determined by the exposure of bare zinc and bare steel test specimens [*10*]. The relative ranking (developed by ASTM Committee B03) showed that the corrosivity at two of the marine sites, Kure Beach and Point Reyes, was about 1.7 to 1.9 times and the industrial Kearny site about 2.6 to 3 3 times that of the rural State College site.

The preceding investigation made no mention of the formation of red-lead oxide on the terne-coated steel panels exposed in the rural and marine atmospheres. It is conceivable, however, that such formation may have occurred before or concurrent with the onset of rust staining associated with substrate corrosion. Recent re-evaluation of the 1964 vintage photographs, particularly that taken at Point Reyes, supports this view.

Additional Findings

Recent examination of archival slides of ASTM B-3/V1 1957 lead test panels showed evidence of "red lead" formation on those exposed at Kure Beach, Point Reyes and State College, but not "New York." In comparison to "New York" exposures, DuRose [*15*] reported greater thickness losses for electrodeposited lead coating on test panels exposed at Kure Beach and Point Reyes, as well as Tela, Honduras.

Copson and LaQue reported in 1963 that…"Lead-coated copper and terne plate are both resistant to the atmosphere and are successfully used as roofing materials. In marine or rural atmospheric exposure, however, bare terne plate does not perform as well as galvanized steel. In industrial atmospheres heavily polluted with sulfur fumes, terne plate has excellent resistance since a protective sulfate layer forms on the surface [*16*]." Copson and LaQue [*16*], nor others [*17-22*] reporting on the atmospheric corrosion behavior of lead and terne plate made no mention of the discoloration or reddish-brown discoloration cited herein. The only specific documentation found was that referenced to Ferrell and Gedge [*7*]. Cook and Smith [*11*], however, did associate red and yellow coloration of stacked lead sheets with rainwater wetting. Pourbaix indicates that powerful oxidizes can convert dissolved plumbous ions (Pb^{++}) into brown PbO_2 oxides [*23*]. A review of a paper by Gradel [*22*] indicates that peroxide may be formed during the lead corrosion process. Whether or not oxidizers such as peroxide or naturally occurring ozone contributed to the discoloration problems cited has not been investigated by the present authors.

Because of the cited cases of discoloration, and degradation, the future use of terne-coated stainless remains uncertain. An alternative material comprising a zinc-tin coating has already been introduced [*24*]. The extent to which this new coating has been field tested is unknown to the authors.

Summary and Conclusions

An overview on metallic roofing with an emphasis on terne (Pb-Sn) coated products including terne-coated stainless steel has been presented. Two case histories reporting on actual roofing applications in non-polluted rural and coastal locations have been reviewed and compared with the performance of similar roofing material exposed in the Kure Beach, NC, atmospheric test site. Under these atmospheric conditions, reddish-brown discoloration associated with the formation of lead oxides has occurred. In the absence of detailed inspection of similar roofing at other locations, the full extent of the observed problem may not be fully identified and appreciated. The patterns developed may in some way be influenced by coil processing and installation crimping operations.

Several references reviewed associate the corrosion resistance of lead with the formation of protective lead sulfate. This formation results from sulfur oxides present in some atmospheres. There is consensus that the absence of sulfur in the atmosphere contributes to degraded performance of terne plate and lead. Except for the earlier work by Ferrell and Gedge [*7*], there appears to be little previously reported concern regarding discoloration that becomes an aesthetic issue for roofing applications.

Presently reported metallographic inspections and test specimen mass loss data have demonstrated that rural and marine exposure of terne-coated stainless steel roofing can result in coating loss due to corrosion and weathering. Because of these occurrences, future use of terne-coated stainless steel may be more restricted.

References

[*1*] Waite, J. G., Gayle, M. and Look, D. W., "Metals in America's Historic Buildings," 2nd edition, U.S. Department of Interior, National Park Service, Cultural Resources, Preservation Assistance, Washington, DC, 1992.

[*2*] LaQue, F. L. and Copson, H. R., "Corrosion Resistance of Metals and Alloys," 2nd edition, Reinhold Publishing Corp., NY, Stainless Steels – Atmospheres, pp. 410-411.

[*3*] Baker, E. A. and Lee, T. S., "Long-Term Atmospheric Corrosion Behavior of Various Grades of Stainless Steel," *Degradation of Metals in the Atmosphere, ASTM STP 965*, S. W. Dean and T. S. Lee, Eds., American Society for Testing and Materials, Philadelphia, PA, 1988, pp. 52-67.

[*4*] Waite, D. S.., "Nineteenth Century Tin Roofing and Its Use at Hyde Hall," Albany: Division for Historical Preservation, New York State Parks and Recreation, State of New York, 1974.

[*5*] Follensbee Steel Company, Follensbee, WV.

[*6*] Pollock, B., "Terne Coated Stainless Choice," *American Metals Market*, January 3, 1992.

[*7*] Farrell, D. and Gedge, G., "Investigation of the Corrosion Performance of Terne Coated Stainless Steel," *Industrial Corrosion*, August/September, 1992, pp. 15-18.

[*8*] Federal Specification QQ-T-201F, Terne plate, for roofing and roofing products, 1973.

[*9*] Copper Development Association, New York, NY, personal communication.

[*10*] Tonini, D. E, "Atmospheric Corrosion Test Results for Metallic-Coated Steel Panels Exposed in 1960," *Atmospheric Corrosion of Metals, ASTM STP 767*, S. W. Dean and E. C. Rhea, Eds., American Society for Testing and Materials, Philadelphia, PA, 1982, pp. 163-185.

[*11*] Cook, A. R. and Smith, R, "Chapter 26 - Atmospheric Corrosion of Lead and Its Alloys," and M. E. Warwick and W. B. Hampshire, "Chapter 36 – Atmospheric Corrosion of Tin and Its Alloys," *Atmospheric Corrosion*, W. H. Ailor, Ed., John Wiley and Sons, NY, 1982, pp. 393-404 and pp. 509-527, respectively.

[*12*] *1998 Annual Book of ASTM Standards, Vol. 03.02 Wear and Erosion; Metal Corrosion*, American Society for Testing and Materials, West Conshohocken, PA, 1998.

[*13*] Doyle, D. P. and Wright, T. E., "Rapid Method for Determining Atmospheric Corrosivity and Corrosion Resistance," *Atmospheric Corrosion*, W. H. Ailor, Ed., Wiley & Sons, NY, 1982, pp. 227-243.

[*14*] Phull, B. S., Pikul, S. J. and Kain, R. M., "Thirty-Eight Years of Atmospheric Corrosivity Monitoring," *Marine Corrosion in Tropical Environments, ASTM STP 1399*, S. W. Dean, G. Hernandez-Duque Delgadillo, and J. B. Bushman, Eds., American Society for Testing and Materials, West Conshohocken, PA, 2000, pp. 60-74.

[*15*] DuRose, A. H., "Atmospheric Exposure of Electroplated Lead Coatings on Steel,"

ASTM STP 197, American Society for Testing and Materials, Philadelphia, PA, 1957, pp. 97-104.

[*16*] LaQue, F. L. and Copson, H. R., *Corrosion Resistance of Metals and Alloys*, 2nd edition, Reinhold Publishing Corp., NY, p. 300.

[*17*] Uhlig, H. H., *Corrosion Handbook*, John Wiley and Sons, NY, 1948, Lead and Lead Alloys, p. 201, and Lead Coatings, p. 845.

[*18*] Rosenfield, I. L., "Lead and Lead Alloys," *Atmospheric Corrosion of Metals*, English Language edition, NACE International, Houston, TX, 1972, p. 185.

[*19*] Shreir, L. L., "Lead and Lead Alloys," *Corrosion*, George Newnes Ltd., London, 1963, pp. 4.59-4.77.

[*20*] Burns, R. M. and Bradley, W. W., *Protective Coatings for Metals*, 2nd edition, Reinhold Publishing Co., New York, 1955, pp. 245-256.

[*21*] McKay, R. J. and Worthington, R., "Corrosion Resistance of Metals and Alloys," *Lead - Atmosphere*, Reinhold Publishing Corp., 1936, pp. 221-222.

[*22*] Gradel, T. E., "Chemical Mechanisms for the Atmospheric Corrosion of Lead," *The Journal of the Electrochemical Society*, Vol. 141, No. 4, 1944, pp. 922-929.

[*23*] Pourbaix, M., et. al., *Atlas of Electrochemical Equilibria in Aqueous Solution*, NACE International, Houston, TX, and Cebelcor, Brussels, Belgium, 1974, p. 488.

[*24*] "Follensbee Environmentals, TCSII, Terne II," Follensbee Steel, Follensbee, WV, September 1998.

M. Morcillo,[1] J.A. González,[1] J. Simancas,[1] and F. Corvo[2]

Outdoor Atmospheric Degradation of Anodic and Paint Coatings on Aluminium in Atmospheres of Ibero-America

Reference: Morcillo M., González J.A., Simancas J., and Corvo F., "**Outdoor Atmospheric Degradation of Anodic and Paint Coatings on Aluminium in Atmospheres of Ibero-America,**" *Outdoor Atmospheric Corrosion, ASTM STP 1421*, H. E. Townsend, Ed., American Society for Testing and Materials International, West Conshohocken, PA, 2002.

Abstract: Though aluminium generally presents good corrosion resistance to the atmosphere, finishing technologies are applied for its protection and decoration in a wide range of applications. These technologies basically consist of anodizing and painting. This paper summarizes the results obtained in Working Group No. 5 "Coatings on Aluminium" in the framework of the PATINA project, a four-year iberoamerican research programme on atmospheric corrosion.

The reported information includes gravimetric and electrochemical impedance spectroscopy determinations for bare and anodized aluminium, and undercutting at the scribe, filiform corrosion, fungal attack and changes in decorative properties (loss of gloss, colour changes and chalking) for the painted specimens. The anodizing and sealing of aluminium avoids the risk of localized corrosion. The degradation of organic coatings refers mainly to their decorative properties: soiling, staining by fungi, colour changes, chalking and loss of gloss; with the last two types of deterioration being dependent upon the intensity of solar radiation at the exposure site.

Keywords: anodized coatings, paint coatings, aluminium, atmospheric degradation, performance

Introduction

In general, aluminium exhibits good resistance to atmospheric corrosion. The reason for this excellent behaviour is an extremely fine oxide passivating layer, of a thickness of about 25 Å [*1*], which forms on the surface of aluminium and its alloys in many natural environments and guarantees the conservation of their properties, including, in many cases, of their appearance.

When corrosion occurs, in the presence of certain pollutants, it is localized and takes the

[1] Head, Researcher, and Technician respectively at Dept. of Materials Engineering, Degradation and Durability. National Centre for Metallurgical Research (CENIM-CSIC). Gregorio del Amo, 8, 28040-Madrid (Spain).

[2] Researcher at Dept. of Corrosion. National Centre for Scientific Research (CNIC) P.O. Box 6412, Havana (Cuba).

form of pitting corrosion, intergranular corrosion, delamination, or stress cracking corrosion; with pitting corrosion being by far the most frequent. In such situations, the maximum penetration of pitting as a function of time follows expressions of semilogarithmic type [*2,3*]. The corrosion attack therefore tends to decrease significantly in time, and if there are no demands relative to the conservation of appearance, aluminium alloys may be used without protection. The best demonstration of the enormous durability of these materials can be seen in a number of applications in public buildings that date from around 100 years ago [*4*].

However, when unprotected surfaces of aluminium and its alloys are exposed to the atmosphere they can acquire a greyish or blackish colouring, especially in industrial atmospheres, which is undesirable from a decorative point of view. In marine atmospheres pitting attack can occur.

For this reason, in many applications aluminium is usually protected when exposed to the atmosphere. The two main protection technologies are anodizing and painting, in the latter case using a broad spectrum of organic coatings which go from conventional oil-type paints to modern polyester powder coatings.

This communication draws on results obtained in the PATINA project [*5*], and refers to the behaviour of aluminium specimens of commercial purity (99.5% Al) protected with anodic films of approximately 7, 18 and 28 μm thickness or with three paint systems: alkyd (130 μm), polyurethane (45 μm) or polyester (110 μm), exposed to a wide variety of climatological conditions representing a broad spectrum of atmospheric corrosivities, from rural-urban to industrial and marine (tropical and polar) environments, with salinity values ranging from negligible up to 684 mg $Cl^{-}\cdot m^{-2}\cdot d^{-1}$.

Experimental Procedure

Painted Specimens

Specimens of 10x15 cm were cut from an aluminium sheet of 2 mm thickness, using alloy L-3350 (UNE 38355) for the alkyd (A) and polyurethane (B) paints and alloy L-3140 (UNE 38314) for the polyester (C) paint. It must be noted the higher copper content of the L-3140 alloy, 3.8%, in relation with the L-3350 alloy, 0.02%.

All of the specimens were subjected to successive degreasing, rinsing and drying operations. Subsequently, the series to be protected with systems A and B was chromatized by immersion for 2 min in a dissolution of $Cr_2O_4Na_2$ 0.1N followed by rinsing and drying. Paint C was applied after a wash primer of polyvinyl butyral-type with zinc chromate catalyzed with phosphoric acid.

The top coats consisted of: a polyester powder coating, of white colour and 120 μm thickness (paint A) with a dry cycle of 20 min. at 190°C; a polyurethane type liquid paint, of white colour and a dry film thickness of 45 μm (paint B) with a curing cycle of 10 min. at 160°C; and a liquid paint of alkyd resin sterified with vegetable oil, also of white colour and an average dry film thickness of 130 μm (paint C) dried at room temperature.

Once the paint coatings had been allowed to cure in the laboratory atmosphere, a scribe of 0.3 mm width and with an angle of attack of 60±15° was made with a knife in the paint coating in the lower zone of the specimens, reaching the base aluminium. Another series of specimens was left without scribes.

Bare and Anodized Aluminium Specimens

In the same network of natural testing stations as the painted specimens (Table 1), aluminium specimens of commercial purity (99.5% Al) were also exposed in bare condition and with anodic coatings of approximately 7, 18 and 28 μm, covering the range of thicknesses habitual in architectural applications. The anodic layers were formed in industrial conditions: 18% weight sulphuric acid bath at 20±1 °C and 1.5 A/dm^2, for the times necessary to achieve the aforementioned thicknesses, followed by sealing with boiling deionized water for 60 min.

Table 1-*Testing stations. Environmental and corrosivity characteristics for aluminium.*

Test site / (Country)	Time of wetness (annual fraction)	Chlorides ($mg\ Cl^-{\cdot}m^{-2}d^{-1}$)	SO_2 ($mg\ SO_2{\cdot}m^{-2}d^{-1}$)	Hours of sunshine / year	First year aluminium corrosion rate ($g{\cdot}m^{-2}y^{-1}$)	ISO 9223 Aluminium Corrosivity Category
Pardo / (Spain)	0.366	3.9	6.4	2700	0.15	C1 (negligible)
Panama / (Panama)	0.629	8.3	4.8	1995	0.36	C2
Esmeraldas / (Ecuador)	0.710	2.1	16.5	1539	0.38	
Cubatao / (Brazil)	0.579	8.1	54.5	1899	0.43	
Lumiar / (Portugal)	0.135	19.6	22.6	2802	0.50	
Jubany / (Argentina, Antarctic)	0.293	-	-	434	0.58	
Tablazo / (Venezuela)	0.504	63.3	6.0	2498	0.65	C3
Punta del Este / (Uruguay)	0.515	147.0	5.0	2338	0.67	
La Voz / (Venezuela)	0.483	374.8	29.9	3073	1.31	
Viriato / (Cuba)	0.500	684.0	31.4	2779	1.78	
Lima / (Peru)	0.589	92.1	14.6	1421	5.21	C5

(-) not available

Techniques Used

Aluminium specimens in bare condition, protected with paints A, B and C, and with the three anodic film thicknesses, were exposed in a variety of atmospheres in the Ibero-American region representing a broad spectrum of atmospheric corrosivities (Table 1). The grid in Figure 1, based on ISO Classification of Corrosivity of Atmospheres (ISO 9223) on the classification of aggressivity as a function of SO_2 and Cl^- (salinity) contents in the atmosphere, situates the different testing stations considered in the study. As can be seen, the coverage of the testing station network goes from the rural-urban atmosphere of El Pardo, with very low SO_2 and Cl^- pollution, to more industrial atmospheres (Cubatao) with much higher SO_2 contents. Marine atmospheres are also considered, from those apparently unpolluted by SO_2 (Jubany in the Antarctic) to others that are more polluted and show very high salinity levels (Viriato and La Voz). Though it was not possible to obtain data on pollutant deposition rates in the Antarctic station of Jubany during the research, previous research [*6*] indicates that the atmosphere surrounding the station is pure marine, without SO_2 pollution.

Aluminium experiences localized corrosion (pitting) but according with ISO 9223 on Classification of Corrosivity of Atmospheres, aluminium corrosion rates were calculated as general corrosion. Pit depth is a better indicator of potential damage, but this characteristic cannot be evaluated after the first year of exposure. Anyway, with the exception of the Lima atmosphere in the rest of the test sites the pit depth is very low.

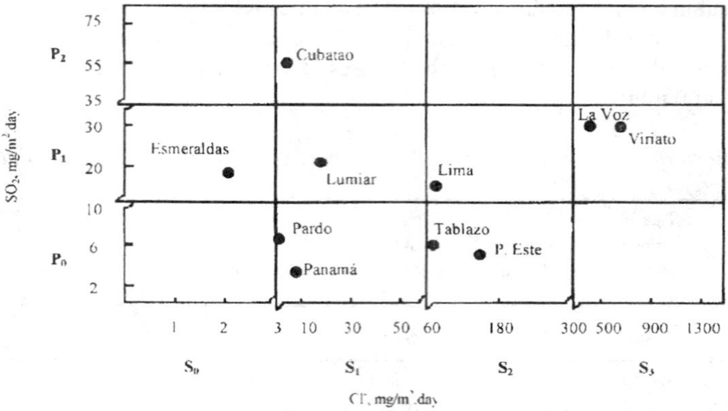

Figure 1 - *Characterization of atmospheres corresponding to the testing stations with regard to SO_2 pollution and salinity, according to ISO 9223.*

Once withdrawn from the testing stations after one, two, three and three and a half years of exposure, a careful inspection of the painted specimens was carried out in the Paints and Coatings Laboratory of INETI* (Lisbon, Portugal), quantifying the changes in colour according to ISO Method for Colour Measurement (ISO 7724/2), loss of gloss according to ISO Method for Measurement of Specular Gloss of Non-metallic Paint Films at 20°, 60° and 85° (ISO 2813) and Rating of Degree of Chalking by Tape Method (ISO 4628: Part 6) of the three types of paints, evaluating the presence of dust, soiling and fungi on the surface, as well as the state of the scribe and the adjacent coating (delamination, blistering, filiform corrosion, etc.).

With regard to the behaviour of bare and anodized aluminium, systematic use was made in the laboratories of CENIM (Madrid, Spain) of traditional gravimetric methods and observation both with the naked eye and with a magnifying glass, as well as standard quality control tests for anodic films and electrochemical impedance spectroscopy (EIS). Use was also occasionally made of optical and electron microscopy.

Experimental Results

With regard to the anticorrosive behaviour of the three paint systems after three years of atmospheric exposure, all have afforded good protection to the aluminium in the different atmospheres. Attention is drawn only to the occurrence of slight undercutting and/or blistering of the paint films in the zones adjacent to the scribe in the cases of paints B and C, the latter in all the atmospheres where it was tested (Table 2). Paint B

---

* Instituto Nacional de Engenharia e Tecnología Industrial

only showed this type of damage in the Lima and La Voz atmospheres. In La Voz station, the blistering at some parts of the specimen edges seemed to show a similar morphology to filiform corrosion, which nevertheless does not develop in a clear way. In the Lima atmosphere, however, filiform corrosion is patently observed on the coating adjacent to the scribe (Figure 2), and the filaments progress notably with exposure time, after an incubation period of approximately 18 months of exposure.

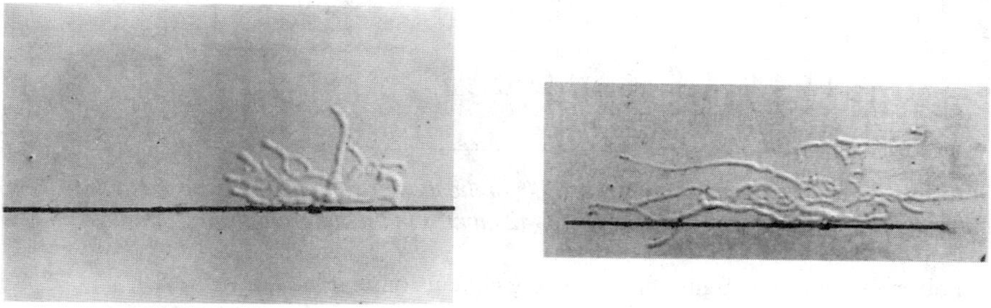

24 months 36 months

Figure. 2 - *Magnified view of filiform corrosion on paint B exposed in the Lima atmosphere. Note the significant progress of filaments with increasing exposure time.*

With regard to the decorative properties of the different coatings, the main changes observed refer to the following aspects: changes in colour, loss of gloss, chalking, presence of fungi on the paint surface and existence of firmly adhered soiling on the paint surface.

Adhered soiling occurred only in the industrial atmospheres considered in the study (Esmeraldas and Cubatao), and on the two paint types (A and B) tested in these atmospheres.

The presence of fungi, which could be seen in some inspections and testing stations, did not constitute very significant damage to the paint films. The slight staining that was produced disappeared with simple washing of the paint surface with water.

Chalking was observed to a greater or lesser degree on all three paint systems in all the testing stations. Qualitative evaluations of the degree of chalking of the three systems were made during the inspections, leading to the following conclusions.

- Paint C is that which presents the highest degree of deterioration due to this phenomenon, and paint B is the most resistant in this respect.
- In general, the degree of chalking increases with exposure time.
- As was foreseeable, the degree of chalking is a function of the number of hours of sunshine per year at the exposure site. Thus, the highest degrees of chalking were found in La Voz, Panama and Lumiar stations, and the lowest in the Lima station.

The changes of colour have been very slight with all three paints. With regard to the loss of gloss experienced by the three paint systems in the different atmospheres, Figure 3 displays the results obtained in the laboratory on specimens exposed for three years in the atmospheres. Data is not presented for Cubatao, since, as has already been mentioned, the specimens suffered heavy soiling in this testing station.

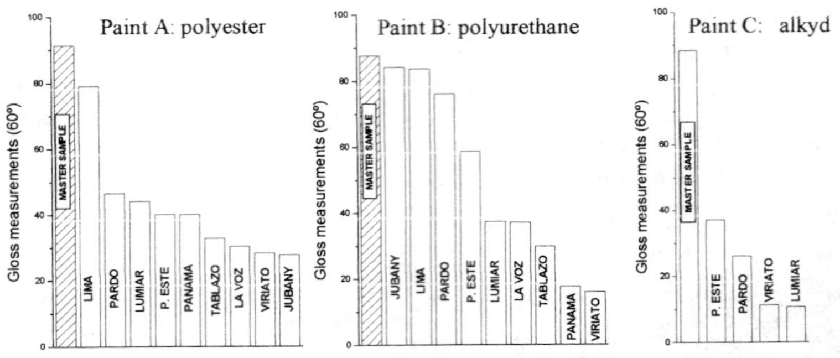

Figure 3 - *Results of gloss measurements (60°) of the master sample (not exposed to the atmosphere) and specimens exposed for 36-42 months in the different testing stations.*

From observation of this figure the following conclusions can be drawn.

- Paint C (alkyd) is that which experienced the greatest loss of gloss and paint B the least.
- A certain relationship is seen between the loss of gloss and the number of hours of sunshine per year at the exposure site. Thus, Viriato (2723 hours of sunshine per year) always registers great losses of gloss, while in Lima (1376 hours of sunshine per year) the losses of gloss are very low.

In summary, in most of the stations there is an evident improvement in terms of the conservation of appearance, compared with bare aluminium. This can be seen, for instance in the case of the Lima station (the most aggressive), by comparing Figure 2, which shows considerable filiform corrosion under the paint, which does not occur on unscribed specimens, with Figure 4, showing an unprotected aluminium covered with blisters, many of which have degenerated into pits.

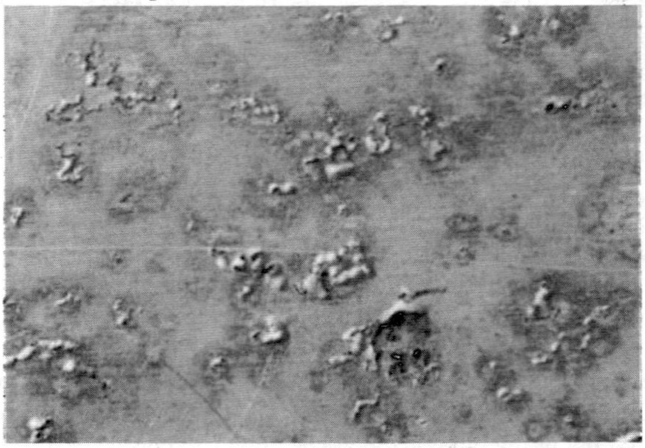

Figure 4 - *Magnified view (x10) of the bare aluminium surface exposed for 42 months in the Lima atmosphere.*

Painted Specimens

In general terms the anticorrosive behaviour of the paint coatings in the different atmospheres may be considered to be good after three years of atmospheric exposure, with some exceptions that are mentioned below. The deterioration observed on some test specimens and in certain atmospheres has mainly affected the decorative characteristics of the specimens more than their anticorrosive properties. Table 2 indicates the deterioration recorded for the three paint systems in the different atmospheres.

Table 2. *Types of damage observed on the three paint systems in the different testing stations.*

Test Site	Paint A	Paint B	Paint C
El Pardo	Fungi	Fungi	Slight undercutting at scribe edges. Fungi.
Panama	Slight adhered soiling.	Slight adhered soiling.	Not tested
Esmeraldas	Adhered soiling (moderate degree) which increases with time.	Adhered soiling (moderate degree) which increases with time.	Not tested
Cubatao	Adhered soiling (severe degree) which increases with time. Fungi.	Adhered soiling (severe degree) which increases with time. Fungi.	Not tested
Lumiar	No damage observed	No damage observed	Very slight undercutting at scribe edges. Slight adhered soiling. Fungi.
Jubany	No damage observed	No damage observed	Not tested
Tablazo	Fungi.	Fungi.	Not tested
Punta del Este	No damage observed	No damage observed	Slight undercutting of the paint adjacent to the scribe, which progresses with time. Fungi.
La Voz	No damage observed	Slight blistering at the scribe and edge area, similar to filiform corrosion.	Slight blistering at the scribe.
Viriato	Fungi, preferentially on the downward facing side.	Fungi, preferentially on the downward facing side.	Fungi, preferentially on the downward facing side. Slight blistering at edges and zones adjacent to the scribes.
Lima	No damage observed	After 18 months, start of filiform corrosion in zones adjacent to scribe, which progress with time.	Not tested

Chalking was observed on all three systems and in all the testing stations, to a greater or lesser degree.

Bare and Anodized Aluminium Specimens

Column 6 in Table 1 shows the mass changes experienced by bare aluminium specimens during the first annual exposure cycle in the different testing stations. Bare aluminium suffers relatively high mass losses in the Lima, Viriato and La Voz stations, fairly significant losses in Tablazo and Punta del Este stations, and very small or insignificant losses in the remaining stations. But even in the least aggressive environments, such as Panama, an ugly staining (also known as blotting) appears, which limits decorative applications (Figure 5).

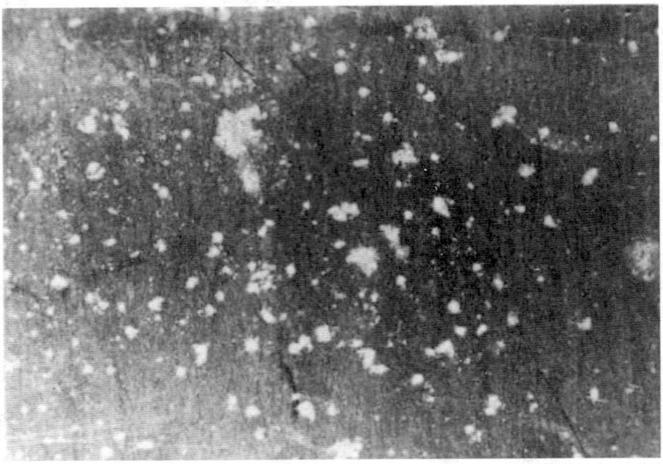

Figure 5 - *Magnified view (x7) of the bare aluminium surface exposed for 12 months in the Panama atmosphere.*

In the case of the anodized specimens, mass losses do not occur and in fact mass gains are recorded. However, these are not due to a corrosion process, since Al in the anodic layers is already in the form of Al^{3+}, as it is found in nature, and these mass gains may be attributed to the filling of the pores during exposure, since they are left practically empty in the sealing operation. This would explain why the mass increases as the thickness of the anodic layers increases. Nevertheless, the greatest mass variation takes place during the first annual cycle, and the mass subsequently stabilizes, within the natural dispersion of results, as the pores become saturated. Figure 6 depicts the mass variations of the 28 μm anodic layers in all the testing stations, as proof of the described behaviour. It can be seen that this behaviour does not depend on the climatological or pollution characteristics of the environment.

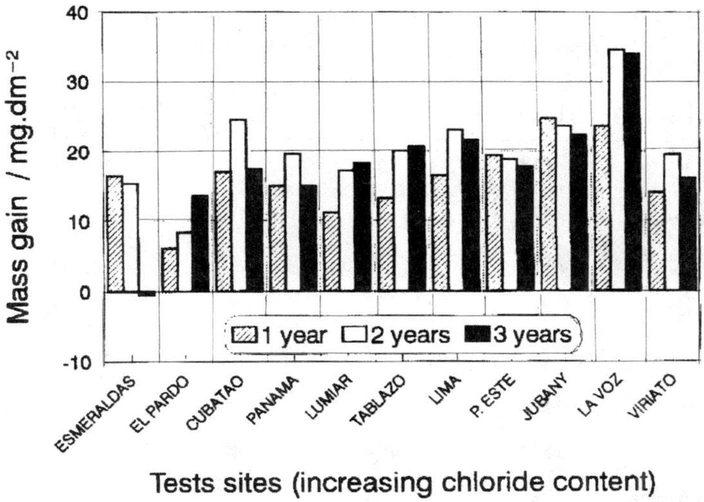

Figure 6 - *Mass gain of 28 μm thickness anodic films in all the testing stations after one, two and three years of exposure.*

Table 3 shows the corrosion symptoms appreciable at up to 60 magnifications on bare aluminium and the three anodic film thicknesses. The symbols indicate the absence of any symptom of localized corrosion or the presence of pitting corrosion, blistering, filiform corrosion or white stains (blotting), which are precursors of pitting. In general terms, anodizing improves the appearance, eliminating any visible symptom of localized corrosion with sufficient thicknesses. The improvement is appreciable even for the lowest of the tested thicknesses, though localized corrosion is not prevented in highly polluted atmospheres.

The physical-chemical characteristics of the anodic films continue to modify during very long time periods, as can easily be seen with EIS, which offers the researcher a very sensitive tool for characterizing the quality and integrity of the barrier and porous layers. EIS permits an estimation of the evolution of the resistance and capacitance of the porous layer (R_p and C_p) and the capacitance of the barrier layer (C_b) from diagrams such as that given in Figure 7, which indicates the responses of the 28 μm thickness anodic film, recently obtained and after one year of exposure in El Pardo and Viriato stations (those with the lowest and highest pollution levels respectively). The behaviour can be seen to be practically independent of the environmental aggressivity, but with a very great increase in impedance with time. Table 4 shows the electrochemical parameters deduced from the impedance diagrams immediately after anodizing and sealing and at the end of the first annual cycle of exposure in the aforementioned stations.

Table 3. *Macroscopic observations of corrosion on bare aluminium and anodized aluminium with different film thicknesses after 42 months of exposure.*

TEST SITE	Bare	Anodized and sealed		
		7 µm	18 µm	28 µm
EL PARDO	✪	○	○	○
ESMERALDAS	✪	○	○	○
PANAMA	✪	○	○	○
CUBATAO	✪	○	○	○
LUMIAR	●	○	○	○
TABLAZO	●	○	○	○
LIMA	● ■	● ς	● ς	○
PUNTA DEL ESTE	●	○	○	○
JUBANY	●	○	○	○
LA VOZ	●	○	○	○
VIRIATO	●	●	○	○

○ Without symptoms of localized attack
✪ Blotting
● Pitting corrosion
■ Blistering
ς Filiform corrosion

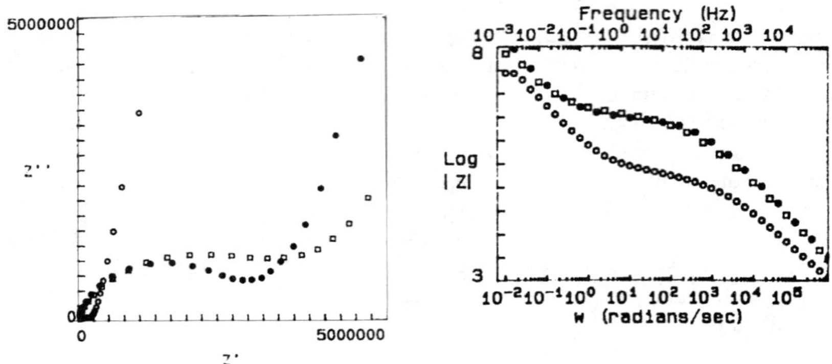

Figure 7 - *Impedance diagrams for the 28 µm anodic layer, recently obtained (○) and after one year of exposure in El Pardo (●) and Viriato (□) testing stations, the least and most polluted atmospheres respectively.*

Table 4 - *Electrochemical parameters of the anodic layers, recently sealed and after the first year of exposure in the test sites of El Pardo and Viriato, the least and most polluted atmospheres respectively.*

TEST SITE	Exposure	R_p ($k\Omega cm^{-2}$)	C_p (nF)	C_b (μF)
LABORATORY (indoor)	7 μm 1 day	248	8.07	0.56
	18 μm 1 day	375	4.60	0.42
	28 μm 1 day	211	28.2	1.37
EL PARDO	7 μm 1 year	1,530	13.3	0.74
	18 μm 1 year	577	7.1	1.21
	28 μm 1 year	3,060	1.4	0.76
VIRIATO	7 μm 1 year	2,990	4.5	---
	18 μm 1 year	4,840	1.1	---
	28 μm 1 year	6,530	1.8	0.72

Discussion

Behaviour of Bare Aluminium

The corrosion data for aluminium shown in Table 1 confirms previous experience [*2,3*], in the sense that it demonstrates the excellent corrosion resistance of aluminium in unpolluted natural environments, as well as its susceptibility to localized corrosion in the presence of depassivating ions such as chlorides in marine atmospheres (Figure 4).

Specifically, bare Al behaves as a passive material and does not show signs of localized attack in any of the atmospheric stations with <30 $mg{\cdot}m^{-2}{\cdot}d^{-1}$ of Cl^- [*7*] pollution during the first year of exposure. Curiously during the second year, although corrosion continues to be negligible, of the order of 0.5 $g{\cdot}m^{-2}{\cdot}y^{-1}$ or less, the small white stains that are observed in Cubatao, Panama and Lumiar stations (Figure 5) gradually give way to tiny blisters which are precursors of pitting, or to very small pits. In the stations with salinity values of ≥50 mg $Cl^-{\cdot}m^{-2}{\cdot}d^{-1}$, localized attack in the form of pitting becomes evident during the first year of exposure, and is more intense the stronger the pollution (Figures 4-6).

Behaviour of Painted Aluminium

Though there is abundant literature on the atmospheric corrosion of aluminium, in particular in European countries and in North America, information about tropical climates, such as those existing in several countries of Central and South America, is scarce, and thus the wide variety of climatological conditions and corrosivities considered (Table 1 and Figure 1) is of great interest for aluminium, particularly when it is anodized or protected with the three paint systems used in this research.

The three paint coatings afford adequate anticorrosive protection to the base aluminium, at least for exposure periods of less than four years. There is a clear contrast between the appearance of bare aluminium (Figure 4) and painted aluminium (Figure 2), excluding the scribe and the area under its influence, in environments of very high aggressivity. Even when painted aluminium subjected to highly corrosive environments (Figure 2) is compared with unprotected aluminium exposed to atmospheres of moderate or low aggressivity (Figure 5), the conservation of its appearance continues to be superior. The delamination which, to a greater or lesser degree, finally affects the scribe edges, and the filiform corrosion which is detected in exceptional cases, despite the previous chromatizing, shows the importance of achieving defect-free protection systems.

The types of damage encountered on specimens without scribes refer exclusively to the conservation of their decorative properties: strongly adhered soiling (industrial atmospheres), slight staining by fungi in many of the atmospheres, which disappears with simple washing (Table 2), and chalking and loss of gloss (Figure 3), the latter two intensifying as the number of hours of sunshine per year at the exposure site increase.

Behaviour of Anodized Aluminium

As is demonstrated by the results obtained, correct anodizing, with an anodic film thickness of ≥18 μm, seems to present an adequate solution to the problems of localized corrosion of aluminium alloys in strongly polluted atmospheres, since numerous points of localized attack have only been detected in Viriato station (which shows the greatest salinity) and the Lima station, in the case of the lowest anodic film thickness (7 μm), and occasionally in the Lima station with the intermediate thickness (Table 3). In the Lima station, symptoms of localized corrosion are already seen after the first year of exposure, while in Viriato station pitting is detected only after the second annual cycle. The number of pits increases with time in both cases.

Though the mass gain due to water diffusion through the pore mouth, which is plugged during sealing, is practically complete after the first year (Figure 7), the changes in the physical-chemical characteristics of the anodic films continue for very long periods of time [*8*]. For instance, the impedance diagrams in Figure 7 show that the resistance of the porous layer (R_p) becomes some 10 times greater during the first year of exposure (Table 4). This parameter has been proposed as a quantitative index for evaluating the advance of the hydration process, or, in other words, the quality of the anodic films: the higher the R_p, the better the sealing quality [*9*]. Except for the coatings of the lowest thickness in the most aggressive environments (Table 3), the quality of the anodic films improves with ageing in all cases, and no significant influence is played either by the coating thickness (within the range habitually employed in architectural applications) or by the level of pollution at the testing stations.

During the 42 months of exposure in this research, the mass variations recorded in 10 of the 11 testing stations involved mass gains, which would continue until the pores become filled. This exposure time is not sufficient to be able to detect the mass losses seen in periods of 10 years or more, which are nevertheless extremely slow, about 0.15 μm/year in atmospheres of moderate aggressivity [*8,10*]. This corrosion rate allows durabilities of 100 years or more to be estimated for the anodic film thicknesses habitual in architectural applications.

Conclusions

- Bare aluminium suffers pitting corrosion in environments with salinity values above 50 mg $Cl^{-}\cdot m^{-2}\cdot d^{-1}$ by the end of the first year. In the rest of the stations, even with salinity values of around 10 mg $Cl^{-}\cdot m^{-2}\cdot d^{-1}$, a white staining (blotting) is showed, which only deteriorates the appearance but gradually becomes transformed into small pits with the pass of the years.

- The three paint coatings offer adequate anticorrosive protection of the base aluminium, at least for time periods of no more than the duration of the project.
- The permanent damage recorded refers to the decorative properties of the coatings, except for those related with the zones adjacent to the scribe.
- The anodizing and sealing of aluminium and its alloys, with a coating thickness of ≥18 μm, is an adequate solution for the problems of localized corrosion in strongly polluted atmospheres, at least for exposure times of less than 42 months. The 7 μm anodic films, on the other hand, already exhibit pitting corrosion during the second year in the Lima and Viriato stations.
- The ageing process supposes an improvement of all the quality indices of the anodic films, irrespective, after a certain critical thickness, of the degree of pollution of the testing stations.

Acknowledgments

The authors would like to thank S. Flores (Peru), E. Almeida (Portugal), M. Sánchez (Panama), J. Peña (Ecuador), F. Fragata (Brazil), S. Rivero (Uruguay), O.T de Rincón (Venezuela) and B. Rosales (Argentina) for exposing the materials in their testing stations.

They also thank Mr Andrés Arranz López and Mr Carlos Paseiro of the Laboratory of the Directorate General of Housing, Planning and Architecture of Madrid for the supply and application of paints A and B, and the Cuban firm VITRAL for the manufacturing and application of paint C.

Finally, they express special gratitude to the Paints and Coatings Laboratory of INETI (Lisbon, Portugal), specifically its Director Elisabete Almeida and María del Rosario Ferreira, for the performance of the infrared spectra of the three paints used, as well as the measurements of the colour and gloss of the specimens aged in the different atmospheres.

Bibliographic References

[1] Goddard, H.P., *Materials Performance*, Vol. 20, No. 7, 1981, pp 9-15.

[2] Feliu, S., and Morcillo, M., "*Corrosión y Protección de los Metales en la Atmósfera*", Bellaterra, Barcelona, 1982.

[3] Morcillo, M., and Feliu, S., "*Mapas de España de Corrosividad Atmosférica*", Programa CYTED, Madrid, 1993.

[4] Gatto, F., and Perrone, A., in "*Atmospheric Corrosion*", W.H. Ailor, Ed., John Wiley and Sons, New York, 1982, pp. 827-839.

[5] Morcillo, M., et al., "Atmospheric Corrosion in Ibero-America: Its Prevention by Coatings (PATINA/CYTED project)", *Proc. 14th International Corrosion Congress*, Cape Town, 1999.

[6] Rosales, B., Fernández, A. and Moriena, G., *Proc. 13th Int. Corrosion Congress*, Melbourne, 1996.

[7] Morcillo, M., Almeida, E. and Rosales, B., *Aluminium*, 76, 2000, pp.610-615.

[8] Lizarbe R., González, J.A., Otero, E. and López, V., *Aluminium*, 69, 1993, p.548.

[9] Hoar ,T.P. and Wood, G.C., *Electrochimica Acta*, 7, 1962, p.333.
[10] López, V., González, J.A., Bautista, A., Otero, E. and Lizarbe, R., *Corrosion Science*, 40, 1998 ,p. 693.

Robert M. Kain,[1] Bopinder S. Phull,[1] and Stanley J. Pikul[1]

1940 'Til Now — Long-Term Marine Atmospheric Corrosion Resistance of Stainless Steel and Other Nickel Containing Alloys

---

Reference: Kain, R. M., Phull, B. S., and Pikul, S. J. **"1940 'Til Now — Long Term Marine Atmospheric Corrosion Resistance of Stainless Steel and Other Nickel Containing Alloys,"** *Outdoor Atmospheric Corrosion, ASTM STP 1421*, H. E. Townsend, Ed., American Society for Testing and Materials International, West Conshohocken, PA, 2002.

Abstract: The history and corrosivity of the LaQue Center for Corrosion Technology, Inc.'s "Kure Beach" atmospheric test site has been reviewed in brief. Stainless steel and other Ni-containing alloy test specimens from a number of earlier investigations remain exposed in the site's "museum area", the oldest of these dates to December 1940. The appearance of these and other 1940, 1950, and 1960 vintage specimens was documented in early 2001. The performance of various test specimens are reviewed, and in some cases compared with that reported previously. Gloss and surface roughness measurements were used as non-subjective tools to assess the current state of reflectivity and luster, and corrosion incurred after periods of up to 60 years in the marine environment. Both alloy composition and original surface finish influenced the performance of different grades. The opinions of earlier researchers appear to have been validated many years later.

Keywords: atmospheric corrosion, Kure Beach, corrosivity, stainless steels, Ni-alloys, gloss measurements, surface roughness

Introduction

"Information on atmospheric corrosion requires years to develop and is more reliable to compare specimens initially exposed at the same time..."[*1*] H. P. Copson was referring to a then newly commissioned International Nickel Company corrosion test program involving simultaneous exposure of 80 materials to the atmosphere in Bayonne, NJ. He and coworkers at their research laboratory in Bayonne and Technical Services Department in New York recognized the need for reliable data in order to provide opinions on the weathering behavior of nickel-containing alloys and competing alloys. Shortly thereafter, in the Spring of 1941, a number of replicate test panels from the "Bayonne Series" were also

---

[1]Corrosion consultant, corrosion consultant, and senior research technologist, respectively, LaQue Center for Corrosion Technology, Inc., 702 Causeway Drive, Wrightsville Beach, NC, 28480.

placed on exposure at a seaside location owned by the Ethyl-Dow Corporation in Kure Beach, NC. Through the earlier efforts of Francis L. LaQue, the International Nickel Company had already helped establish a seawater test site on the Ethyl-Dow property and by 1940 space about 800 feet inland from the oceanfront was set aside for atmospheric testing.

The Ethyl-Dow property was a large tract of land in southeast North Carolina, in the vicinity of Federal Point, which extended from the Atlantic Ocean to the Cape Fear River on its western boundary. The Kure Beach site had been developed in 1933/1934 to facilitate construction of an ethylene dibromide plant that utilized seawater as a source of bromine ions from the raw material. The original atmospheric corrosion test site was located between the oceanfront and a large man-made lagoon that supplied seawater to the actual extraction plant further west near the river bank. Ethylene dibromide was used as an anti-knock additive for aviation fuel. Reportedly, production peaked at 16 000 pounds per day during the height of World War II. The bromine extraction plant closed and was dismantled in 1946, but corrosion testing continued under the direction of Frank LaQue, who by then was in charge of the International Nickel Company's corrosion engineering section. Ultimately, a section of the Ethyl-Dow tract was purchased by the International Nickel Company (later known as INCO) in order to perpetuate marine atmospheric corrosion testing at this location. Although under new ownership since 1998, a portion of the original property continues to be used for marine atmospheric corrosion and weathering test by the LaQue Center for Corrosion Technology, Inc.

This paper has been prepared primarily to document the condition of a number of stainless steel test panels that have remained on continuous exposure since 1940/41. Also noted is the behavior of other specimens that were exposed between 1940/41 and 1967 and still remain in a dedicated "museum section" of the test site. It is beyond the scope of this paper to report on all of the long-term specimens at the Kure Beach test site.

Test Site and Its Corrosivity

The coordinates for the Kure Beach atmospheric test site are latitude 34°N and longitude 77.5°W.

As recently described by Phull et al. [*2*], the site comprises two newly designated areas for testing, the oceanfront lot and the near-ocean lot. The "museum section" is located in the latter. Previously, these areas were known as the 25-meter lot and 250-meter lot, respectively. Earlier references cite them as the 80-foot lot and 800-foot lot (original area). The corrosivities of these two test sites have been extensively monitored by various means for the past 38 years [*2*]. Earlier monitoring, commencing in 1949, utilized iron calibration specimens [*3*].

A recent review of historical files found no mention of concerns about any possible deleterious effect associated with the proximity of the bromine plant during the initial years (1940-1946) of atmospheric testing at Kure Beach. By the account of one recently interviewed plant engineer, "...there was little or no atmospheric release due to the fact that the process operated under negative pressure". Moreover, a site calibration reported earlier by Baker [*3*] indicated that the winds across the Kure Beach test lots blow more predominantly from directions other than from west-to-east. This geographical trend continues to this day.

Because of setback distances from the ocean, chloride levels in the near-ocean lot are historically lower than on the oceanfront. Over the recent 30 years for which data have been reported, chloride levels in the near-ocean lot have been in the 100 to 200 mg/sq m/day (monthly average) range, while those at the oceanfront range from 200 to 700 mg/sq m/day. Based on wire-on-bolt monitoring, the Marine Corrosivity Index [*4*] (MCI) for the oceanfront lot may vary from moderately severe to very severe, while that in the near-ocean lot ranges from moderate to moderately severe. Over a recent nine-year period, the corrosion rate for A-36 and 1008 carbon steel in the less aggressive environment ranged from 17 to 34 μm/yr. Rates of 62 to 576 μm/yr were recorded in the oceanfront lot for the same period.

All test panels described herein were boldly exposed on a 30° incline, to the horizontal, facing due south. Over the past 60 years, the test site has encountered weather conditions associated with numerous hurricanes as well as the March 1993 "storm of the century".

Current Evaluation Procedure

It is common practice to prepare and expose multiple specimens of each material and surface condition of interest so that from time-to-time one or more may be removed for thorough evaluation. In the case of the subject 1941 series, for example, duplicate 4-in. x 6-in. (100 mm x 150 mm) panels of most materials were exposed. In 1956, following 15 years' exposure, one specimen each was removed, cleaned, reweighed and microscopically examined for evidence of surface pitting. The appearance of the 300 series stainless steels following one-year and 15 years' exposure at Kure Beach was previously reported, along with 15-year corrosion rates and pit depths [*5*]. It is hoped that the remaining specimens in the atmospheric corrosion "museum" will remain on test for many more years. Accordingly, cleaning was not contemplated for the present evaluation. However, the condition of many of the stainless steel test panels from the May 1941 exposure were weighed in their present condition. The weighing was actually performed about 48 hours following a rainstorm that is expected to have removed, from the skyward-facing surfaces, most deposits other than adherent stain and surface rust.

The corrosion resistance of stainless steel specimens exposed to atmospheric conditions is often reported in terms of the percent surface area covered by rust stain. The previously mentioned 15-year report, for example, cited slight to moderate rust stain covering 10 to 25% of the surfaces on the various grades of 300 series stainless steel. Such reporting can be somewhat subjective, particularly when the actual inspections are performed by different individuals, and/or without prior photographs or other pictorial standard. While there are ASTM standards covering the assessment of coated and plated panels, none exist presently for bare stainless steel, per se.

In an attempt at providing some quantitative measure of performance, not biased by subjective observations and experience, two somewhat unconventional methods of assessment were employed. Since the manufacturers, specifiers, and users of stainless steel are likely to be more interested in how much surface reflectivity or luster is retained over many years' exposure, in lieu of the amount of staining developed, it seemed appropriate to gloss measurements for which a Minolta multi-gloss Model 268 reflectometer set at an 85° angle of measurement was utilized. The reflectometer reading is related to a black gloss standard with a defined index of refraction which equals 100 units. The second method selected was surface profile measurements using a Taylor-Holson Surtronic 10 model surface roughness indicator. Both instruments provide simple digital readouts. Measurements were performed in-situ at five randomly selected areas on the skyward facing surfaces, and the values obtained averaged. Except where indicated, the gloss meter was placed directly on the test panel with its long dimension parallel to the specimen length. The surface roughness indicator was positioned so that its stylus traversed across the width of the panel; each 1 cm traverse yielded a R_a value in μm or μin. All data were collected originally in English units.

Unfortunately, no unexposed replicates of the actual test materials were available for "baseline" calibration and comparison. Accordingly, it is suggested that the present gloss measurement represent some averaged indication of the original and diminished reflectivity due to surface corrosion or staining. Likewise, the surface roughness measurements are indicative of the original profile and possible changes due to any localized pitting and buildup of rust.

Preliminary Measurements

Gloss measurements were performed on the reflective, corrosion-free surface of a highly polished Hastelloy®[2] C alloy test panel on exposure since 1942, and on the 100% rust covered surface of a 9% Ni-steel test panel on exposure since 1960. The average gloss values for these extreme conditions were 121 and 2, respectively. As will be demonstrated, measurements on the various stainless steel panels discussed fell within these extreme values.

The relationship between gloss and surface roughness measurements for exposed stainless steel (Type 304) with an original 2B finish is shown in Figure 1. The data provided are for two 100 x 150 mm test panels that were exposed in the Kure Beach near-ocean and oceanfront test lots in December 1991. Measurements were performed in February 2001. As can be seen, higher gloss and lower R_a values were found for the groundward facing surfaces than for those facing skyward. Moreover, the specimen exposed in the near-ocean site consistently exhibited higher gloss and lower R_a values than the corresponding specimen in the oceanfront exposure lot.

[2]Hastelloy® is the registered trademark of Haynes International, Inc., Kokomo, IN.

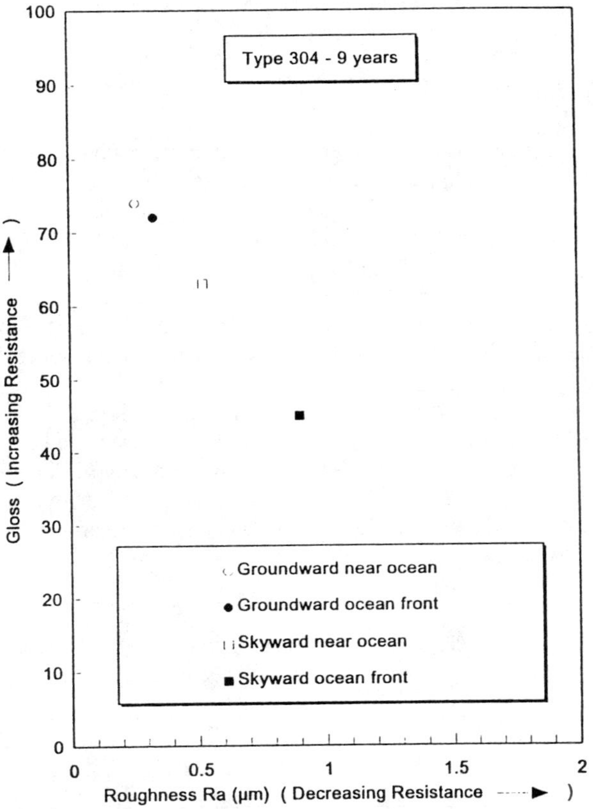

Figure 1 – *Surface gloss vs. roughness plot for groundward and skyward surfaces of Type 304 test specimens exposed in Kure Beach atmospheric test lots.*

Surface Finishes

The following provides brief general description of surface finishes cited.

No. 1 finish - Hot or cold rolled, annealed and pickled; characteristically dull matte to fairly reflective.

No. 2 finish - As above plus light cold rolled pass with polished rolls; more reflective than No. 1 finish.

No. 2B finish - Bright cold rolled after annealing and descaling.

No. 4 finish - General purpose polished finish; typically 120 to 150 mesh abrasive.

No. 6 finish - A dull satin finish having lower reflectivity than No. 4 finish.

No. 7 finish - Highly reflective buffed finish but with retained "grit" lines from prior fine grinding.

No. 8 finish - Most reflective finish; obtained by buffing with fine rouges

(Source: Reference [*6*])

Identification of Specimens and Their Performance in Long-Term Marine Atmospheric Corrosion Tests at Kure Beach, NC

September 1967 Series

Architectural usage of stainless steel often requires material with different surface finishes to provide contrasting appearances or to minimize glare. In addition, forming with simple and compound bends is typically required. In 1967, a series of convoluted test specimens with 90° and 180° bends and with five different surface finishes were exposed at Kure Beach. The grades represented were Type 302, Type 316 and Type 430. One specimen of each was exposed in the following finishes: No. 2, No. 4, electropolished, 2B and No. 8; the latter two being designated as "British" finishes. While it could not be confirmed at the time of writing this paper, it is quite conceivable that more than one heat of each alloy was used for specimens stock.

The performance of these specimens after 13 years was documented in an earlier technical bulletin published in color [7]. Subsequent reports commented on the condition of the subject specimens following 25 years [8], and later 28 years [9] of continuous exposure. As might be expected, Type 316 reportedly performed best overall. Based on visual appearance, all materials benefited from electropolishing in terms of enhanced corrosion resistance.

Table 1 – *Thirty-three years' exposure: gloss meter measurements average 85° angle measurement for September 1967 series.*

Surface Finish	Type 430 Set 1[1]	Set 2[1]	Avg.	Type 302 Set 1	Set 2	Avg.	Type 316 Set 1	Set 2	Avg.
No. 2	28	29	29	69	77	73	64	72	68
No. 4	13	14	14	40	40	40	71	80	76
El-Pol.[2]	36	36	36	91	98	95	77	75	76
2B[3]	35	38	37	54	57	56	55	51	53
No. 8[3]	33	35	34	67	67	67	68	76	72

[1] Five locations each set
2 El-Pol. = Electropolished
[3] British finishes

Table 1 compares gloss values obtained in February 2001, after >33 years exposure. Two sets of data were collected by the same investigator on two separate occasions. For any given alloy-surface finish combination, the difference between the two sets of measurements varied from 0 to 10%. Type 430 specimens, which exhibited the most visible rust stain overall, consistently exhibited the lowest average gloss values and with the least variability between sets. The most distinguishable trend observed as a function of stainless steel composition is from the data for specimens with the No. 4 finish. As can be seen in Table 1, gloss results for Type 302 and Type 316 were similar for specimens in the No. 2, 2B and No. 8 finishes. Somewhat surprisingly, the highest gloss value was determined for the Type 302 in the electropolished condition; this despite the fact that visual observations indicated less staining on the electropolished Type 316 specimen. This anomaly may be the result of differences in the original surface condition of the specimens. Type 302 generally has a higher manganese-sulfide inclusion content than other 300 grades. It is conceivable that preferential dissolution of inclusions at the surface, during electropolishing, created "facets" that yielded a more reflective glossy surface originally, and that is probably retained to this day.

1950s Series

Numerous test programs were initiated during the 1950s. Specimens of the associated specimens are retained in the museum area and others elsewhere in the test lots. Of particular note are a large number of spot-welded lap joined stainless steel test panels from a 1958 ASTM sponsored program. Thirteen different grades of material are represented. While the condition of some has been documented from time-to-time, the last comprehensive reports were written by Baker and Lee [*10*] and Baker and Kirk [*11*] following 26 years of testing. The now 42-year-old test panels remain in both the oceanfront and near-ocean test lots for an indefinite period.

Another 1950s vintage series comprises six grades of stainless steel with three different surface finished (2B, No. 4 and No. 6). These 100 mm x 150 mm test panels were originally exposed in 1957 and are presently located in the museum area. Table 2 gives the average 85° angle gloss measurement after more than 43 years' exposure.

Table 2 - *Average 85° gloss measurements for 1957 series stainless steel.*

Surface Finish	Type 201	Type 202	Type 302	Type 316	Type 430	Type 442
2B	70	58	73	54	41	71
No. 4	41	57	68	33	61	80
No. 6	41	64	59	63	63	65

For this particular series, there does not appear to be any consistent trend regarding surface finish among the six grades. Some grades, for example, exhibited similar gloss values for test panels with No. 2B and No. 4 finishes, while others showed similarities between the No. 4 and No. 6 finishes. Type 430 with a 2B finish exhibited a gloss reading identical to Type 201 with the No. 4 and No. 6 finishes. Visually, the least stained specimen during the February 2001 inspection was the Type 316 test panel with the No. 4 finish; yet it showed the lowest gloss value. Rotating the gloss meter 90°, so that it was parallel to the highly visible No. 4 grinding pattern, resulted in a new average reading of 93. Correspondingly, the average R_a changed from 22 μin (0.6 μm) against the grinding marks to 7 μin (0.2 μm) with the grinding marks. Table 3 compares gloss and R_a data for the six materials in the 2B surface condition. Considering the limited number of measurements for each, there appears to be a reasonable correlation between gloss and R_a for the various materials. Type 302, for example, exhibited the highest gloss number and the lowest R_a, while Type 430 exhibited the lowest gloss value and the highest R_a. Considering an average R_a of 18 μin (0.5 μm) for all six materials, Types 316, 201 and 302 exhibited roughness below this average while R_a's for Types 442, 202 and 430 were above the average.

Table 3 - *Comparison of gloss and R_a measurements for 1957 series stainless steels with 2B finishes.*

Parameter	Type 430	Type 202	Type 302	Type 201	Type 316	Type 442
Gloss	41	58	73	70	54	71
R_a (μin)	26	20	13	14	16	19
R_a (μm)	0.7	0.5	0.3	0.4	0.4	0.5

Single test panels of Type 302 and Type 304, both in the 2B condition, have been exposed since July 1956. After more than 44.5 years' exposure, the Type 304 panel exhibited an 85° gloss measurement of 87 and a R_a of 8 μin (0.2 μm). The Type 302 panel, which exhibited somewhat more general rust staining, exhibited corresponding values of 63 and R_a 21 μin (0.5 μm). Because of the narrow width of these two specimens, the gloss meter was rotated 90°.

Ni-Base Alloys

The museum area also contains several test panels of Ni-base INCONEL® and INCOLOY®[3] type alloys exposed in 1952 and 1953. Included, for example, are panels of alloys 600, 800 and 825. The oldest INCONEL alloy 625 test panel dates to 1965. As would be expected, those alloys containing molybdenum, e.g., alloys 825 and 625, remain rust free, while the Mo-free alloys 600 and 800 show very light to moderate staining as they approach 50 years' exposure. Earlier, Baker [*12*] cited corrosion rates of 1 μm for these and a number of other high nickel alloys.

[3]INCONEL® and INCOLOY® are registered trademarks of Special Metals Family of Companies, Huntington, WV.

1940s Series

A significant number of specimens remain from exposures commencing in 1948 and between 1940 and 1942. The remainder of this review will focus on these, the oldest of the continuously exposed stainless steel.

December 1940 Series (60 Years) - The oldest series comprises martenisitic, ferritic, and austenitic (stabilized and non-stabilized grades) stainless steels with a variety of surface finishes. No. 1 and No. 2 finishes were common to all seven alloys. Test panels of Type 304 were exposed with six difference finishes, the finest being a No. 7. Gloss rating values and surface roughness (μm) averages are given in Tables 4 and 5, respectively.

Table 4 - *Sixty years' exposure: gloss meter measurement for 12/13/40 series of stainless steels.*

	Average 85° Angle Measurement for Indicated Grade[1]						
Surface Finish	Type 410	Type 416	Type 430	Type 304	Type 316	Type 321	Type 347
No. 1	2	2	10	11	77	64	16
No. 2	3	2	4	16	62	33	45
No. 2B	--	--	45	69	81	14	--
No. 4	--	--	60	75	108	49	--
No. 6	--	--	--	85	--	--	--
No. 7	--	--	--	115	--	--	--

[1] Average of five locations.

Table 5 - *60 Years' Exposure: Surface Roughness Measurements for 12/13/40 Series of Stainless Steels*

	Average R_a for Indicated Grade[1]													
Surface	Type 410		Type 416		Type 430		Type 304		Type 316		Type 321		Type 347	
Finish	μin	μm	μin	μm	μin	μm	μin	μm	μin	μm	μin	μm	μin	μm
No. 1	282	7.2	320	8.1	176	4.5	106	2.7	23	0.6	19	0.5	33	0.8
No. 2	205	5.2	304	7.7	168	4.3	70	1.8	20	0.5	33	0.8	86	2.1
No. 2B	--	--	--	--	76	1.9	32	0.8	14	0.3	38	1.0	--	--
No. 4	--	--	--	--	29	0.7	13	0.3	7	0.2	25	0.0	--	--
No. 6	--	--	--	--	--	--	23	0.6	--	--	--	--	--	--
No. 7	--	--	--	--	--	--	7	0.2	--	--	---	--	--	--

[1] Average of five locations.

Consistent with the general appearance of the various grades, Type 410 and Type 416 exhibited the lowest gloss values (less retained luster) and the roughest surfaces (fully rusted). Improvement is observed with Type 430, but not to the extent quantified for the four austenitic grades with comparable surface finishes.

As can be seen from Tables 4 and 5, gloss values generally increased and surface roughness values decreased on going from No. 1 and No. 2 finishes to the higher finishes indicated. Test results suggest that one alloy with a given surface finish (e.g., Type 304 with No. 7) can exhibit an appearance similar to another alloy with a different surface finish (e.g., Type 316 with No. 4).

May 1941 Series (~60 Years) - As previously mentioned in the introduction, a group of specimens replicating others exposed in Bayonne, NJ, in 1940 were exposed to the Kure Beach atmosphere in May 1941. Non-stainless type alloys for this series which remain in the museum area include: "Nickel-Silver", "Everdur", "Constantan", "70-30 CuNi", "Hastelloy A", and "Hastelloy B". Also remaining are single test panels (100 mm x 150 mm) of nine 300 series stainless steels and Type 430 stainless steel. While not part of the "Bayonne Series", test panels of Type 201 and Type 202 stainless steel were exposed along with the above on 5/14/41. Actual heat analyses for the 300 series and 430 grade are presented in Table 6. Table 7 lists the nominal composition for Type 201 and Type 202 grades.

Table 6 - *Reported chemical analyses for stainless steels exposed at Kure Beach on 5/14/41.*

Alloy	Weight Percent Composition							
	C	Ni	Cr	Other	Si	Mn	S	P
Type 301	0.11	8.14	17.74	--	0.47	1.40	0.014	0.015
Type 302	0.10	10.05	18.61	--	0.41	0.39	0.003[1]	0.020
Type 304	0.07	8.92	18.39	--	0.38	0.41	0.013	0.010
Type 308	0.07	10.74	20.38	--	0.38	0.63	0.012	0.020
Type 309	0.09	13.60	23.62	--	0.38	1.15	0.017	0.020
Type 310	0.07	19.77	24.12	--	0.39	1.46	0.008	0.018
Type 316	0.08	13.16	17.82	2.81 Mo	0.39	1.52	0.016	0.017
Type 317	0.05	14.13	18.55	3.5 Mo	0.39	1.70	0.018	0.027
Type 321	0.05	9.66	18.65	0.48 Ti	0.53	0.54	0.015	0.015
Type 347	0.07	11.23	18.64	0.78 Cb	0.24	0.56	--	--
Type 430	0.05	0.32	17.10	--	0.31	0.30	0.018	0.018

[1]Considering it unlikely that 1940's commercial melting practice could produce a stainless steel heat with this low a sulfur content, the authors suspect a typographical error in the original report from which this data was obtained.

Table 7 - *200 series nominal compositions (wt%).*

Alloy	C	Mn	Cr	Ni	P[1]	S[1]	Si[1]	N[1]
Type 201	0.015	5.5-7.5	16-18	3.5-5.5	0.60	0.30	1.00	0.25
Type 202	0.15	7.5-10.0	17-19	4.0-6.0	0.60	0.30	1.00	0.25

[1] Maximum

All 12 stainless steel grades represented in this series were exposed with a 2B finish. Historical records indicate that at least four different mills may have been involved with production of the sheet products.

Table 8 lists average gloss and surface roughness values. Also included are the visual observations from the earlier 15-year evaluation [*5*] of the originally exposed duplicate. As can be seen, there appears to be good correlation between results for the austenitic alloys; the best performance being indicated for Types 316, 310 and 317.

Based on gloss results, Table 8 also shows a clear distinction between the performance of the 300 series grades with that for Type 430 and Type 201. Type 202, on the other hand, appears comparable with Type 301 and Type 304 after nearly 60 years in the marine atmosphere. A similar observation is noted in the Figure 2 plot of gloss versus R_a data.

Table 8 - *Performance of 1941 series stainless steels.*

Stainless Grade	60 Year Gloss At 85°	Visual Appearance After 15 Years [*5*]
Type 301	54	Light rust and rust stain on 20% of surface
Type 302	- no panel[1] -	Spotted with rust stain on 10% of surface
Type 304	41	Spotted with slight rust stain on 20% of surface
Type 308	80	Spotted by rust stain on 25% of surface
Type 309	51	Spotted by slight rust stain on 25% of surface
Type 310	81	Spotted by slight rust stain on 20% of surface
Type 316	87	Extremely slight rust stain on 15% of surface
Type 317	72	Extremely slight rust stain on 20% of surface
Type 321	67	Spotted with slight rust stain on 15% of surface
Type 347	48	Spotted with moderate rust stain on 20% of surface
Type 430	20	Not previously reported in Reference 5
Type 201	25	Not previously reported in Reference 5
Type 202	43	Not previously reported in Reference 5

[1] No Type 302 test panels remained after the 1956 (15-year) removals.

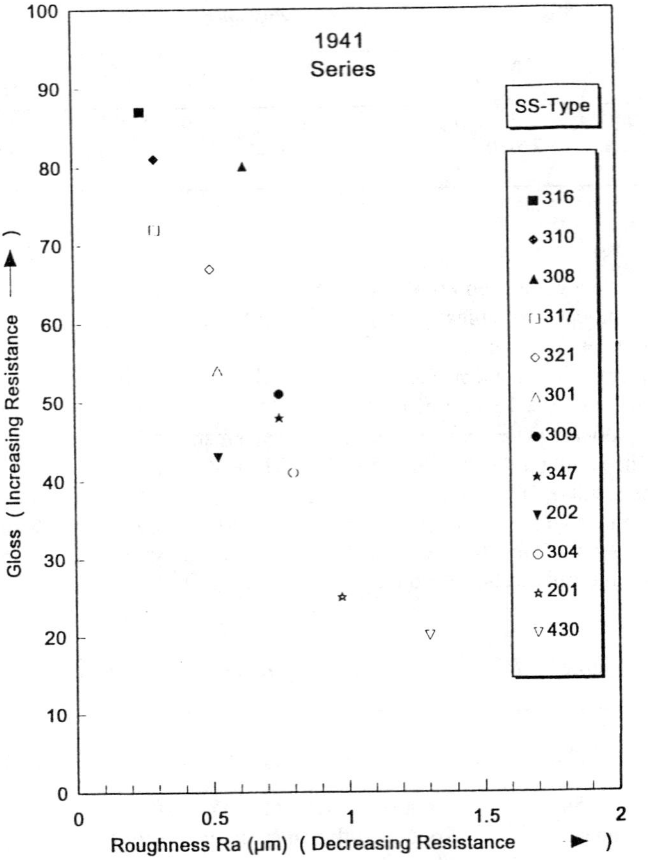

Figure 2 - *Surface Gloss vs. Roughness Plot for Various 2B Finish Stainless Steels After 60 Years' Exposure at Kure Beach, NC*

After 15 years' exposure, the corrosion rates determined from mass loss for the austenitic grades were <0.001 mpy (2.5 x 10^{-5} mm/y), except for Type 347 which was slightly higher (0.001 mpy). Type 347 also exhibited the highest average depth of pitting (3.4 mils or 86 μm). Table 9 ranks the austenitic grades by the 15-year pit depth results [*5*]. Also shown are the present 60-year R_a roughness values and the present uncleaned mass loss results. When these three parameters are compared, Types 316, 310 and 317 again exhibit the best overall performance, followed closely by Type 309. Such behavior would be expected from materials with the Cr and Cr+Mo contents listed in Table 6.

Table 9 - *Ranking of austenitic stainless steels according to 15-year pit depths.*

Stainless Grade	Average Pit Depth Mils	Average Pit Depth μm	Average R_a @ 60 Years μin	Average R_a @ 60 Years μm	60-Year Mass Change
Type 347	3.4	86	30	0.8	-0.06
Type 321	2.6	66	20	0.5	-0.06
Type 308	1.6	41	25	0.6	no data
Type 301	1.6	41	21	0.5	-0.03
Type 302	1.2	31	------- no specimen -------		
Type 304	1.1	28	32	0.8	-0.07
Type 309	1.1	28	30	0.8	-0.02
Type 317	1.1	28	12	0.3	-0.03
Type 316	1.0	25	10	0.3	-0.01
Type 310	0.4	10	12	0.3	-0.03

November 1941 Series (~60 Years) - In addition to the above, a 4-inch x 12-inch (100 mm x 300 mm) panel of conventional "18Cr-8Ni-1.5Mn" stainless steel and a "18Cr-4Ni-4Mn" variety, both with a 2B finish, were also exposed in 1941. After nearly 60 years in the marine atmosphere, the latter material exhibits somewhat less rust staining overall. Recent gloss and surface roughness measurements substantiate this observation. For example, while the gloss for the 18Cr-8Ni-1.5Mn alloy averaged 67, that for the 18Cr-4Ni-4Mn averaged 97. Surface roughness values for the two averaged 17 μin (0.5 μm) and 11 μin (0.3 μm), respectively. Note, however, that the above values differ significantly from those for Type 301/304 and Type 201 shown in Table 8. These differences are likely related to surface area and spacing between measurements on the larger and small size specimens.

Summary and Conclusions

A brief history of the Kure Beach atmospheric test site and its corrosivity has been presented as a backdrop to the long-term testing of various stainless steels and other nickel-containing alloys. Attention has been focused primarily on the performance of various grades of stainless steel on continuous exposure for periods now exceeding 60 years. Some shorter term experience, e.g., 34 to 44 years, has also been described. Present methods of evaluation included surface gloss measurement and surface profile measurements. It is suggested that these techniques provide some average measure of the original condition of the test materials and changes resulting from decades of exposure in a natural marine environment.

Results have clearly demonstrated the inherent resistance associated with increased chromium content of the alloys and the presence of molybdenum. However, in some cases, differences in surface finish appeared to have an equal or greater influence on the performance. As demonstrated by gloss measurements of the 1967 series of exposed stainless steel test panels, increasing the number of measurements may or may not influence the average value.

Long-term testing has demonstrated the durability of stainless steels in moderate to moderately-severe marine environments. The tolerance for rust staining will depend on the specific application. Resistance to staining can be maximized by selection of appropriate alloy and surface finish combinations. All specimens described were subjected to only natural rain washing and never benefited from deliberate, periodic cleaning as might be performed in actual use in certain cases.

Post Script - Following the 15-year inspection of the 1941 series, LaQue and co-worker T. P. May offered the following comment concerning exposures in the "800 foot lot": "It has been our experience that during the first few months of exposure, the stainless steels will develop a certain degree of rusting. This seems to approach its limiting maximum after about six months and from that time on the staining does not progress very severely [*13*]." This may be true for some grades of 300 series, but not necessarily all stainless steels.

Acknowledgment

The authors recognize the contribution of earlier researchers associated with these exposures, and thank the present ownership of the LaQue Center for Corrosion Technology, Inc. for its support in preparing this paper. Also recognized is the Nickel Development Institute for its financial and technical support of the current inspections.

References

[1] Copson, H. R., Internal File Memorandum, LaQue Center for Corrosion Technology, Inc., Wrightsville Beach, NC, 1940.

[2] Phull, B. S., Pikul, S. J. and Kain, R. M., "Thirty-Eight Years of Atmospheric Corrosivity Monitoring," *Marine Corrosion in Tropical Environment, ASTM STP 1399*, S. W. Dean, G. Hernandež-Duque Delgadillo, and J. B. Bushman, Eds., American Society for Testing and Materials, West Conshohocken, PA, 2000, pp. 60-74.

[3] Baker, E. A. and Lee, T. S., "Calibration of Atmospheric Corrosion Test Sites," *Atmospheric Corrosion of Metals, ASTM STP 767*, S. W. Dean, Jr. and E. C. Rhea, Eds., American Society for Testing and Materials, West Conshohocken, PA, 1982, pp. 250-266.

[4] Doyle, D. P. and Wright, T. E., "Rapid Method for Determining Atmospheric Corrosivity and Corrosion Resistance," *Atmospheric Corrosion*, W. H. Ailor, Editor, Wiley & Sons, NY, 1982, pp. 227-243.

[5] "Corrosion Resistance of the Austenitic Chromium-Nickel Stainless Steels in Atmospheric Environments," The International Nickel Company, Inc., 2nd Edition, 1983, Nickel Development Institute Reference No. 318.

[6] "Standard Forms and Finishes for Stainless Steels," The International Nickel Company, 1963, Nickel Development Institute Reference No. 959.

[7] "Marine Atmospheric Corrosion," The International Nickel Company, Inc., 1978.

[8] Kain, R. M., "Performance and Limitations of Engineering Alloys in Seawater and the Marine Atmosphere," *Proceedings of the International Conference on Innovative Marine Materials and Technologies*, LaSpezia, Italy, 1992.

[9] Phull, B. S., Melton, D. G., and Kain, R. M., "Engineering Alloy Behavior in Marine Marine Environments," presented at NACE Northeast Regional Meeting, Atlantic City, NJ, October 1992.

[10] Baker, E. A. and Lee, T. S., "Long-Term Atmospheric Corrosion Behavior of Various Various Grades of Stainless Steel," *Degradation of Metals in the Atmosphere, ASTM STP 965*, S. W. Dean and T. S. Lee, Eds., American Society for Testing and Materials, West Conshohocken, PA, 1988, pp. 52-67.

[11] Baker, E. A. and Kirk, W. W., "Long-Term Atmospheric Corrosion Behavior of Various Grades of Stainless Steel in Rural, Industrial, and Marine Environments," *Corrosion Testing and Evaluation: Silver Anniversary Volume, ASTM STP 1000*, R. Baboian and S. W. Dean, Eds., American Society for Testing and Materials, West Conshohocken, PA, 1990, pp. 177-190.

[12] Baker, E. A., "Long-Term Corrosion Behavior of Materials in the Marine Atmosphere," *Degradation of Metals in the Atmosphere, ASTM STP 965*, S. W. Dean and T. S. Lee, Eds., American Society for Testing and Materials, West Conshohocken, PA, 1988, pp. 125-144.

[13] LaQue, F. L. and May, T. P., Internal File Memorandum, LaQue Center for Corrosion Technology, Inc., Wrightsville Beach, NC, 1957.

Caifeng Liang[1] and Wentai Hou[1]

Twelve Year Atmospheric Exposure Study of Stainless Steels in China

---

Reference: Liang, C., & Hou, W., **"Twelve Year Atmospheric Exposure Study of Stainless Steels in China,"** *Outdoor Atmospheric Corrosion, ASTM STP 1421,* H. E. Townsend, Ed., American Society for Testing and Materials International, West Conshohocken, PA, 2002.

Abstract: Corrosion data of 12 years of atmospheric exposure of five types of stainless steel at six sites in China were obtained. The exposed steels are of commonly used stainless steels. The test sites cover typical environments of temperate and subtropical, industrial, marine and rural. Generally, stainless steels are corrosion resistant in the atmosphere. Corrosion occurred mostly in the form of pitting. Of the main environmental factors that have influence on atmospheric corrosivity of stainless steels, the most influential one is Cl pollution. If high humidity and heat are combined with Cl pollution, extra high corrosion will be the result, a phenomena not being noticed before. Sulphur dioxide has very little effect on the atmospheric corrosion of stainless steels. Deposition of dust and surface defects are the main starting points of atmospheric corrosion for stainless steels. 13% Chromium content is the basis for steel to be stainless. When Chromium content is over 18% stainless steels become excellent in atmospheric corrosion resistance. Molybdenum raises corrosion resistance of stainless steels, especially in marine environments. Extra low carbon raises their corrosion resistance. Even in humid and hot marine environments of high corrosivity, the steels with extra low carbon and some molybdenum are not corroded after 12 years of exposure.

Key Words: atmospheric corrosion, exposure test, stainless steel

Stainless steels have widespread use in various fields. Because of their good performance in atmosphere, people's attention has been moved to the corrosion of stainless steels in other mediums that are much more corrosive.

---

[1] Research engineer and Professor, Qingdao Marine Corrosion Institute, Central Iron and Steel Institute, Wheat Island, Qingdao, 266071, P. R. China.

However, the corrosion of stainless steels in the atmosphere is not as simple as people thought. The environment has a big effect on the corrosion of stainless steel. In addition, the application of stainless steels in the atmosphere is mostly for decorative cladding in architecture. Slight corrosion will cause serious damage to the look of the buildings. Therefore, the study of corrosion of stainless steels in the atmosphere has not stopped in recent years. Khanna et al. [*1*] exposed type 304 and 316 stainless steels in India and studied the effect of thermal treatment. Button [*2*] et al. studied the corrosion of 316 stainless steel exposed for 4 to 30 years in the United Kingdom. It was observed that the deposition of particulate matter could facilitate corrosion pitting and rusting. Kearns [*3*] made exposure studies of 38 stainless steels at two sites in the United States. A spectrocolorimeter was used to measure the rusting of the coupons. Molybdenum combined with nitrogen was observed to improve corrosion resistance significantly in the atmosphere. More studies were given to the practical application of stainless steels, especially in the field of architecture [*4*]. For corrosion mechanism, Johnson [*5, 6*] studied the relation between inclusions in stainless steels and pitting in atmospheric corrosion, and concluded that inclusions do not have a certain relation with pitting. A synergistic effect was found with sulphur dioxide and sodium chloride on pitting of stainless steels. Azzerri [*7*] measured the variation of corrosion potential with exposure time. The development of passive film was discussed.

To obtain comprehensive and systematic corrosion data of materials in natural environments, a national corrosion station network was established in China. A 20-year exposure program was started from 1983. Stainless steel is a part of it. Corrosion data of stainless steels exposed for 12 years has been obtained. This paper summarizes the results.

Exposure Tests

Five stainless steels were exposed. The first is one of the most commonly used austenitic stainless steel 1Cr18Ni9Ti (AISI 321, UNS S32100). It is used in various atmospheres for many applications, such as column cladding etc. The second is a feritic stainless steel 00Cr17TiAl (AISI 439LL, UNS S43908). It is used in less corrosive atmospheres. The third is a martensitic stainless steel 2Cr13 (AISI 420, UNS S42000). It is one of the cheapest stainless steel used in low request situations. The other two are a low carbon austenitic stainless steel 00Cr19Ni10 (AISI 304L, UNS S30403) and an extra low carbon feritic stainless steel 00Cr18Mo2 (a modified AISI 444, UNS S44400 with extra low carbon). The chemical compositions of the steels are listed in Table 1.

Coupons with a size of 200mm×100mm×6mm were taken from commercial cold rolled plates. The rolled surface was kept as the exposure surface. Four parallel coupons were tested for each exposure term, three for weight loss measurement and one for rust layer analysis. The coupons were exposed facing south at an angle of 45 degrees to the ground.

Table.1- *Chemical composition of the tested steels* □□□

STEEL	C	Mn	Si	S	P	Ni	Cr	Mo	Ti	Al	AISI
1Cr18Ni9Ti	0.078	0.73	0.65	0.007	0.014	8.39	17.60		0.57		321
00Cr17TiAl	0.009	0.53	0.21	0.016	0.036	0.22	16.73	0.09	0.24	0.03	439LL
2Cr13	0.199	0.37	0.35	0.009	0.030		13.79				420
00Cr19Ni10	0.010	0.29	0.59	0.006	0.030	9.28	18.90				304L
00Cr18Mo2	0.003	<0.1	0.11	0.015	<0.0		19.92	1.92			444

All the coupons were uniformly prepared and exposed on the same day, October 20, 1983 at six exposure sites in China. The brief meteorological and pollution data of the sites are given in Table 2. Details of the site can be found elsewhere [*8*].

Table 2 *Meteorological and pollution data of the atmosphere at the exposure sites.*

Site	Temp.	Humidity	Pollution			Rain		Dist.
	Aver.	Aver.	Cl^-	SO_2	PH	Cl^-	SO_4^{2-}	to sea
	(°C)	(%)	(mg/100cm^2/d)			[mg/m^3]	[mg/m^3]	(m)
Beijing (BJ)	12.0	57	0.049	0.442	5.5			
Qingdao (QD)	12.5	71	0.250	0.704	6.1	11044	81654	34
Wuhan (WH)	16.9	77	0.011	0.272	6.5	917	11390	
Jiangjin (JJ)	18.4	81	0.006	0.667	4.4	1994	31642	
Guangzhou(GZ)	22.4	78	0.024	0.107	5.8	353	13702	
Wanning (WN)	24.6	86	0.387	0.060	5.0	11229	3552	350

The coupons were withdrawn after 1, 3, 6, 12 years. Weight losses were obtained after cleaning the coupons in an inhibitor borne 10% nitric acid solution. All the procedures are in accordance with ISO-DIS 8565.2, "Metals and alloys - Atmospheric corrosion testing - General requirement for field test" (1991).

Results and Discussion

Weight Loss and Appearance Change of the Coupons

Figures 1 to 5 show the weight loss of the steels after twelve years of exposure at various sites. They have shown a complete different atmospheric corrosion behavior from the low alloy steels [*8*]. The data scattering of the parallel coupons is much higher with more chance of abnormality, especially for the four latter steels of high corrosion resistance. However, the difference in corrosion resistance of various steels and the effect of the environment can be seen clearly. It means the weight loss can reflect the corrosion behavior of stainless steels, in agreement with the study of Johnson [*5*]. It was found in his study that the results of weight loss agreed with the results of pitting depth and pitting size measurement.

Although the data scatter is also present in the development of weight loss over time, we may see that all steels had nearly no corrosion at all exposure sites in their first year of exposure. Corrosion developed only in the later years, and the corrosion loss stopped increasing after three years of exposure. This is true for all the specifications of the tested steels. This means that the corrosion product of stainless steel in the atmosphere has a protective property.

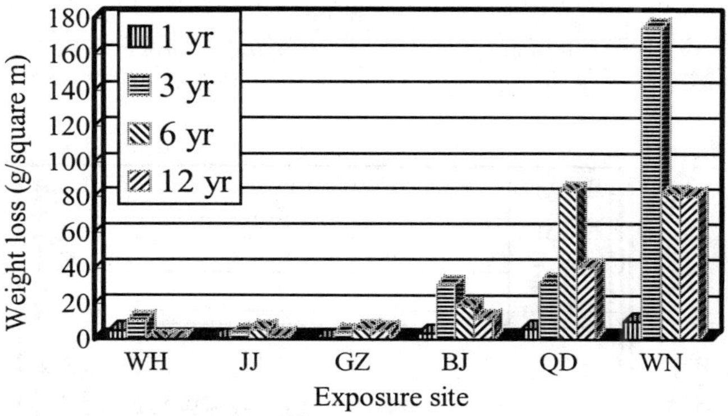

Figure 1 - *Weight loss of 2Cr13 stainless steel after twelve years of exposure at various sites.*

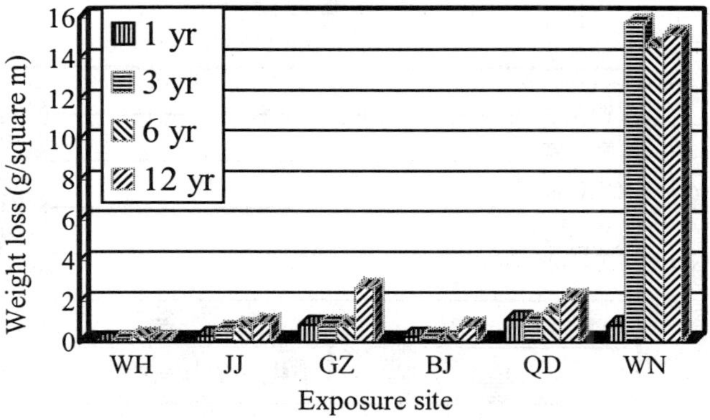

Figure 2 - *Weight loss of 00Cr17AlTi stainless steel after twelve years of exposure at various sites.*

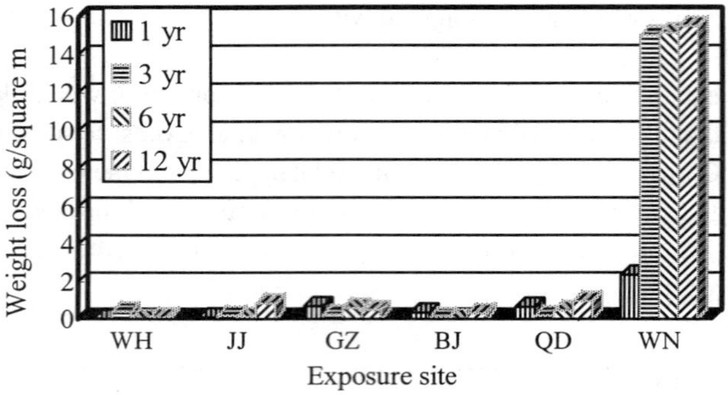

Figure 3 . *Weight loss of 1Cr18Ni9Ti stainless steel after twelve years of exposure at various sites.*

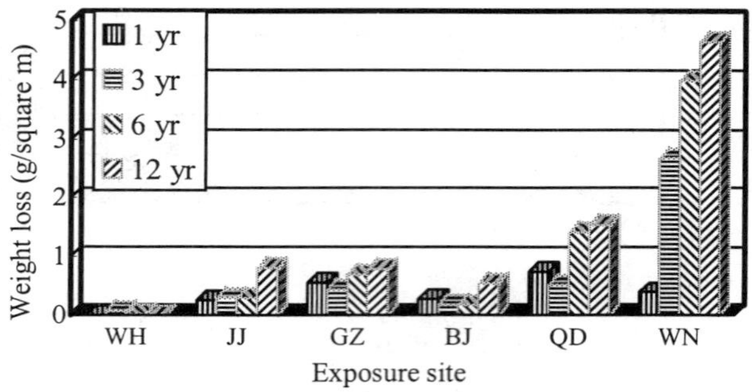

Figure 4 - *Weight loss of 00Cr19Ni10 stainless steel after twelve years of exposure at various sites.*

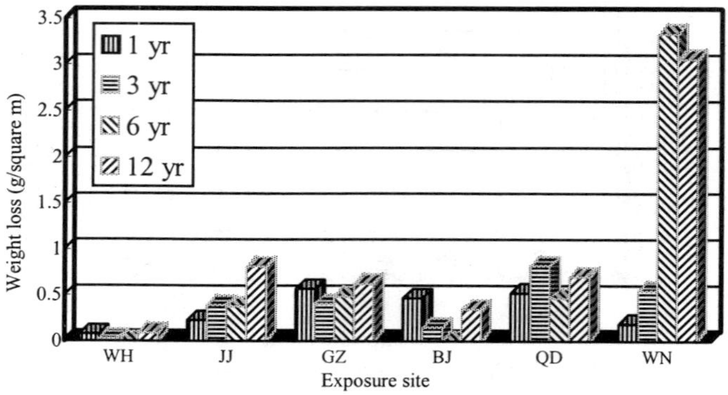

Figure 5 - *Weight loss of 00Cr18Mo2 stainless steel after twelve years of exposure at various sites.*

Differences in atmospheric corrosion resistance of various types of stainless steels were obvious. 2Cr13 has the lowest corrosion resistance. 1Cr18Ni9Ti, 00Cr17AlTi, 00Cr18Ni10 all have good resistance. 00Cr18Mo2 has the best.

For steel 2Cr13, the lowest alloy containing steel, there was no serious weight loss after exposure at Wuhan, Guangzhou and Jiangjin. The weight loss was about $0.2 g/m^2$ after three years of exposure. The weight loss was $11 g/m^2$ at Beijing, a bit higher. While at Wanning and Qingdao, scattered rust had occurred on all the coupons even after one year of exposure. The rust had covered all coupons after the third year of exposure. The weight loss reached $40 g/m^2$ after three years of exposure at Qingdao. It reached $80 g/m^2$ at Wanning.

1Cr18Ni9Ti, 00Cr17AlTi, 00Cr18Ni10 have a much higher corrosion resistance. There were only occasionally slight rust spots after three years of exposure at Beijing, Wuhan, Guangzhou Jiangjin and Qingdao. The weight loss was nearly zero. There was no rust after a one year exposure at Wanning. Occasional rust occurred only after three years of exposure. The weight loss was $15 g/m^2$ after three years of exposure.

00Cr18Mo2 steel exhibited the best corrosion resistance in the tested steels. The weight loss was nearly zero at almost all exposure sites after three years. A few rust areas occurred after three years of exposure only at Wanning. The weight loss was $3 g/m^2$ after twelve years.

All rust started from dust deposits or surface defects and developed with the form of shallow pitting. Pitting rust occurred more in quantity and dispersion and developed less with time on the coupon surface facing up. Meanwhile, it occurred less in quantity and dispersion on the coupon surface facing down. It was concentrated more at the corners and the edges and they developed more in size. This phenomenon was shown more clearly on the rust cleaned coupons.

Effect of Alloying on the Atmospheric Corrosion Resistance of Stainless Steels

Stainless steels rely on the passive film formed on its surface to resist corrosion. Passivation ability comes mainly from chromium. Therefore, chromium is the base of any stainless steel. The stability of the passive film increases with the chromium content.

To see the effect of alloying on the atmospheric corrosion resistance of stainless steels more clearly, we put the corrosion data of various steels at one site together in one figure for Qingdao and Wanning, the two most corrosive sites. They are shown in Figures 6 and 7.

A minimum content of 13.79% chromium may make the stainless steel form passive film spontaneously in ordinary atmosphere. However, this content is not enough for the steel to have a sufficient self-rehabilitation ability in a corrosive environment when the passive film is destroyed. It was covered with yellow rust after one year in Qingdao and Wanning.

To give a stainless steel sufficient self-rehabilitation ability in atmosphere, 17%

chromium is needed. Nickel and molybdenum can improve the passive film stability of steels. The effect of molybdenum is more significant in marine environments.

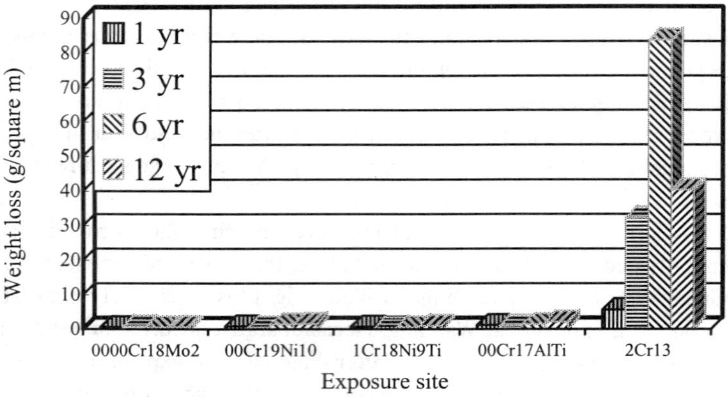

Figure 6 - *Weight loss of various stainless steels after 12 years of exposure at Qingdao.*

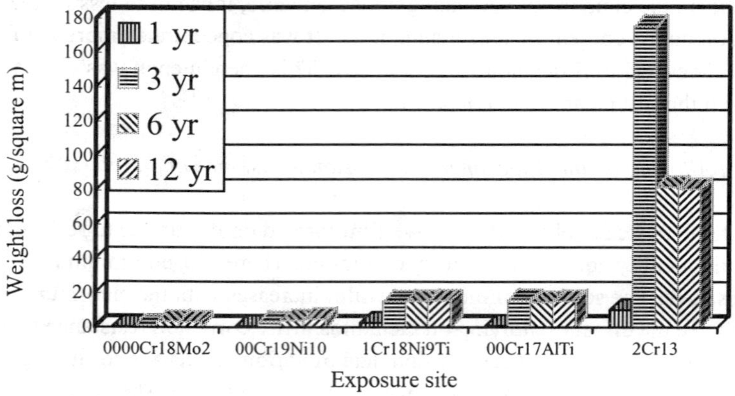

Figure 7 - *Weight loss of various stainless steels after 12 years of exposure at Wanning.*

Another problem is that chromium carbide can form in stainless steels at the grain boundaries when the steel is cooled down from the rolling temperature. The carbide will lower the chromium content there and cause intergranular corrosion.

To counter intergranular corrosion, strong carbide former alloying elements such as Al, Ti, and V were added. Carbon will then be taken into their carbide and dispersed in the matrix. 1Cr18Ni9Ti is the typical steel with 18% chromium and 9% nickel and an addition of titanium. It is nearly free from rust in all atmospheres.

The second way to counter intergranular corrosion is lowering the carbon content as much as possible. With the development of steel making technology, extra low carbon can be obtained with little difficulty. The 00Cr17AlTi steel, which has only 16% Cr, has a similar corrosion resistance to the 1Cr18Ni9Ti. Furthermore, extra low carbon combined with addition of molybdenum gaves the steel 00Cr18Mo2 a very good performance in any atmosphere. It stays rust free even in the most corrosive condition of Wanning.

Atmospheric Corrosivity for Stainless Steels

Atmospheric corrosion is an electrochemical reaction under a water film on the surface of materials. With the sole factor of surface water film in an ordinary atmosphere, the passive film of a stainless steel will not be destroyed. Even destroyed, it can self-rehabilitate easily. However, when dust deposits exist, crevices formed under the deposit. The water film will then be kept easily and the supply of oxygen will be blocked. When a passive film is dissolved under certain pollution conditions and corrosion starts under the deposit, the concentration of the dissolved oxygen there will decrease. Then a secluded cell forms that will facilitate the corrosion progress.

Our data proved this generally accepted mechanism.

All steels had nearly no corrosion at all exposure sites in their first year of exposure. The coupons were smooth and clean when they were newly exposed. Dust was not easy to deposit on the coupons with the wash of rain. After the loss of gloss with time, dusts started to be easy to deposit on the coupons. This resulted in the sudden growth of corrosion loss after the first year exposure. Corrosion was always observed at dust deposits and was in a form of pitting.

Another proof for the function of dust deposits is that the rust spots are more by amount but less developed on the upper side of the coupons. This means the chance is greater for dust deposits on the upper surface of the coupons, but they are more easily cleaned by the rain and dried by sunshine. It is difficult for dust to deposit on the lower surface of the coupons, but once deposited, it is difficult too to be cleaned and dried.

Beijing is notorious for being dusty. However, there had been nearly no corrosion during the whole 12 exposure years for the tested steels except 2Cr13. This is the result of another character of Beijing, dryness. Dust has to be accompanied with certain humidity to facilitate the start of corrosion in high grade stainless steels. But for 2Cr13, once the corrosion started, it developed quickly. Beijing is more corrosive than Wuhan, Guangzhou, and Jiangjin for this steel.

With the function of dust deposits, corrosion occurred rather randomly. This resulted in the irregularity of the corrosion development of stainless steels.

From Figure 1 to 5, the corrosion rate at Qingdao and Wanning is much higher than at other sites for all the stainless steels. Marine atmosphere is the character of these two sites. This shows the strong effect of chloride ions on intensifying the atmospheric corrosion of stainless steels.

However, compared with Qingdao, Wanning has an extra high atmospheric corrosivity. Another character of Wanning is its high humidity and heat. This means pollution with chloride ion under high humidity and heat condition creates extra high corrosion of stainless steel.

SO_2 pollution has a very bad effect on atmospheric corrosion of low alloy steels [*7*]. However, it has little effect on the stainless steels. The pollution character of Jiangjin is SO_2 pollution. Its corrosivity is the lowest together with Guangzhou, Wuhan, and Beijing.

This may be explained by the known fact that iron may be passivated by sulfate in a neutral solution. During the development of crevice corrosion, the corrosion product of sulfate may accumulate and reach saturation. Then, sulfate crystal may precipitate and prevent further corrosion and promote the recovery of passive film.

This seems to be in contradictory to the results of Johnson [*6*]. Johnson measured the corrosion of stainless steel under various concentrations of SO_2. It was concluded that the corrosion of stainless steel increased with increasing SO_2 concentration. However, their study of the effect of concentration of SO_2 was under the condition of coexistence of chloride ions. SO_2 might be effective only through chloride ions. Meanwhile, their study was of a short period (4 months). It might be different in a longer run.

Conclusion

- Stainless steels are corrosion resistant in the atmosphere. They exhibit excellent corrosion resistance when Chromium content is over 18%. Molybdenum raises their corrosion resistance, as does extra low carbon. Their effect is obvious even in the most corrosive environment, the high humidity and heat marine atmosphere.
- The most damaging pollutant to the atmospheric corrosion resistance of stainless steels is the chloride ion. With addition of high humidity and heat, extraordinary corrosion may be incurred. Pollution of sulfide has little effect.
- The sites of dust deposits and surface defects are the starting points of atmospheric corrosion.

Acknowledgement

The project was sponsored by the China National Natural Science Foundation. The data was the result of united effort of Beijing 621 Institute, Qingdao Marine Corrosion Institute, Wuhan Materials Protection Institute, Chongqing 59 Institute, Guangzhou Electro-appliances Institute.

References:

[1] Khanna A. S., and Gnannmoorthy T. B.,: Atmospheric Corrosion, W. H. Ailor Ed., John Wiley and Sons Inc., New York, NY, 1982, p.461.

[2] Button H. E., and Simm D. W., "The Influence of Particulate Matter on the Corrosion Behavior of Type 316 Stainless Steel," Anti-Corrosion, June 1985, p. 8.

[3] Kearns J. R., and Barkowski A. L. R., "Appearance and Corrosion of Stainless Steels in the Atmosphere," Materials Performance, 27(2), 1988, p. 47.

[4] Tochihara M., etal., "Atmospheric Corrosion of Stainless Steels Used for the Eaves of Buildings," Materials Performance, Vol.35, No10, 1996, p. 58.

[5] Johnson K. E., "Role of Inclusions in the Atmospheric Pitting of Stainless Steel," British Corrosion Journal, Vol. 15, No 3, 1980, p. 123.

[6] Johnson K. E., "Airborne Contaminants and the Pitting of Stainless Steels in the Atmosphere," Corrosion Science, Vol. 22, 1982, p. 175.

[7] Azzerri N., Aging of Passive Surfaces of Stainless Steels in an Urban-Industrial Atmosphere," Corrosion Science Vol. 22, 1982, p. 867.

[8] Hou w., and Liang C., "Eight Year Atmospheric Exposure of Steels in China," Corrosion, Vol. 55, No.1, 1999, p. 65.

Wentai Hou[1] and Caifeng Liang[1]

Effects of Alloying on Atmospheric Corrosion of Steels

Reference: Hou, W. and Liang, C., "**Effects of Alloying on Atmospheric Corrosion of Steels,**" *Outdoor Atmospheric Corrosion, ASTM STP 1421,* H. E. Townsend, Ed., American Society for Testing and Materials International, West Conshohocken, PA, 2002.

Abstract: The result of a previous eight-year atmospheric exposure of steels at seven sites in China showed significant effect of variation of Cu, P, and S content within the allowance in carbon steels and copper-free low-alloy steels in corrosive environments. To verify this phenomenon, steels with 0.035% phosphorous and 0.2% copper were prepared and tested. Accelerated corrosion tests and four year, three period atmospheric exposure at Beijing, Qingdao, Wuhan, Jiangjin, Guangzhou, Qionghai and Wanning, seven typical environmental sites in China, were made on these steels together with ordinary mild steels. The results showed that the new steels were more corrosion resistant than ordinary steels at all exposure sites and in the alternating immersion accelerated corrosion tests. The effect is bigger at Qingdao, Jiangjin, characterized by serious pollution of marine salt and acid rain. The effects were most significant at Wanning, a site of high humidity and heat in addition to marine salt pollution. Alternate immersion accelerated corrosion tests showed an obvious advantage of the steels over the common mild steels. Suggestion was made of an introduction of a minimum content of phosphorous and copper in the standard mild steels. Combined with a low sulphur content, the routine request now for steels, it would make any ordinary steel a weathering steel—economic weathering steel—without additional cost, a way of avoiding enormous amounts of corrosion loss for the whole world.

Keywords: corrosion, steel, atmospheric exposure, weathering steel, alloying element

---

[1] Professor and research engineer, respectively, Qingdao Marine Corrosion Institute, Central Iron and Steel Institute, Wheat Island, Qingdao, 266071, P.R. China.

Mild steel and ordinary low alloy steels will corrode in the atmosphere. But, for their excellent mechanical properties and low price, no other materials can replace them completely. To improve their corrosion resistance, atmospheric corrosion study of steels has become one of the main subjects in the study of corrosion since the beginning of the century.

On the basis of large-scale exposures, effects of alloy elements were determined and low alloy weathering steels were developed. The most popular weathering steel is ASTM A242 with a high content of Cu & P. and ASTM A588 with alloying of Cr, Mn and Cu. They are now widely used for containers, vehicles, bridges and other structures. The effect of alloy composition on their atmospheric corrosion resistance was further studied in later years [*1,2*]. Standard guidelines for estimating the atmospheric corrosion resistance of low alloy steels were also stipulated in the ASTM " Standard Guide for Estimating the Atmospheric Corrosion Resistance of Low Alloy Steels" (G101-94, 1995)

As for the variation of Cu, P, and S content within the allowance of carbon steel and copper free steels, it was assumed that there was no effect on the corrosion resistance [*3*]. ISO/DIS9226 Corrosion of metals and alloys - Corrosivity of atmospheres - Method of determination of corrosion rate of standard specimens for evaluation of corrosivity (1988) stipulates that low carbon steel can be exposed to determine corrosivity of atmosphere. The chemical composition request is only P$\leqslant$0.08% and Cu$\leqslant$0.3%. This means that within this area there is no large effect of chemical composition variation on the atmospheric corrosion resistance.

In China, weathering steel study started in 1960s. A few specifications were developed and used widely. A 20-year exposure program of various materials was started in 1984. A national network was organized under the National Science and Technology Commission and National Natural Science Foundation of China to carry out the program. Eight-year exposure data of four periods (1,2,4,8 years) of steels at seven sites in China were obtained [*4-6*].

The results showed significant effect of Cu, P, and S content within the allowance in carbon steels and copper free low alloy steels in corrosive environments. When P content reaches 0.035% and Cu content reaches 0.07% from the ordinary 0.02% P and no Cu , the corrosion resistance can be raised considerably. A low P content under 0.01% might result in extremely high corrosion. Meanwhile, when S content is lowered to 0.01% instead of the usual 0.02%, the corrosion resistance of ordinary carbon steel can be raised to the level of A588, and the corrosion resistance of A588 can be raised to the level of A242.

The corrosion performance of various steels at Wanning, where it is high humidity, heat, and marine polluted, is shown in Figure 1. The chemical composition of the steels is shown in Table 1. A242, A588 does have lower corrosion rate. However, the mild steel C, although a plain carbon steel only, has a lower corrosion rate than mild steel B and a much lower corrosion rate than mild steel A. The only difference in the chemical composition of these three steels is a little variation in phosphorus and copper content.

For the ordinary carbon steels, the chemical composition standard is S≤0.05%, P≤ 0.05%, Cu ≤0.2% without a low limit. The actual chemical composition of the commercial carbon steels is S≈0.2% and P≈0.2%. Cu varies with the source of scraps or ores. We may define a new specification as follows:

P: 0.03% -- 0.05%,
S: 0.01 max., and
Cu: 0.07% -- 0.2%.

This is similar to the condition of mild steel C in Figure 1. This steel will be easy to melt with modern melting technology and with the copper containing scraps. It will give a corrosion resistance as good as A242 without extra expense of alloying. We call it economic weathering steel. With this specification, the economic weathering steel will have as low a corrosion rate as one half of ordinary mild steels and one fourteenth of low phosphorous mild steel at Wanning.

Table 1 - *Chemical composition of the steels in Figure 1*

	C	S	P	Mn	Si	Cu	Cr	Ni	Type
1	0.08	0.025	0.008	0.4	0.1				Mild steel A
2	0.16	0.025	0.009	1.4	0.4				Mild steel B
3	0.17	0.023	0.030	1.4	0.3	0.07			Mild steel C
4	0.13	0.022	0.011	0.8	0.3	0.20	0.6		A588
5	0.08	0.023	0.070	0.4	0.5	0.41	0.6	0.3	A242

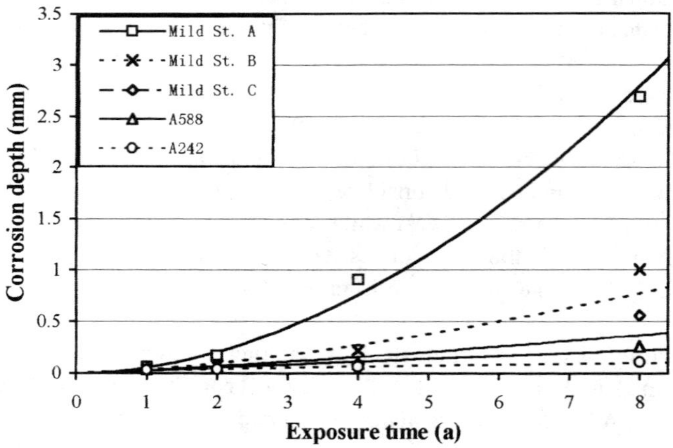

Figure 1 - *Corrosion of various steels at Wanning, Hainan Island, P. R. China*

Study of Economic Weathering Steel

To confirm the effect of variation of Cu and P content within the allowance in carbon steels on the corrosion resistance of the steels and to develop a new type of weathering steel, steels were melted and hot rolled with equipments for industrial production.

The content of P and Cu of the steels is 0.035% and 0.16%. The other elements were controlled according to GB 700-88 for Q235, a relevant steel of ASTM A570. The mechanical properties are all within the request of the standard.

Two batches of steel coupons were made. Both outdoor exposure and alternate immersion accelerated corrosion tests were carried out for the corrosion resistance tests.

Alternate Immersion Accelerated Corrosion Test

Alternate immersion testing machine FL-65 was used. Coupons with a size of 100mm × 50mm × 4mm were grounded on the two plain surfaces. Three periods of 165 hrs tests were made with three parallel coupons for each period. The chemical compositions of the steels are shown in Table 2. Testing results are shown in Figure 2.

Figure 2 shows that the corrosion rate of all the testing steels is lower than the steels for comparison from the second period. The corrosion rate of the testing steels is only one half of the steels for comparison after the third period. Addition of a rare earth elements has no apparent influence on corrosion performance for both ordinary mild steels and economic weathering steels.

Atmospheric Exposure Tests and Results:

Two batches of exposure were made. The exposures were of four years with three periods (1, 2 and 4 year) at seven sites of Beijing, Qingdao, Wuhan, Jiangjin, Guangzhou, Qionghai and Wanning. The meteorological and pollution conditions have been reported in other papers [*4,6*].

For the first batch of three heats of steel plates (2 mm in thickness), the rolled surface was taken as exposure surface after pickling, cleaning, and drying. They were put into exposure on November 15, 1992. Two mild steels with ordinary chemical composition were also exposed at same time for comparison.

The second batch of three heats was put into exposure on December 1, 1993, together with two mild steels of ordinary chemical composition. For this batch of coupons the two large surfaces were grounded as the exposure surface.
The Coupon size was 200mm*100mm*4mm. Four coupons were tested for each exposure period, three for weight loss measurement and one for rust analysis. The coupons were fixed with ceramic cylinders on an aluminum frame and exposed facing south at an angle of 45 degrees to the ground (facing sunshine). Exposed coupons were

cleaned with inhibited hydrochloric acid. The exposure and treatment process is in accordance with ISO-DIS 8565.2 Metals and alloys - Atmospheric corrosion testing - General requirement for field test (1991).

Table 2 - *Chemical composition of steels for alternate immersion accelerated corrosion test (%)*

No.	Steel	C	Si	Mn	S	P	Cu	RE	Remark
1	Mild St. 1	0.14	0.16	0.50	0.024	0.008	0.03	0.02	For
2	Mild St. 2	0.10	0.19	0.38	0.023	0.025	0.04		comparison
3	Testing St. 1	0.17	0.36	1.37	0.024	0.028	0.15	0.04	
4	Testing St. 2	0.11	0.16	0.58	0.025	0.035	0.17		
5	Testing St. 3	0.15	0.33	1.26	0.020	0.027	0.16	0.03	
6	Testing St. 4	0.12	0.19	0.58	0.025	0.040	0.17	0.01	
7	Testing St. 5	0.11	0.13	0.52	0.026	0.041	0.18	0.02	
8	Testing St. 6	0.12	0.13	0.48	0.020	0.018	0.17	0.03	

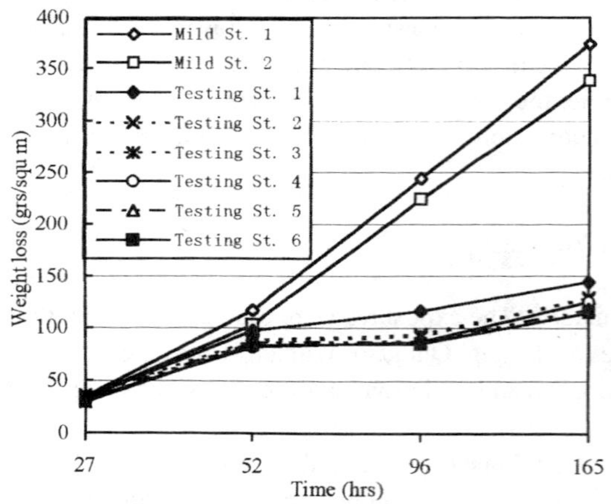

Alternate immersion circle: 60 Minutes
Solution: 0.01M $NaHSO_3$
pH: 4.4~4.8
Temperature: 45℃
Relative humidity of drying space: 70%
Max. temp. of specimens after drying : 70℃
Type of testing machine: FL-65

Figure 2 - *Alternate immersion accelerated corrosion test*

The chemical compositions of the first batch of steels are given in Table 3. Corrosion rate after exposure for four years at seven sites are given in Table 4.

Table 4 shows that the corrosion rate of all the testing steels are obviously lower than the steels for comparison at all exposure sites. The difference in corrosion resistance was less at Beijing (BJ), where pollution is little. The corrosion rate of the testing steels was 20% lower than that of the comparison steels at Qingdao (QD), where pollution by chloride ion was serious. It became 40% lower at Jiangjin, where pollution by sulphur dioxide was serious. Further more, it became 60% lower at Wanning (WN), where pollution by chloride ion was serious in addition to the high humidity and heat. Addition of rare earth element has no obvious influence in consideration of corrosion for both ordinary mild steels and economic weathering steels.

Table 3 - *Chemical composition of first batch of steels for outdoor exposure (%)*

No.	Steel	C	Si	Mn	S	P	Cu	RE	Remark
1	Mild St. 1	0.12	0.16	0.52	0.022	0.014	0.03		For
2	Mild St. 2	0.09	0.22	0.37	0.022	0.013	0.03	0.025	comparison
3	Testing St. 1	0.10	0.12	0.52	0.023	0.035	0.17	0.027	
4	Testing St. 2	0.13	0.14	0.54	0.024	0.046	0.16		
5	Testing St. 3	0.13	0.17	0.60	0.024	0.050	0.17	0.023	

Table 4 - *Corrosion rate after four year exposure of the first batch of steels (mm/a)*

		BJ	QD	WH	JJ	GZ	QH	WN
1	Mild St. 1	0.020	0.052	0.015	0.066	0.031	0.023	0.094
2	Mild St. 2	0.018	0.047	0.014	0.053	0.031	0.022	0.086
3	Testing St. 1	0.016	0.041	0.013	0.034	0.025	0.019	0.046
4	Testing St. 2		0.039		0.034		0.019	0.033
5	Testing St. 3	0.016	0.036	0.012	0.032	0.025	0.018	0.032

The chemical composition and exposure results of the second batch of steels are given in Table 5 and Table 6. The result was similar as that of the first batch of steels.

Table 5 - *Chemical composition of the second batch of steels for outdoor exposure (%)*

	Steel	C	Si	Mn	S	P	Cu	RE	Remark
6	Mild St. 1	0.14	0.15	0.50	0.024	0.008	0.02	0.020	For
7	Mild St. 2	0.10	0.19	0.37	0.020	0.018	0.03		comparison
8	Testing St. 1	0.14	0.32	1.26	0.025	0.040	0.16	0.025	
9	Testing St. 2	0.11	0.12	0.47	0.026	0.041	0.17	0.027	
10	Testing St. 3	0.11	0.12	0.51	0.025	0.035	0.17	0.021	

Table 6 - *Corrosion rate after four years of exposure of the second batch of steels (mm/a).*

		WH	QH	BJ	GZ	JJ	QD	WN
6	Mild St. 1		0.019	0.019	0.026	0.038	0.052	0.146
7	Mild St. 2	0.015	0.020	0.018	0.026	0.032	0.051	0.127
8	Testing St. 1	0.013	0.018	0.019	0.024	0.034	0.042	0.082
9	Testing St. 2	0.011	0.016	0.018	0.024	0.029	0.038	0.078
10	Testing St. 3	0.012	0.017	0.017	0.023	0.032	0.043	0.060

To see the difference more clearly, corrosion development of various steels at

Wanning is shown in Figure 3 and Figure 4. The curves are the regression results by the empirical formulae $D=At^n$, where D: Corrosion depth in mm, t: Exposure time in years. The advantage of economical weathering steels is obvious.

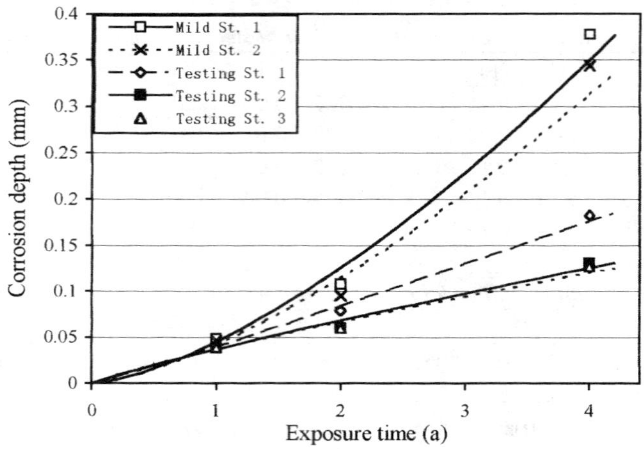

Figure 3 - *Atmospheric corrosion of various steels of 1st batch exposed at Wanning*

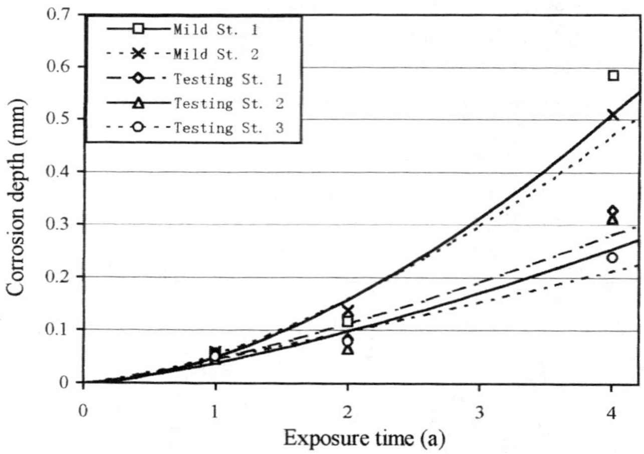

Figure 4 - *Atmospheric corrosion of the second batch of steels exposed at Wanning*

It has been mentioned above that for the first batch tests, the rolling surface was taken as exposure surface after pickling, cleaning and drying, while, for the second batch tests, the two large surfaces of the coupons were grounded as the exposure surfaces. The different condition of surfaces did not give an obvious difference in the results of exposure. This means that machining of the coupon surface does not affect the atmospheric corrosion behavior of carbon steels.

A further confirmation study has been started. The effect of alloying on the atmospheric corrosion resistance of steels by controlling content of all three alloy elements P, Cu, and S together with Si is studied. Its first year data showed a positive result. A steel with a phosphorous content of 0.035%, a sulphur content of 0.005%, a copper content of 0.15% and a silicon content of 0.3% gives as good corrosion resistance as that of A242 weathering steel.

Discussion

From the point of decoration, all steel structures are painted. However, painted steel is not immune to corrosion. Water and some ions can permeate the paint film steel substrate. Meanwhile, defects always exist in paint films. On the other hand, there are always locations where painting is difficult to reach or to be maintained, such as inner walls, and crevices. Here, corrosion starts from the inside. If the steel substrate is not corrosion resistant, once the corrosion starts, it develops even more quickly than that without painting and makes the paints delaminate. There are proofs that weathering steels increase the age of painting. Therefore, weathering steel is still important for structures even with painting.

However, weathering steels are not used widely for structures in open air. The problem of knowledge about corrosion and protection is one of the reasons. The more important reasons are financial and technological.

A242 has been the dominant choice for weathering steel in the world at the moment. Its high phosphorous, copper, and chromium content raises its atmospheric corrosion resistance. Meanwhile, high phosphorous brings along low temperature fragility and weld crack; and high copper increases its tendency to hot fissuring. In addition, high phosphorous made the melting plant scared of contaminating the furnace. All these led to a higher price of steel with lower bendability and lower weldability than carbon steel, and then led to the reluctance to use weathering steels.

With the suggested new type of economic weathering steels, all these problems will be solved. This steel will be easy to melt with modern melting technology, and it will give a corrosion resistance as good as A242 without the extra expense of alloying and without the faults of low bendability and low weldability. We call it economic weathering steel. It certainly will benefit both manufacturers and customers.

This kind of weathering steel can be applied to any type of steels. We take steel for bridges as an example: To get enough strength, we define the carbon content as 0.15%

and manganese content as 1.6%. To guarantee its atmospheric corrosion resistance, we give the steel an additional request of S, P, and Cu as shown above, and define the specification with a suffix of W (weathering). Then, it will become the weathering steel for bridges. Meanwhile, its price will be kept nearly the same as the ordinary steel for bridges.

Ideally, all structural steels, except for especially high request for low temperature toughness, should be given a specification of phosphorous, sulphur, silicon and copper content as follows:

P: 0.03% -- 0.05%,
S: 0.01 max.,
Si: 0.3% -- 0.5%, and
Cu: 0.07% -- 0.2%.

Then, all structural steels will become weathering steels without extra cost.

By our data, the corrosion age of this kind of steels will last 1.3 times longer than ordinary structural steels in average. In addition, it will avoid the occurrence of steels with extreme low corrosion resistance as the mild steel A shown in Figure 1.

The whole world now turns out over 400 million tons of steels every year. Steels for open-air structures and vehicles take two thirds of the total. If we can produce them according to the suggested economic weathering steel specification, that will prolong their length of service. Being conservative, according to the data obtained from the tests, we estimate it prolongs the service life by one third. Thus, 100 Mt of steels will be saved. That will be about 50 billion dollars every year. What a big savings of natural resources will it be for our world. This does not even consider the indirect spending from incidents caused by corrosion failure and the savings from a longer service life.

Conclusions

- Variation of S, P, and Cu contents within the allowance may alter the corrosion resistance considerably.
- Combining 0.035% P with 0.15% Cu decreased the ordinary copper free carbon steel' s corrosion rate at all exposure sites in China. The decrease was 60% lower at Wanning, where pollution by chloride ion was serious in addition to high humidity and heat.
- The corrosion rate of the tested steels becomes only one half of that of the steels for comparison after third period in an alternate immersion accelerated corrosion test.
- Addition of rare earth element has no obvious influence in consideration of corrosion for both ordinary mild steels and economic weathering steels.

Acknowledgement

The project was sponsored by the China National Natural Science Foundation. The data was the result of united effort of Beijing 621 Institute, Qingdao Marine Corrosion Institute, Wuhan Materials Protection Institute, Chongqing 59 Institute, and Guangzhou Electro-appliances Institute.

References

[1] Legault R.A., and Leckie H. P., " Effect of Alloy Composition on Atmospheric Corrosion Behavior of Steels Based on a Statistical Analysis of Larabee-Coburn data set," *Corrosion in Natural Environments, ASTM STP 558*, American Society for Testing and Materials, West Conshohocken, PA, 1974, pp. 334-347.

[2] Townsend H.E., " Effect of Silicon and Nickel Contents on the Atmospheric Resistance of ASTM A588 Weathering Steel," *Atmospheric Corrosion, ASTM STP 1239,* W. W. Kirk and H. H. Lawson, Eds., American Society for Testing and Materials, West Conshohocken, PA, 1995, pp. 85–100.

[3] Chandler K.A, and Killcullen M.B., " Corrosion-Resistant Low-Alloy Steels: A Review with Particular Reference to Atmospheric Conditions in the United Kingdom," British Corrosion Journal, vol. 5, January 1974, p. 24.

[4] Hou W., Yu J., and Liang C., " Atmospheric Corrosion of Carbon and Low Alloy Steels," Journal of Chinese Society for Corrosion and Protection, Vol. 12, 1993, p. 291.

[5] Liang C., and Hou W., " Atmospheric Exposure of Carbon and Low Alloy Steels," Corrosion Science and Corrosion Protection Technology, Vol. 7, No. 3, 1995, p. 182.

[6] Hou w. and Liang C., " Eight Year Atmospheric Exposure of Steels in China," Corrosion, Vol. 55, No.1, 1999, p. 65

Author Index

Subject Index

Y

Z